APPLIED VEGETATION SCIENCE

Applied Vegetation Science

Editor:

Dr. Hazem Fouda, Ph.D.

Delve Publishing

www.delvepublishing.com

Applied Vegetation Science

Editor: **Dr. Hazem Fouda, Ph.D.**

708 3rd Avenue, 6th Floor
New York, NY 10017
United States of America
www.delvepublishing.com

ISBN: 978-1-68095-779-2

Library of Congress Control Number: 2016951535

About the Editor

Dr. Hazem Fouda, Ph.D.

Hazem Shawky Fouda has a PhD. In Agriculture Sciences, obtained his PhD. From the Faculty of Agriculture, Alexandria University in 2008, He is working in Cotton Arbitration & Testing General Organization (CATGO).

Contents

Section II

Section III

Section IV

List of Contributors

Teegalapalli Karthik
Nature Conservation Foundation, 3076/5, IV cross, Gokulam Park, Mysore, India - 570002

Gopi Govindhan Veeraswami
Wildlife Institute of India, Chandrabani, Dehradun, India - 248001

Prasanna Kumar Samal
G B Pant Institute of Himalayan Environment & Development, North East Unit, Vivek Vihar, Itanagar, Arunachal Pradesh, India – 791113

A. Nurbekov
International Center for Agricultural Research in the Dry Areas (ICARDA) Central Asia and the Caucasus Regional Office, Tashkent, Uzbekistan

A. Akramkhanov
International Center for Agricultural Research in the Dry Areas (ICARDA) Central Asia and the Caucasus Regional Office, Tashkent, Uzbekistan

A. Kassam
School of Agriculture, Policy and Development, University of Reading, United Kingdom

D. Sydyk
South-Western Research Institute of Livestock and Crop Production, Chimkent, Kazakhstan

Z. Ziyadaullaev
Kashkadarya Research Institute of Breeding and Seed Production of Cereals, Karshi, Uzbekistan

J.P.A. Lamers
Center for Development Research, Bonn, Germany

W. Mupangwa
CIMMYT Southern Africa Regional Office, Mount Pleasant, Harare, Zimbabwe

I. Nyagumbo
CIMMYT Southern Africa Regional Office, Mount Pleasant, Harare, Zimbabwe

E. Mutsamba
CIMMYT Southern Africa Regional Office, Mount Pleasant, Harare, Zimbabwe

Oriol Grau,
Department of Plant Biology and Institute for Research on Biodiversity, University of Barcelona, Av. Diagonal 643, ES-08028 Barcelona, Catalonia, Spain

Josep M. Ninot
Department of Plant Biology and Institute for Research on Biodiversity, University of Barcelona, Av. Diagonal 643, ES-08028 Barcelona, Catalonia, Spain

Aaron Pérez-Haase
Department of Plant Biology and Institute for Research on Biodiversity, University of Barcelona, Av. Diagonal 643, ES-08028 Barcelona, Catalonia, Spain

Terry V. Callaghan
Royal Swedish Academy of Sciences, Lilla Frescativägen 4A, SE-114 18, Stockholm, Sweden
Department of Animal and Plant Sciences, University of Sheffield, S10 2TN Sheffield, UK

Marcos Jiménez
Remote Sensing Area, National Institute of Aerospace Technologies (INTA), Ctra. Ajalvir s/n, Torrejón de Ardoz, 28850 Madrid, Spain

Ricardo Díaz-Delgado
Remote Sensing and GIS Lab., Doñana Biological Station, CSIC, Avda. Americo Vespucio, 41092 Sevilla, Spain

Davies, A.L.
School of Biological and Environmental Sciences, University of Stirling, Stirling, Scotland FK9 4LA, UK

Bunting, M.J.
Department of Geography, University of Hull, Cottingham Road, Hull HU6 7RX, UK

Brian W. Miller
Natural Resource Ecology Laboratory, Colorado State University, Fort Collins, CO, 80523-1499, USA
Department of the Interior North Central Climate Science Center, Fort Collins, CO, 80523-1499, USA

Leonardo Frid
Apex Resource Management Solutions Ltd., Bowen Island, BC V0N 1G1, Canada

Tony Chang
Montana State University, Bozeman, MT, 59717, USA

Nathan Piekielek
The Pennsylvania State University, University Park, PA, 16802, USA

Andrew J. Hansen
Montana State University, Bozeman, MT, 59717, USA

Jeffrey T. Morisette
U.S. Geological Survey, Department of the Interior North Central Climate Science Center, Fort Collins, CO, 80523-1499, USA

Raquel Benavides
Department Biogeografı´a y Cambio Global, Museo Nacional de Ciencias Naturales-CSIC, Madrid, Spain

Sonia G. Rabasa
Department Biogeografı´a y Cambio Global, Museo Nacional de Ciencias Naturales-CSIC, Madrid, Spain

Elena Granda
Department Biogeografı´a y Cambio Global, Museo Nacional de Ciencias Naturales-CSIC, Madrid, Spain

Adria´n Escudero
A´ rea de Biodiversidad y Conservacio´n, Universidad Rey Juan Carlos, Mo´stoles, Madrid, Spain

Jose´ A. Ho´dar
Departamento Ecologı´a, Universidad de Granada, Granada, Spain

Jordi Martı´nezVilalta
CREAF, Cerdanyola del Valle`s, Barcelona, Spain
Universitat Auto`noma Barcelona, Cerdanyola del Valle`s, Barcelona, Spain

Ana M. Rinco´n
Departamento Proteccio´n Vegetal, Instituto de Ciencias Agrarias-CSIC, Madrid, Spain

Regino Zamora
A´ rea de Biodiversidad y Conservacio´n, Universidad Rey Juan Carlos, Mo´stoles, Madrid, Spain

Fernando Valladares
Department Biogeografı´a y Cambio Global, Museo Nacional de Ciencias Naturales-CSIC, Madrid, Spain

A. V. Lozhkin
Northeast Interdisciplinary Scientific Research Institute, Far East Branch, Russian Academy of Sciences, 16 Portovaya Street, Magadan, 685000, Russia

P. M. Anderson
Earth & Space Sciences and Quaternary Research Center, University of Washington, Seattle, 98195-1310, USA

Richard D. Alward
University of Nebraska–Lincoln

James K. Detling
Colorado State University - Fort Collins

Daniel G. Milchunas
Colorado State University - Fort Collins

Jan Willem Erisman
Louis Bolk Institute, Hoofdstraat 24, 3972 LA Driebergen, The Netherlands
VU Amsterdam, Department of Earth sciences, De Boelenlaan 1085, Amsterdam, The Netherlands

Nick van Eekeren
Louis Bolk Institute, Hoofdstraat 24, 3972 LA Driebergen, The Netherlands

Jan de Wit
Louis Bolk Institute, Hoofdstraat 24, 3972 LA Driebergen, The Netherlands

Chris Koopmans
Louis Bolk Institute, Hoofdstraat 24, 3972 LA Driebergen, The Netherlands

Willemijn Cuijpers
Louis Bolk Institute, Hoofdstraat 24, 3972 LA Driebergen, The Netherlands

Natasja Oerlemans
WWF Netherlands, Zeist, The Netherlands

Ben J. Koks
Dutch Montagu's Harrier Foundation, Scheemda, The Netherlands

P. Biber
Chair for Forest Growth and Yield, Department Ecology and Ecosystem Management, Technische Universität München, Freising, Germany

S. Seifert
Chair for Forest Growth and Yield, Department Ecology and Ecosystem Management, Technische Universität München, Freising, Germany

M. K. Zaplata
Geobotany, Department Ecology and Ecosystem Management, Technische Universität München, Freising, Germany

W. Schaaf
Soil Protection and Recultivation, Brandenburg University of Technology, Cottbus, Germany

H. Pretzsch
Chair for Forest Growth and Yield, Department Ecology and Ecosystem Management, Technische Universität München, Freising, Germany

A. Fischer
Geobotany, Department Ecology and Ecosystem Management, Technische Universität München, Freising, Germany

Korehisa Kaneko
Ecosystem Conservation Society-Japan, Tokyo, Japan

Seiich Nohara
Center for Environmental Biology and Ecosystem Studies, National Institute for Environmental Studies, Ibaraki, Japan

Muhammad Kamal
Biophysical Remote Sensing Group, School of Geography, Planning and Environmental Management, The University of Queensland, Brisbane, QLD 4072, Australia
Cartography and Remote Sensing Study Program, Faculty of Geography, Gadjah Mada University, Bulaksumur, Yogyakarta 55281, Indonesia

Stuart Phinn
Biophysical Remote Sensing Group, School of Geography, Planning and Environmental Management, The University of Queensland, Brisbane, QLD 4072, Australia

Kasper Johansen
Biophysical Remote Sensing Group, School of Geography, Planning and Environmental Management, The University of Queensland, Brisbane, QLD 4072, Australia

Maurizio Quartieri
Department of Agricultural Sciences, University of Bologna, viale G. Fanin 46, 40127 Bologna, Italy

Elena Baldi
Department of Agricultural Sciences, University of Bologna, viale G. Fanin 46, 40127 Bologna, Italy

Giovambattista Sorrenti
Department of Agricultural Sciences, University of Bologna, viale G. Fanin 46, 40127 Bologna, Italy

Graziella Marcolini
Department of Agricultural Sciences, University of Bologna, viale G. Fanin 46, 40127 Bologna, Italy

Moreno Toselli
Department of Agricultural Sciences, University of Bologna, viale G. Fanin 46, 40127 Bologna, Italy

Abha Mishra
Asian Center of Innovation for Sustainable Agriculture Intensification, Asian Institute of Technology, Klong Luang, Pathumthani 12120, Thailand

Prabhat Kumar
Asian Center of Innovation for Sustainable Agriculture Intensification, Asian Institute of Technology, Klong Luang, Pathumthani 12120, Thailand

Jan Willem Ketelaar
FAO Inter-Country IPM/Pesticide Risk Reduction Programme, FAO Regional Office for Asia and the Pacific, 39 Phra Athit Road, Bangkok 10200, Thailand

Theodor Friedrich
FAO representative in Cuba, Cuba

Amir Kassam
Reading University, UK

K.R. Fowler
Dept. of Mathematics, Clarkson University, 8 Clarkson Avenue, Potsdam, NY 13699, USA

E.W. Jenkins
Dept. of Mathematical Sciences, Clemson University, O105 Martin Hall, Clemson, SC 29634, USA

M. Parno
Dept. of Aeronautics and Astronautics, Massachusetts Institute of Technology, 77 Massachusetts Avenue, Cambridge, MA 02139, USA

J.C. Chrispell
Dept. of Mathematics, Indiana University of Pennsylvania, 210 South 10th St., Indiana, PA 15705 USA

A.I. Colón
Dept. of Mathematics, Clarkson University, 8 Clarkson Avenue, Potsdam, NY 13699, USA

R.T. Hanson
California Water Science Center, U.S. Geological Survey, 4165 Spruance Road, San Diego, CA 92140, USA
Nature Conservation Foundation, 3076/5, IV cross, Gokulam Park, Mysore, India - 570002

Preface

Applied Vegetation Science focuses on community-level topics relevant to human interaction with vegetation, including global change, nature conservation, nature management, restoration of plant communities and of natural habitats, and the planning of semi-natural and urban landscapes.

The book in the Section Chapter is talking about forest recovery following shifting cultivation, conservation agriculture for combating land degradation in Central Asia, Effect of different mulching materials on maize growth and yield in conservation agriculture systems of sub-humid Zimbabwe and plant co-existence patterns and High-Arctic vegetation composition in three common plant communities in north-east Greenland and in Section Two it talks about anticipating the effects of climate changes, Section Three talks about global warming, Section Four about agriculture and biodiversity and Section Five talks about food security and rural development. I hope this book will inspire people, I like to think that this could be a start of human race to understand the ecosystem, how applied vegetation science can help us restoring and preserving our planet for healthy environment for us and for the next generations.

Editor
Dr. Hazem Fouda, Ph.D.

SECTION I

CHAPTER 1

FOREST RECOVERY FOLLOWING SHIFTING CULTIVATION: AN OVERVIEW OF EXISTING RESEARCH

Teegalapalli Karthik[1], Gopi Govindhan Veeraswami[2] and Prasanna Kumar Samal[3]

[1] Nature Conservation Foundation, 3076/5, IV cross, Gokulam Park, Mysore, India - 570002

[2] Wildlife Institute of India, Chandrabani, Dehradun, India - 248001

[3] G B Pant Institute of Himalayan Environment & Development, North East Unit, Vivek Vihar, Itanagar, Arunachal Pradesh, India – 791113

ABSTRACT

Shifting cultivation is a predominant practice in the majority of tropical hilly tracts. Relatively few studies have examined forest recovery following shifting cultivation and we have reviewed these studies to identify and synthesize general recovery patterns. Most studies report that, although pioneer tree species recover relatively faster, woody biomass of mature forest trees recovers several decades after suspension of cultivation. Analysis of bird-species inventories in 10 studies revealed that up to 70% of mature forest birds may recover in successional sites and that species composition is about 55% similar to that in mature forests within 25 years. Results of mammal studies were likely to have been confounded by hunting effects and therefore the effect of time since suspension of cultivation and hunting impacts need to be investigated separately. As shown by a few studies, we conclude that community recovery can be expected to be accelerated when relatively large forest tracts adjoin a shifting cultivation landscape, in comparison with recovery in sites with shorter fallow cycles in the absence of contiguous forests, which act as sources for recolonization of fauna and vegetation.

INTRODUCTION

Shifting cultivation, also referred to as forest agrarian system, has been widely practiced by hill communities in Asia, Africa, and Latin America since the Neolithic period (13,000 to 3,000 BC) [1-2]. The practice is characterized by a cultivation phase, which involves clearing of primary or secondary forest and crop cultivation for one to three years, followed by a fallow phase, during which cultivation is suspended to allow recovery of soil fertility [3]. Conklin [3] defined integral and partial forms of shifting cultivation. The integral form of cultivation, which is primarily practiced for farmers' subsistence, includes pioneer swidden farming in which primary or climax forest is cleared annually as well as established swidden farming which involves clearing mostly of secondary forests. On the other hand, partial shifting cultivation is undertaken as a supplementary or temporary activity [3].

Shifting cultivation (hereafter, SC) is considered as a major driver of deforestation: globally, until the year 1991, SC had accounted for 61% of overall tropical forest destruction [4]. Nevertheless, the practice persists since it provides subsistence livelihoods to at least 300 to 500 million people worldwide [5]

Citation: Teegalapalli K., Gopi, G. V. and Prasanna K. Samal. 2009. Forest recovery following shifting cultivation: an overview of existing research. Tropical Conservation Science Vol.2 (4):374-387.

and is intricately linked to cultural, ecological, and economic aspects of communities [6]. While certain ecologists question the sustainability of the practice, since it involves clearing of primary and secondary forests [e.g. 4, 7], others appreciate the existence of the practice for several millennia and acknowledge the fact that timber-felling, monoculture plantations, and other such economic-oriented objectives are also critical drivers of deforestation [e.g. 8-9]. However, when fallow cycles drop below a critical time period due to increased human population leading to unavailability of land, the productivity of the plot as well as forest regeneration are negatively affected [2, 6, 10]. Whitmore [11] suggested that the density of human population SC can sustain is 7 per km^2, which is considerably lower than present densities in SC landscapes in the tropics. Human population density in the year 1995 even in several of the biodiversity hotpots in the tropics was much higher than the world average of 42 people per km2 [12].

Shifting cultivation throughout the tropics is largely a subsistence activity practiced in areas with few alternative options and is therefore a practice that is likely to continue. Secondary forests formed following logging and SC cover more than 600 million hectares and play an important role in biodiversity conservation in the tropics [13]. In this context, studies that research biodiversity in secondary forests formed following SC are of significant value. Studies that have been carried out in various tropical and sub-tropical sites differ in spatial scales of sampling, taxa focused on, and sampling methods used. We therefore found it relevant to review existing research on plants, birds and mammals, which are the groups that have been most researched in SC landscapes and specifically, a) find generalizations in patterns recorded from these studies; b) identify factors possibly responsible for variance in study-results, and c) make recommendations for future studies.

Literature Review

We searched for published studies using the ISI web of knowledge (1985-2008) with the combination of keywords shifting/swidden/slash-and-burn cultivation and vegetation/ forest recovery, avifaunal/bird recovery, faunal/mammal recovery. Studies on vegetation and mammals in SC landscapes were reviewed qualitatively whereas species inventories in avifaunal studies were analyzed to compare bird species composition in successional sites with that in mature forests. While we realize that avifaunal studies vary in geographic locations, sampling methods, spatial scale of sampling, and features such as extent of anthropogenic disturbances and quality of surrounding forest (mature/ near-mature/ old secondary forest) used for comparison in studies, we have nevertheless undertaken meta-data analyses based on species inventories to generalize recovery patterns. A species list was obtained from other research papers when data from the control forest were not available for comparison (species list was obtained from Karr [14] for Blankespoor [15], which the latter author used for species comparison). We included a study that focused on spatial heterogeneity and its effect on raptor diversity in a SC landscape since a SC mosaic was recorded as spatially more heterogeneous than primary forest, and therefore a direct comparison of species composition in the two habitats was possible [16]. We grouped 10 studies available on recovery of birds following SC in the following time periods since abandonment to track recovery of species over time: 1 to < 5 years, 5 to < 11 years, 11 to 25 years, and sites older than 25 years. These categories were chosen to arrive at approximately equal number of studies in each temporal group.

We calculated Jaccard's index of similarity between bird species recorded from different successional stages and from mature forests in respective studies. We also investigated changes in proportions of mature forest and disturbed habitat species in successional fallows over time. Species that were encountered in the control forest were considered mature forest species, whereas disturbance-associated birds were species that were recorded from 1–5 year fallows. Average values for Jaccard's indices and proportion of mature and disturbance-associated species were calculated when different methods were used for sampling birds in the same study [17].

We summarized results of 33 studies on plants, birds, and mammals in SC landscapes and four review articles on birds and vegetation in anthropogenically disturbed habitats. A majority of studies were from South and Central American sites (17), with comparatively fewer studies from African (4) and Asian sites (12). Few studies (6) have focused on mammals in comparison with studies on avifauna (10) and vegetation (17) in SC landscapes. Further, only six studies have been carried out in site adjoining tropical dry and semi-deciduous forests and most of the understanding of recovery following SC comes from sites adjoining humid and moist forests.

Vegetation Recovery

Factors that significantly affected the vegetation recovery trajectory in clearings formed following suspension of cultivation were existing seed bank disturbance, presence of resprouters, and poor seed dispersal rates from surrounding forests [18-22]. In general, succession of pioneer vegetation following suspension of cultivation occurred rapidly in the early stages of regeneration, followed by delayed recovery of woody biomass [10, 21, 23].

Plant species richness has been reported to recover to values comparable with those in mature forest in 20-40 years; however, species composition recovery has been reported to take several decades [20, 24-26]. Endemic species, particularly, have been reported to have not fully recovered even after a period of 50-60 years [25]. In contrast, fallows adjoining tropical dry forests in Bolivia accumulated 75 percent of mature forest species within five years of active cultivation, although basal area was comparable with that in mature forests only in 50-year old fallows [27]. Mature forests sampled for comparison also had a disturbance history of fires, some of which may have been anthropogenic, which may have led to underestimates of recovery time from successional to mature forest [27].

The soil seed bank following suspension of cultivation consisted of predominantly pioneer species, herbs, and grasses: about half the seedlings that germinated from the soil seed bank from five-year fallows in Amazonia were of pioneer tree species [21]. Sprouting of tree stumps, roots, and rhizomes played an important role in regeneration in early successional stages [19, 22-23, 27]. In 2- to 15-yearold successional fallows in Paraguay, the number of species recorded resprouting was twice that of mature forest for individuals in the height range

30-130 cm [19]. Regeneration strategies of plants differed with successional stages. Early successional herbaceous species that colonize disturbed habitats were r-strategists that were primarily comprised of wind-dispersed or small-seeded vertebrate-dispersed species, whereas late successional primary forest species were k-strategists that were large-seeded vertebrate-dispersed species [18, 28-29]. Vertebrate seed disperser avoidance of open habitats prevented the dispersal of relatively large seeds into clearings in early successional stages [21, 30].

The number of years of cultivation as well as number of cultivation cycles affected the recovery process [21, 31]. Vegetation parameters such as canopy cover, tree species richness, percentage of forest species in regeneration, and biomass were considerably higher in plots cultivated for three years in comparison with sites cultivated continuously for six years [21]. Species density and evenness declined considerably with the increase in number of cultivation cycles in SC successional sites in Indonesia [31]. Factors such as harsh microclimatic conditions in clearings, competition with existing vegetation, lack of soil nutrients due to several years of cultivation, and high rates of seed predation and seedling herbivory further impeded plant regeneration [21, 28, 30, 32]. The vegetation recovery period was also affected by the crop cultivated: fields abandoned after cultivation of rice recovered tree species diversity faster than fields cultivated with poppy [33]. Presence of microhabitats such as remnant trees and tall shrubs, which provide sites for seedling regeneration and may improve bat- and bird-dispersed seed arrival, were also identified as important factors in clearings [21, 30].

Fallow cycles as short as 4–5 years in northeast India resulted in arrested succession, since weedy species were not succeeded by pioneer woody species, and over time the soil seed bank was replaced with seeds of weedy shrubs [34]. Fallows as old as 10 years in the region were dominated by bamboo cover [10, 35]. However, early colonizers such as bamboo, with relatively faster growth rates in comparison with woody tree species, may have facilitated soil-nutrient recovery and provided microhabitats for regeneration of shade-loving species [10, 35].

Response of Mammals

Generalist mammal species that are adapted to habitat disturbance have been reported to be abundant in successional sites whereas specialist frugivore-foliovore species have been reported to be restricted to mature or primary forest [36-39]. A majority of studies undertaken report that "garden hunting," a term coined by Linares [40] to refer to hunting of terrestrial mammals that are relatively more abundant close to agricultural habitats and are sources of protein for communities, is common in SC landscapes.

The impact of SC on mammals is relatively less understood and the few studies (6) have been only recently undertaken. Studies of large mammals (size > 2 kg) concluded that these animals may not persist in secondary forests and such sites may therefore harbor populations of fast-reproducing "weedy" disturbance-tolerant species [36-37, 39]. In 13- to 18-year-old secondary forests regenerating after mechanical clearing, mammal abundances were comparable to those in primary forest [37]. However, the composition differed between these forest types, with generalist species occurring in the secondary forest and specialist frugivore species restricted to primary forest [37]. Studies from northeast India reported that species feeding on young leaves and seeds (e.g., Phayre's leaf monkey, Trachypithecus phayrei) persisted in 10-year and older secondary sites to some extent [41]. However frugivore-foliovore species (e.g., Hoolock gibbon, Hoolock hoolock, and capped langur, Trachypithecus pileatus) rarely occurred far from primary or relatively undisturbed forests [38]. In Peru, the proximity to primary or relatively undisturbed forest was an important determinant of the number of large mammal species and average body mass of wildlife visiting agricultural fields [36].

Disturbance-tolerant small mammal species were also abundant in SC fallows [38]. The hoary-bellied squirrel, Calloscurius pygerythrus, a species known to colonize disturbed habitats in India, was frequently encountered in recently abandoned fallows, whereas forest-dependent species such as Malayan Giant squirrel, Ratufa bicolor, and the red-bellied squirrel, C. erythraeus, were only observed in 25-year and older secondary sites [38]. In contrast with findings of other studies, Medellin & Equihua [42] found that small (approx. 1–3 ha) and relatively young abandoned agricultural fallows (six years old), interspersed with mature forest supported forest-dependent as well as open-habitat small- and medium-sized mammal species. Further, only about a quarter of species sampled in the study were found exclusively in mature forest.

Wilkie and Finn [43] found few significant differences in mammal abundances between successional forests (15–20 years since cultivation suspension) and mature forests in Africa. However, large, wide-ranging species such as leopards, Panthera pardus, and okapis, Okapia johnstoni, were virtually absent in the former. Both successional and climax sites that were sampled in the study were within 2–6 km from a village and the authors mention that the areas were exploited by local hunters; therefore it is quite likely that hunting may have affected mammal abundances recorded in the sampled sites. Hunting has been reported as a practice often associated with SC [35, 39-40]. In Panama, Smith [27] found that about half the biomass of mammals hunted by the Buglé community was from agricultural fields. This included about a third each from active gardens, fallows younger than five years, and fallows older than five years [39].

Findings of studies that have covered mammal abundances in agricultural habitats suggest that effects of hunting by human communities override the factor of time since abandonment of a field [36, 42]. Therefore, further research on mammal recovery following SC, which incorporates effects of hunting, can contribute significantly to understanding the value of secondary habitats for mammals. It is also critical that studies on the effects of SC on mammals examine more explicitly, the effect of landscape factors such as distance to mature forest, contiguity and the size of patches on species diversity; richness and abundance, which are factors that may affect different groups of mammals differently [36].

Response of Avifauna

Avifaunal research in secondary habitats has produced varied results; studies have documented an increase or no effect [16-17, 43-45] as well as a decrease [10, 15, 46-48] in species diversity resulting from converting homogeneous forests into heterogeneous patches of fields and successional fallows interspersed with primary forest. Mean (SE) value of Jaccard's index of similarity increased from 0.26 (0.08) in 1–5 year fallows to 0.55 (0.12) in fallows younger than 25 years (Fig. 1). Within 25 years since suspension of cultivation, the proportion of mature forest species increased from 0.39 (0.14) to 0.71 (0.16), whereas the proportion of disturbance-associated species decreased from 0.58 (0.16) to 0.25 (0.11).

Our analysis of results of 10 studies revealed that birds accumulated relatively rapidly in the first 11 years, and the similarity between birds recorded from SC fallows and mature forest increased from about 25 to 55 percent within 25 years after suspension of cultivation (Figure 1). Within 25 years, only about 25 percent of disturbance-associated species were found in secondary sites and about 70 percent of mature forest species had recovered. Dunn [49], in a review of studies spanning recovery of faunal communities in secondary forests, reports that although species richness in a secondary site may recover within 20–40 years after abandonment, recovery of species composition may take a longer period. Guilds such as large-bodied canopy frugivores and understory insectivores are adversely affected by such habitat modification and are confined to primary or relatively undisturbed forests [10, 43, 45-46]. The type of agriculture system being practiced also affected bird species diversity in a SC landscape: the traditional SC mosaic of fallows and forest supported higher diversity and species evenness than a non-traditional practice with a fallow cycle of only four years and a cultivation period of up to six years [48].

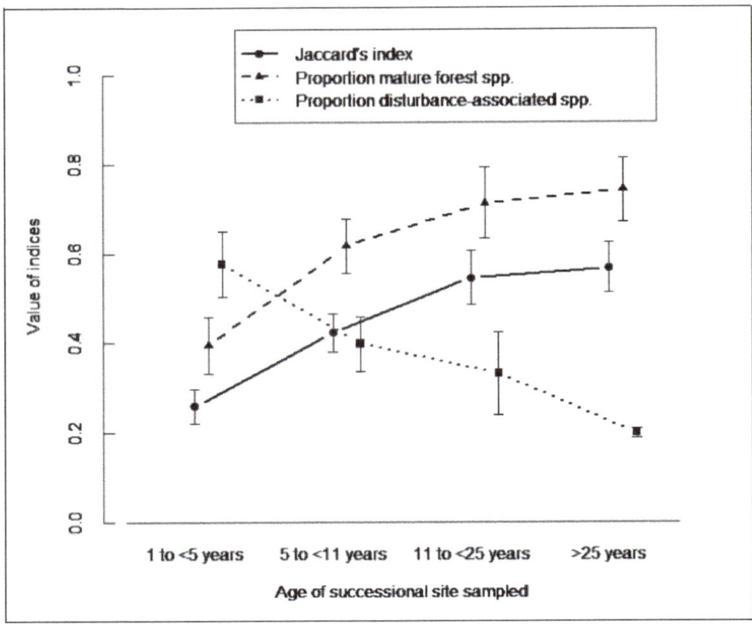

Fig. 1. Jaccard's index of similarity between bird species composition in successional sites and mature forests, proportion of mature forest species, and proportion of disturbance-associated species in different-aged successional sites. Each point represents average value (bars represent standard error) of these indices calculated from species inventories from 10 studies.

Hill and Hamer [50] suggest that results of studies that have recorded a decrease, no effect, or an increase in avifaunal diversity with conversion of homogeneous forest into a heterogeneous SC mosaic are often spatial scale-dependent. Studies that have sampled at smaller scales, mostly using transects or point counts [10 (5 ha), 47 (0.4 ha)], have reported decreased diversity whereas studies that sampled at larger scales generally using mistnets [17 (52 ha), 43 (60 ha)] have reported increased diversity. Barlow and others [51] identify other factors affecting study results; edge effects in a secondary site from surrounding habitat may bias results and near-primary forest considered as a control and assumed to be equivalent to mature undisturbed forest may have been disturbed, leading to an inaccurate comparison of habitats. In the specific case of impact of SC on birds, spatial scales of sampling for most studies are by default small, since across the world the practice is undertaken at relatively small scales in comparison with other land uses such as plantations that are often larger. When sampling scales are large, studies will incorporate effects of surrounding habitat and therefore reflect bird diversity in the agriculture-fallow-forest matrix rather than the diversity in each habitat specifically [17, 43].

IMPLICATIONS FOR CONSERVATION

Shifting cultivation fallows have been shown to recover vegetation faster in terms of basal area, regeneration, and accumulation of species than other human-modified and subsequently abandoned lands such as pastures, agroforestry sites, and plantations [52]. One factor responsible for this is likely to be that the practice is undertaken across the world at small scales in comparison with pastures (for instance, SC fallows in Meghalaya, north-east India ranged from 1–2.5 ha [6], whereas in Amazonia, pastures extended over hundreds to thousands of ha [21]). Another factor is that often SC fallows are interspersed with primary/near-primary forests [53].

Land-use conversions of forest for timber logging, monoculture cash crops, and infrastructural projects such as hydro-electric projects, usually benefit stakeholders not resident in the particular landscape, in comparison with SC which is often a subsistence activity [8]. Alternatives to the practice often recommended, such as intensive sedentary cultivation, monocultures, or agroforestry, have been suggested to possibly have long-term detrimental effect for both farmers and forest [52, 54-55].

SC can be considered as a small-scale high-intensity disturbance when relatively large tracts of primary forest adjoin the mosaic of fallows and active farms [17, 43]. Deleterious effects of SC such as forest and biodiversity loss, soil erosion, and similar environmental impacts are often consequences of shortened fallow cycles which are likely an offshoot of increasing human population. The practice of SC can be sustainable and can also support biodiversity in the overall landscape if large tracts of relatively mature forests are preserved, few trees are retained during cultivation, and fallow cycles are relatively long [21, 26]. Further, regulation of hunting of relatively rare species can ensure retention of high mammal diversity in a SC landscape [36]. Mature forests can act as critical sources for recolonization of plants, mammals, birds, and other groups in secondary sites and therefore can maintain biodiversity in the larger landscape.

ACKNOWLEDGEMENTS

We thank Dr. Aparajita Datta for critical comments and Dr. Brian Finegan, Dr. Divya Mudappa, and Kulbushansingh Suryawanshi for suggesting improvements to the manuscript. We also thank the Director, GBPIHED, for logistic support and the Ministry of Environment & Forests for funding a project monitoring recovery of vegetation and fauna in the Eastern Himalayas. We thank two anonymous reviewers for helpful comments on early drafts of this manuscript.

REFERENCES

1. Mazoyer, M. and Roudart, L. Eds. 2006. A history of world agriculture from the neolithic to the current crisis, pp. 101-102, Earthscan publishers, London.
2. Spencer, J. E. 1966. Shifting cultivation in southeastern Asia, pp. 7-8, University of California Press (republished by Bishen Singh Mahendra Pal Singh publications, Dehradun).
3. Conklin, H. C. 1961. The study of shifting cultivation. Current Anthropology 2: 27-61.
4. Myers, N. Ed. 1991. Tropical forests and climate, pp. 3-21, Kluwer Publishers, London, UK.
5. Brady, N. C. 1996. Alternatives to slash-and-burn: A global imperative. Agricultural Ecosystems and Environment 58: 3-11.
6. Ramakrishnan, P. S. 1992. Shifting agriculture and sustainable development: An interdisciplinary study from North-eastern India, UNESO-MAP series, Paris (republished by Oxford University Press, New Delhi, India).
7. Bandy, D. E., Garrity, D. P. and Sanchez, P. A. 1993. The worldwide problem of slash-and-burn agriculture. Agroforestry Today 5: 2-6.
8. Dressler, W. 2005. Disentangling Tagbanua lifeways, swidden and conservation on Palawan island. Human Ecology 12: 21-29.
9. Ickowitz, A. 2006. Shifting cultivation and deforestation in tropical Africa: critical reflections. Development and Change 37: 599-626.
10. Raman, T. R. S., Rawat, G. S. and Johnsingh, A. J. T. 1998. Recovery of tropical rainforest avifauna in relation to vegetation succession following shifting cultivation in Mizoram, NorthEast India. Journal of Applied Ecology 35: 214-231.
11. Whitmore, T. C. 1984. Tropical rainforests of the Far East, pp. 276-278, Oxford Science publications, Clarendon Press, Oxford.
12. Cincotta, R. P., Wisnewski, J. and Engelman, R. 2000. Human population in the biodiversity hotspots. Nature 404: 990-992.
13. Brown, S. and Lugo, A. E. 1990. Tropical secondary forests, Journal of Tropical Ecology 6: 1-21.
14. Karr, J. R. 1976. Within- and between-habitat avian diversity in African and Neotropical lowland habitats. Ecological Monographs 46: 457-481.
15. Blankespoor, G. W. 1991. Slash-and-burn shifting agriculture and bird communities in Liberia, West Africa. Biological Conservation 57: 41-71.
16. Anderson, D. L. 2001. Landscape heterogeneity and diurnal raptor diversity in Honduras: the role of shifting cultivation. Biotropica 33: 511-519.
17. Blake, J. G. and Loiselle, B. A. 2001. Bird assemblages in second-growth and old-growth forests, Costa Rica: perspectives from mist nets and point counts. The Auk 118: 304-326.
18. Del Castillo, R. F. and Rios, M. A. P. 2008. Changes in seed rain during secondary succession in a tropical montane cloud forest region in Oaxaca, Mexico. Journal of Tropical Ecology 24: 433-444.
19. Kammesheidt, L. 1999. Forest recovery by root suckers and above-ground sprouts after slashand-burn agriculture, fire and logging in Paraguay and Venezuela. Journal of Tropical Ecology 15: 143-157.
20. Finegan, B. 1996. Pattern and process in neotropical secondary rain forests: the first 100 years of succession. TREE 11: 119-124.
21. Uhl, C. 1987. Factors controlling succession following slash-and-burn agriculture in Amazonia. Journal of Ecology 75: 377-407.
22. Vieira, I. C. G. and Proctor, J. 2007. Mechanisms of plant regeneration during succession after shifting cultivation in eastern Amazonia. Plant Ecology 192: 1573-5052.
23. Toky, O. P. and Ramakrishnan, P. S. 1983. Secondary succession following slash and burn agriculture in north-eastern India: I. Biomass, litterfall and productivity. Journal of Ecology 71: 735-745.
24. Saldarriaga, J. G., West, D. C., Tharp, M. L. and Uhl, C. 1988. Long-term chronosequence of forest succession in the upper Rio Negro of Columbia and Venezuela. Journal of Ecology 76: 938- 958.
25. Van Gemerden, B. S., Shu, G. N. and Olff, H. 2003. Recovery of conservation values in central African rainforest after logging and shifting cultivation. Biodiversity and Conservation 12: 1553- 1570.
26. N'Dja, K. Justin, K. and Guillaume, D. 2008. Successional patterns of plant species and community diversity in a semi-deciduous tropical forest under shifting cultivation. Journal of Vegetation Science 19: 809-820.
27. Kennard, D. K. 2002. Secondary forest succession in a tropical dry forest: patterns of development across a

50-year chronosequence in lowland Bolivia. Journal of Tropical Ecology 18: 53-66.

28. Ingle, N. R. 2003. Seed dispersal by wind, birds, and bats between Philippines montane rainforest and successional vegetation. Oecologia 134: 251-261.

29. Tabarelli, M. and Peres, C. A. 2002. Abiotic and vertebrate seed dispersal in the Brazilian Atlantic forest: implications for forest regeneration. Biological Conservation 106: 165-176.

30. Aide, T. M. and Cavelier, J. 1994. Barriers to lowland tropical forest restoration in the Sierra Nevada de Santa Marta, Columbia. Restoration Ecology 2: 219-229.

31. Lawrence, D. 2004. Erosion of tree diversity during 200 years of shifting cultivation in Bornean rainforest. Ecological Applications 14: 1855-1869.

32. Hammond, D. S. 1995. Post-dispersal seed and seedling mortality of tropical dry forest trees after shifting agriculture, Chiapas, Mexico. Journal of Tropical Ecology 11: 295-313.

33. Fukushima, M., Kanzaki, M., Hara, M., Okhubo, T., Preechapanya, P. and Choocharoen, C. 2008. Secondary forest succession after the cessation of swidden cultivation in the montane forest area in northern Thailand. Forest Ecology and Management 255: 1994-2006.

34. Saxena, K. G. and Ramakrishnan, P. S. 1984. Herbaceous vegetation development and weed potential in slash and burn agriculture (jhum) in N.E. India. Weed Research 24: 135-142.

35. Rao, K. S., and Ramakrishnan, P. S. 1989. Role of bamboos in nutrient conservation during secondary succession following slash and burn agriculture (jhum) in north-east India. Journal of Applied Ecology 26: 625-633.

36. Naughton-Treves, L., Mena, J. L., Treves, A., Alvarez, N. and Radeloff, V. C. 2003. Wildlife survival beyond park boundaries: The impact of slash-and-burn agriculture and hunting on mammals in Tambopata, Brazil. Conservation Biology 17: 1106-1117.

37. Parry, L., Barlow, J. and Peres, C. A. 2007. Large vertebrate assemblages of primary and secondary forests in Brazilian Amazon. Journal of Tropical Ecology 23: 653-662.

38. Raman, T. R. S. 1996. Impact of shifting cultivation on diurnal squirrels and primates in Mizoram, northeast India: A preliminary study. Current Science 70: 747-750.

39. Smith, D. A. 2005. Garden game: shifting cultivation, indigenous hunting and wildlife ecology in western Panama. Human Ecology 33: 505-537.

40. Linares, O. 1976. "Garden Hunting" in the American Tropics. Human Ecology 4: 331-349.

41. Gupta, A. K. and Kumar, A. 1994. Feeding ecology and conservation of the Phayre's leaf monkey Presbytis phayrei in Northeast India. Biological Conservation 69: 301-306.

42. Medellin, R. A. and Equihua, M. 1998. Mammal species richness and habitat use in rainforest and abandoned agricultural fields in Chiapas, Mexico. Conservation Biology 35: 13-23.

43. Wilkie, D. S. and Finn, J. T. 1990. Slash-burn cultivation and mammal abundance in the Ituri forest, Zaire. Biotropica 22: 90-99.

44. Andrade, G. I. and Rubio-Torgler, H. 1994. Sustainable use of the tropical rainforest: evidence from the avifauna in a shifting-cultivation habitat mosaic in the Columbian Amazon. Conservation Biology 8: 545-554.

45. Naidoo, R. 2004. Species richness and community composition of songbirds in a tropical forestagricultural landscape. Animal Conservation 7: 93-105.

46. Smith, A. L., Ortiz, J. S. and Robertson, R. J. 2001. Distribution patterns of migrant and resident birds in successional forests of the Yucatan Peninsula, Mexico. Biotropica 33: 153-170.

47. Borges, S. H. 2007. Bird assemblages in secondary forests developing after slash-and-burn agriculture in the Brazilian Amazon. Journal of Tropical Ecology 23: 469-477.

48. Bowman, D., Woinarski, J. C. Z. Sands, D. P. A. Wells, A. and McShane, V. J. 1990. Slash-and-burn agriculture in the wet coastal lowlands of Papua New Guinea: Response of birds, butterflies and reptiles. Journal of Biogeography 17: 227-239.

49. Zhijun, W. and Young, S. S. 2003. Differences in bird diversity between two swidden agricultural sites in mountainous terrain, Xishuangbanna, Yunnan, China. Biological Conservation 110: 231- 243.

50. Dunn, R. R. 2004. Recovery of faunal communities during tropical forest regeneration. Conservation Biology 18: 302-309.

51. Hill, J. K. and Hamer, K. C. 2004. Determining the impacts of habitat modification on diversity of tropical forest fauna: the importance of spatial scale. Journal of Applied Ecology 41: 744-754.

52. Barlow, J., Mestre, L. A. M., Gardner, T. A. and Peres, C. A. 2007. The value of primary, secondary and plantation forests for Amazonian birds. Biological Conservation 136: 212-231.

53. Ferguson, B. G., Vandeermeer, J., Morales, H. and Griffith, D. M. 2001. Post-agricultural succession in El-Petén, Guatemala. Conservation Biology 17: 818-828.

54. Finegan, B. and Nasi, R. 2004. The biodiversity and conservation potential of shifting cultivation landscapes. In: Agroforestry and biodiversity conservation in tropical landscapes. Schroth, G., da Fonseca, G. A. B. Harvey, C. A., Gascon, C., Vasconcelos, H. L. and Izac, A. M. N. (Eds), pp. 153- 197. Island press, Washington, D. C.

55. Robichaud, W. G., Sinclair, A. R. E., Odarkor-Lanquaye, N and Klinkenberg, B. 2009. Stable forest cover under increasing populations of swidden cultivators in central Laos: the roles of intrinsic culture and extrinsic wildlife trade. Ecology and Society 14: 33.

56. Teegalapalli, K. 2008. Shifting livelihood options and changing attitudes of communities in the Garo hills, western Meghalaya. Current Conservation 2.3: 18-21.

CHAPTER 2

CONSERVATION AGRICULTURE FOR COMBATING LAND DEGRADATION IN CENTRAL ASIA: A SYNTHESIS

A. Nurbekov[1], A. Akramkhanov[1], A. Kassam[2], D. Sydyk[3], Z. Ziyadaullaev[4], J.P.A. Lamers[5]

[1] International Center for Agricultural Research in the Dry Areas (ICARDA) Central Asia and the Caucasus Regional Office, Tashkent, Uzbekistan

[2] School of Agriculture, Policy and Development, University of Reading, United Kingdom

[3] South-Western Research Institute of Livestock and Crop Production, Chimkent, Kazakhstan

[4] Kashkadarya Research Institute of Breeding and Seed Production of Cereals, Karshi, Uzbekistan

[5] Center for Development Research, Bonn, Germany

ABSTRACT

This manuscript reviews scientific findings on agricultural systems, associated land degradation and selected remedies such as Conservation Agricultural (CA) practices to counterbalance these. In particular, this review addresses the research findings on CA practices conducted in the rainfed and irrigated systems in Central Asia. The arid and semi-arid croplands in this region are vulnerable to different types of soil and environmental degradation, and particularly to degradation caused by intensive tillage, irrigation water mismanagement, and cropping practices, especially in the Aral Sea Basin. Overall, the evidence shows that various CA elements, such as permanent beds, seems to be technically suitable for the major cropping systems and despite the heterogeneous conditions in the region. CA practices can contribute to combating on-going land degradation. No-till seeding along with the maintenance of a permanent soil coverage e.g. by residue retention, reduces wind and water erosion, increases water infiltration and storage which can reduce crop water stress, improve soil quality and increase soil organic matter. Further, CA practices can lead to similar or even higher crop yields while reducing production resource needs and costs considerably, including fuel, seeds, agrochemicals, water and labor. Nevertheless, the growing research evidence on the productivity, economic and environmental benefits that can be harnessed with CA, still is from a limited number of studies and hence more research at local scale is needed.

INTRODUCTION

Central Asia comprises the five independent republics Kazakhstan, Kyrgyzstan, Tajikistan, Turkmenistan and Uzbekistan. Six of the agro-climatic zones in this vast region occupy about 90% of the entire region. Two of the agro-climatic zones (Semi-arid-cold winter-warm summer, and Arid-cold winter-warm summer) alone occupy up to 60% (Table 1). The climate in Central Asia is mostly arid and semi-arid, and strongly continental, with long, hot summers and short, cold winters. Average annual precipitation, which is concentrated in winter and spring, is about 270 mm, but varies from 80-150 mm in the arid regions to 600 to 800 mm in the semi-arid mountainous zones. The land area of the five countries covers about 393 M ha (Table 2).

Citation: A. Nurbekov, A. Akramkhanov, A. Kassam, D. Sydyk, Z. Ziyadaullaev, J.P.A. Lamers. Conservation Agriculture for combating land degradation in Central Asia: a synthesis. AIMS Agriculture and Food, 2016, 1(2): 144-156. doi: 10.3934/agrfood.2016.2.144.

Table 1. Main agro-climatic zones and extent of land area under CA. (Adapted from [4])

Agro-climatic zones§	**Total area, ha**	**Area planted with elements of CA, ha**	**Description of the elements**	**Source**
Semi-arid, cold winter, warm summer	151,387,760	13,700,000§	Including 2,100,000 ha of no-tillage, i.e. direct seeding of spring wheat and barley	Ministry of Agriculture of Kazakhstan
Semi-arid, cold winter	*26,419,800*			
Sub-humid, cold winter	23,617,700			
Arid, cold winter, warm summer	*123,027,520*	< 700,000§	Conservation tillage, sowing of winter wheat into standing cotton	Ministry of Agriculture and Water Resources of Uzbekistan, Ministry of Agriculture of Tajikistan and Ministry of Agriculture and Melioration of the Kyrgyz Republic
Arid, cool winter, warm summer	*19,572,560*			
Semi-arid, cool winter, warm summer	*5,991,600*			
§for detailed description of the agro-climatic zones, see [4] and [5].				

Table 2. Land resources, population and various agricultural indicators of five Central Asian countries

Country	**Total territory (M ha)**	**Land area (M ha)**	**Cropland (M ha)**	**Irrigated land (M ha)**	**Rainfed land (M ha)**	**Population (M)**	**% rural population**	**Per capita cropland (ha)**	**% Agric. GDP**
Kazakhstan	272.5	269.7	24	1.6	22.1	17.01	42.8	1.41	5.3
Kyrgyzstan	20.0	19.2	1.4	1.1	0.3	5.93	63.4	0.23	25.8
Tajikistan	14.2	14.0	0.9	0.7	0.2	8.61	71.4	0.10	19.8
Turkmenistan	48.8	47.0	1.8	1.8	0.0	5.41	72.0	0.33	22.1
Uzbekistan	44.7	42.5	4.9	4.3	0.5	31.00	63.5	0.15	19.4
Total	400.3	392.7	33	9.5	23.1	68.05			
Mean							62.6	0.48	9.9

Mikhalev and Reimov postulated that Central Asia's drylands are to be regarded as dry steppe, semi-desert, desert, and salt marshes, which are known to be vulnerable to different kinds of degradation including soil degradation (here defined as a loss of fertility, or increase in salinization or waterlogging), degradation of pastures (due to overgrazing and excessive agricultural and firewood harvest), degradation of forests (due to illegal logging, fires, grazing, erosion), and erosion, landslides, and mudflows [1]. Extensive and intensive land use during the Soviet Union period (1924-1990) and disorganized land management systems introduced after the collapse of the Soviet Union (1991) worsened land degradation in Central Asia. Mono-cropping and a production strategy aimed at increasing the production of agricultural commodities without considering environmental consequences have been listed as major causes of past and on-going land degradation [2,3].

Gupta et al. argued that during the post-Soviet period, the three primary causes of land degradation included the (i) mismanagement and over-use of natural resources, (ii) insufficiency of economic infrastructures and market mechanisms, and (iii) insufficient development of capacity and weak inter-sector coordination [11]. Despite the control of areas by governmental agencies during Soviet reign, the on-going land degradation could not be stopped and therefore remained high on the political agendas of the countries in Central Asia after independence.

Areas under land degradation are wide spread in the arid and semi-arid zones of the Central Asian countries, and comprising over 80% of the agricultural area [12]. Some 68% of the agricultural land in the region is degraded due to erosion and increased salinity [3] (Table 3). The degradation of agricultural land in Kazakhstan and Uzbekistan has amounted to 73% and 44%, respectively, mainly caused by increased soil salinity, erosion and loss of vegetation cover. Most of the land resources in Kyrgyzstan and Tajikistan are prone to erosion due to the high proportion (above 90%) of mountainous areas in these two countries (Table 3). The types of degradation within a country vary according to land use type. However, the largest portion of the degraded land is in response to improper farming practices [3]. In particular, the numerous soil tillage practices, which invert the soil using heavy machinery with high ground pressures, agricultural practices that neglect to protect the soil surface, and the insufficient supply of organic material to the soil has resulted in increased soil erosion, decrease of topsoil depth and increase in salinity, with consequent losses in soil fertility and land value. In addition, poor irrigation management have resulted in soil degradation due to waterlogging (Table 3).

Table 3. Soil degradation in Central Asia, %.

Type of soil degradation	Percent of Agricultural land				
	Kazakhstan	Kyrgyzstan	Tajikistan	Turkmenistan	Uzbekistan
Erosion	17.2*	85.0*	75.0*	8.9*	11.6*
Waterlogging	8.7**	1.0**	3.6**	3.2**	8.7**
Salinity	47.7**	0.5**	5.0**	19.1**	24.4**
Total	73.6**	86.5**	83.6**	31.2**	44.7**

The on-going degradation of soil resources in Central Asia is not only widespread, but represents a direct threat to the productive capacity and sustainability of the agricultural production base in the region. Substantial improvements in soil management are therefore direly needed to counter these threats.

Conservation Agriculture (CA) has the potential to provide various tools to combat soil degradation as well as raise productivity and resilience, and reduce production costs [5,14,15,16]. Conservation Agriculture rests on three interlinked principles:

- no or minimal mechanical disturbance of the soil through no-till direct seeding to maintain the quality and productivity of the soil, which is at the base of all CA-based farming practices and systems;
- maintenance of a permanent soil mulch cover with for instance plant residues including stubbles or cover crops to improve infiltration, reduce water loss and erosion, protect the soil from harsh climate extremes, and serve as a substrate for soil microorganisms and fauna; and
- diversified cropping systems over time (rotations, sequences) and space (associations) to further strengthen the systems' resilience against biotic and abiotic threats [16,17,18]. In this way, CA practices provide important benefits to the environment and the land user alike [18,19,20].

COMBATING SOIL DEGRADATION WITH CA

Originally, CA practices were promoted to combat soil degradation and erosion resulting from tillage that caused the destruction of soil structure and aggregate stability, deplete soil organic matter and soil biological health. In later years, CA helped to reduce production costs, and raise productivity (yield and efficiency) [17]. During the past 20 years, CA has spread across all continents and most agro-ecological zones, particularly in North and South America and in Australia, but more recently also in Asia, Africa and Europe. In 2013, CA was used globally on 155 M ha of annual cropland, corresponding to about 11% of global annual cropland [21,22]. About 50% of the CA area is located in developing countries.

In Central Asia, the research on crop residue management under no-till and its effect on soil erosion is still in its infancy. Yet, a review of a wealth of literature from outside Central Asia, illustrates numerous benefits. Hence, CA has been shown to be an innovative approach that helps in reducing soil erosion, improving water use efficiency as well as soil quality and helps in increasing soil organic matter, decreasing energy use and above all improving crop and land productivity and in turn the income of (resource poor) farmers [22,24]. For instance, soil erosion in Brazil decreased from 3.4-8.0 t ha^{-1}under conventional tillage to 0.4 t ha^{-1}under CA, while water loss decreased from approximately 990 to 170 t ha^{-1} [25,26]. The reduction in soil erosion led to enhanced surface and ground water quality whilst crop residues retention on the surface helps in holding soil particles in place and keeping any applied plant nutrients and pesticides on the field.

The overarching experimental evidence from the many different production environments worldwide demonstrate that CA-based management can have both immediate (e.g. reduced production costs, reduced erosion, stabilized crop yield, and improved water productivity) and long-term benefits (e.g. higher soil organic matter contents and improved soil structure), although the magnitude of these benefits tends to be site and year specific depending on the nature of the initial status of land degradation and the prevailing yield level [27,28,29,30]. Increasingly, CA is considered to be climate-smart also, because of its better adaptability to climate change, and as a means to reach a sustainable intensification of agricultural production with minimum negative impacts on the environment [31,32]. As such, CA is a means for the integration of ecological management with modern, scientific, agricultural production practices. This holistic embrace of knowledge, as well as the capacity of farmers to apply this knowledge, innovate and adjust to evolving local conditions, ensures the sustainability of those who practice CA. A major strength of CA is furthermore the option of a step-wise implementation by farmers of complementary, synergetic soil husbandry practices that build to a robust, cheaper, more productive and environmentally friendly farming system. Therefore, CA practices have an important role to fulfill in the production systems of Central Asia.

EFFECT OF CA ON SOIL QUALITY AND LAND DEGRADATION PROBLEMS

Soil organic matter (SOM) dynamics: Worldwide evidence has also shown the main benefits of no-till or low soil disturbance tillage on soil organic matter and soil carbon (C) interactions. The maintenance of these important parameters for soil quality depends on a permanent soil mulch cover developed through crop residue retention or cover crops, which is one of the three main principles of CA. In CA in Central Asia, the use of cover crops is not fully developed yet, but crop residues including stubble are retained on the soil surface after harvest where they benefit soil properties and crops, as shown by numerous field investigations [33,34]. Keeping crop residues on the soil surface reduces soil losses, protects the soil from water and wind erosion, and adds organic matter to the soil both in the rainfed and irrigated agriculture conditions worldwide [35,36].

In CA, no-till, direct seed drilling is the only mechanical operation causing disturbance to the soil surface. All other operations that are normally employed under "conventional tillage agriculture" in the rainfed areas of Kazakhstan such as sweep tillage, disking and harrowing, are thus not included in CA [2]. Intensive experiences with CA practices in Central Asia date from the year 2000 onward. The concept of CA within the irrigated areas of CA has taken some time to become accepted, which has delayed the experimentation and the documentation thereof.

Organic matter is one of the major indicators of soil quality and biological health, which affects, among other factors, crop yield and the ability of soils to resist erosion. A number of researchers have investigated the impact of different tillage systems on soil organic matter (SOM). There is general agreement that no-till can increase SOM as shown in arid and semi-arid regions in and outside the Central Asia region. Hernanz et al.[37], for instance, conducted a long term experiment in a semi-arid area of Spain using different tillage methods and reported that under no-till with mulch cover, the SOM at a depth of 0-10 and 20 cm had higher organic contents compared to conventional tillage [37]. Numerous results from the irrigated areas of Central Asia showed that crop residue retention improves SOM and soil N content [38,39,40]. The CA practices examined in Central Asia increased SOM significantly with corresponding improvements in soil structure and greater soil moisture holding capacities [38,39].

Most beneficial effects on soil physical properties reported due to plant residue retention were the positive influences on soil quality, decreasing soil bulk density, increasing soil moisture retention, and increasing biological activity of the soil. That is why a general preservation of crop residues, irrespective of its make-up, improved physiological and biological properties of the soil, which in turn significantly increased soil fertility [41]. More recently, the positive impact of no tillage and crop residue management on properties of a silty loam soil under irrigation in Uzbekistan was reported for a rotation of winter wheat and maize for two years followed by cotton for another two years [42].

Soil salinization: The on-going soil salinization in the irrigated areas of Central Asia is predominantly caused by the capillary rise of the ground water. This is the major cause of the on-going cropland degradation, especially in the Aral Sea basin [43,44]. A mulching experiment with crop residues decreased soil salinity under the irrigated conditions of Uzbekistan [45]. Pulatov et al. [39] reported that after four years, a no-till CA system had the lowest soil salinity level of all practices tested, i.e., no tillage and residue retention, which influenced also the location and accumulation of salts by reducing evaporation and the upward salt transport in the soil [39].

Soil erosion: Increasing the SOM content and maintaining crop residues on the soil surface also reduced wind erosion [47]. Depending on the amount of crop residues retained on the soil surface, soil erosion could be reduced to insignificant levels compared to the unprotected, intensively tilled exposed fields [48] and this benefit can be harnessed in Central Asia as well. Water erosion too enhances soil degradation in Central Asia, especially on hilly areas and under irrigated conditions. The effective CA practices showed for many years to constitute a promising set of improved and financially feasible methods of crop production, which concurrently reduced wind and water erosion [49]. The regularly occurring wet springs in much of northern Kazakhstan resulted in severe soil erosion of exposed soil surface in fallowed fields [50]. Although information about the effect of slopes is lacking, in general where they are considered to be long, they resulted in water accumulation in the lower parts and in increasing the velocities of runoff water. However, with crop residue retention, soil erosion could be reduced drastically on the cropped areas [50]. Nevertheless, water and wind erosion studies remain rare in Central Asia despite having been acknowledged as being a core reason for on-going soil degradation (Table 3). Based on the research findings and lessons learned from different agro-climatic regions, several remedies could be examined for adoption in the region.

CROP YIELD UNDER CONSERVATION AGRICULTURE PRACTICES

Early research from similar semi-arid environments showed the yield enhancing effects on barley of reduced and zero tillage systems compared to conventional practices [51]. Crop yields after four years of permanent bed planting in North-western Uzbekistan was 20% higher with zero tillage system compared to the conventional tillage methods [52]. Although results from numerous findings of CA practices on crop yields have been mixed, in the end crop yield is a critical assessment criterion for farmers. Hence, more research needs to be directed towards yield and its parameters under irrigated agriculture. Similarly, for the rainfed areas, results have been promising, but still are sparse. For example, from 1992 to 1995, minimum soil disturbance tillage techniques were introduced and tested particularly in the northern, rainfed parts of Kazakhstan. Excellent results were obtained throughout the areas cultivated with minimized soil disturbance, resulting in both economic savings and increased crop yields [53].

Many research results from the irrigated areas in Central Asia indicated that bed planting practices improved wheat yields, increased fertilizer efficiency, reduced herbicide use, saved seeds, reduced water demands (on average 30%), and reduced production costs by 25-35%[54,55,56]. According to the Ministry of Agriculture of Kazakhstan, CA and conservation tillage practices were applied on some 11.7 M ha (Table 3), which is 70% of the total area sown to wheat in Kazakhstan [57] (Figure 1). Consequently, the country harvested a record gross output of grain of 20 M t, corresponding to a yield of 1.7 t ha^{-1} [57]. Hence, CA practices may have contributed to these increased yields and output, although the area under full CA in Kazakhstan is only 2.1 M ha. Results from Tajikistan and Kyrgyzstan showed 25-38% higher wheat yields under raised bed and no-till planting conditions compared to the traditional, tillage-driven planting [54,56].

Figure 1. Rainfed, no-till winter wheat in Kazakhstan (2008). Photo by Aziz Nurbekov.

In addition to yield increases, seeding rates under CA in Kyrgyzstan could be reduced by 50% while irrigation water requirement could be lowered by 27% [54]. Similar results were reported in the irrigated conditions of Tajikistan [56]. On the other hand, Nurbekov et al. reported that the application rate of N had no significant effect on winter wheat yields in no-till and conventional systems in Uzbekistan [40]. The yields with 120 kg N ha^{-1} rates turned out to be as good as with 140 kg N ha^{-1} under conventional practices using mouldboard ploughs, while with no-till practices a slight increase in grain yields was observed with the higher N rates. Nurbekov et al. reported that winter wheat yields increased with no-till compared to conventional tillage system [40]. Sanginov and Khalikov, carriying out research on the planting of winter wheat before the harvest of cotton in Yavan and Gozimalik districts of Tajikistan, reported that wheat growth and development under no-tillage system resulted in savings of seed quantity and in increased yield [58]. The adoption of CA methods could thus bring about significant productivity and environmental benefits [41].

So far, only Kazakhstan has issued supportive policies to introduce and spread CA practices and this has increased the area under CA-based practices from virtually none in 2001 to 2.1 M ha in 2013. The other four countries in Central Asia are only gradually moving towards the adoption of supportive policies on CA and in general, a wide-spread adoption of CA is still pending and would need more extension and research support [60].

Permanent raised bed planting in Uzbekistan consists of raised beds that have been prepared and used during a previous season and subsequently used for growing the next crop (Figure 2). Over the last 20 years, Uzbekistan has been researching different ways of introducing grain crops into the existing crop rotations, which included cotton and alfalfa mainly, albeit predominantly during the Soviet Union epoch. However, since 1990s, winter wheat, previously grown under rainfed conditions only in Central Asia, is being cultivated also under irrigation. Research findings showed that a timely, no-till planting of winter wheat in standing cotton is a promising relay cropping practice. As a consequence, the area under this cotton-winter wheat relay cropping has now reached some 600,000 ha annually [59]. Several development projects in Central Asia, supported by the international donor community, currently include the promotion of CA with permanent raised beds system as part of their priority activities, but according to many these efforts need to be intensified [60].

Figure 2. Permanent bed planted winter wheat in Uzbekistan (2012). Photo by Aziz Nurbekov.

Hence, despite the numerous positive research results, CA is still not widely practices among the farming population in the irrigated areas of Central Asia. This is partly due to a predetermined mindset but also due to the relative complexity of CA practices compared to conventional tillage agricultural practices.

CONCLUSIONS

Current research evidence from the rainfed areas of Central Asia, shows that CA practices are promising to combat a series of flaws in the existing cropping systems. However, much less research evidence exists for the irrigated areas even though such research has introduced in all five Central Asian countries and while covering the heterogeneous local conditions. These preliminary research results, albeit limited to a few locations, show the potential for achieving similar, or even higher crop yields over time. The CA practices favored, such as permanent no-till beds, showed their effectiveness in lowering the rate of land degradation caused by soil salinization. Research on CA practices and its role in combating the on-going water and wind erosion have not been placed high on the research agendas yet. However, the maintenance of a soil coverage by residues reduces wind and water erosion, increases water infiltration and storage capacity, which helps reducing crop water stress, improves soil quality and increases organic matter. These benefits are promising to the scientists in the first place, but not yet to farmers! The findings underscored furthermore that CA is not a single, uniformly applicable technology that can be immediately applied anywhere and in a standard manner. Rather, it represents a set of principles that encourage the formulation of locally adapted practices, approaches and methods, which need to be tested, evaluated and then adopted or implemented not only under various climatic setting but also while considering the socio-economic conditions. Hence, also socio-economic research has to be promoted for instance when addressing the residue management component since this needs to be packaged into an easily adoptable technology, acceptable to farmers. Finally, as is evidenced in Kazakhstan, encouraging policies are needed as well as an effective and functioning agricultural extension system, which only is in its infancy in most Central Asia countries.

Further research is needed across the agro-climatic zones that should address in detail the effects of various types of CA crop rotations and mulch covers on weed management, on nutrients, pests and water management, on residue levels, sowing depths, dates and density, and on fertilizer and irrigation rates. Needless to repeat the importance of an impact assessment on livelihoods and environmental conditions including the potential of integrating trees and timber production, pastures and livestock into CA farming systems particularly with small-scale farmers.

REFERENCES

1. Mikhalev V, Reimov A. Land Degradation in Central Asia. 2008. Available from: http://www.developmentandtransition.net/
2. Pender J, Mirzabaev A, Kato E (2010) Central Asian Countries Initiative for Land Management Multicountry Partnership Framework Support Project: Economic Analysis of Sustainable Land Management Options in Central Asia. Asian Development Bank, Manila, Philippines.169.
3. Qushimov B, Rustamova IM, Haitov B(2007) Land degradation by agricultural activities in Central Asia. In: Rattan Lal et al. (Eds.). *Climate change and terrestrial carbon sequestration in Central Asia.*Taylor & Francis, London.
4. De Pauw E(2008) ICARDA regional GIS datasets for Central Asia: Explanatory notes. GIS Unit Technical Bulletin. International Centre for Agricultural Research in the Dry Areas (ICARDA).
5. Nurbekov A, Akramkhanov A, Lamers J, et al.(2013) Conservation Agriculture in Central Asia. In: Jat RA, Sahrawat KL, Kassam AH (Eds.).*Conservation Agriculture: Global Prospects and Challenges*,Wallingford:CABI, 223-247.
6. Kazakhstan State committee on Statistics. 2015. Available from:http://www.stat.gov.kz/faces/wcnav_externalId/publicationsPage?_afrLoop=30154916533175647#%40%3F_afrLoop%3D30154916533175647%26_adf.ctrl-state%3D5am21wla0_70
7. Kyrgyzstan State Agency. Statistical Yearbook of the Kyrgyz Republic. 2016. Available from: http://www.stat.kg/en/publications/statisticheskij-ezhegodnik-kyrgyzskoj-respubliki/
8. Tajikistan Statistical Committee. Tajikistan in Figures. 2015. Available from: http://www.stat.tj/en/img/695c206e2b1ce86f333f33fdc268a469_1439617381.pdf
9. Turkmenistan State Agency on statistics. 2015. Available from: http://www.stat.gov.tm/assets/plan/stat_work_2016_ru.pdf
10. The State Committee of the Republic of Uzbekistan on statistics (2015) "O'zbekiston raqamlarda". Tashkent. 211.
11. Gupta R, Kienzler K, Martius C, Mirzabaev A, et al. (2009) Research Prospectus: A Vision for Sustainable Land Management Research in Central Asia. ICARDA Central Asia and Caucasus Program. Sustainable Agriculture in Central Asia and the Caucasus Series 1. CGIAR-PFU. Tashkent. Uzbekistan, 84.
12. Bekturova G, Romanova O (2007) Traditional Land Management Knowledge in Central Asia: Resource pack. Almaty: S-Print. 2007, 86.
13. Islamic Research and Information Center. 2006. Available from:http://www.iric.org
14. Karabayev M (2008) Improvement of soil and water management in Kazakhstan: Conservation Agriculture for wheat production and crop diversification. In: Proceedings of the International Technical Workshop on Investing in Sustainable Crop Intensification: The Case for Improving Soil Health. 22-24 July 2008. Integrated Crop Management Vol. 6. FAO, Rome.
15. Fileccia T (2009) Conservation Agriculture and Food Security in Kazakhstan. Working Paper. FAO Investment Centre, Rome.
16. Kassam AH, Friedrich T, Derpsch R, et al (2012) Conservation Agriculture in the dry Mediterranean climate. *Field Crops Res* 132: 7-17.
17. Kassam AH, Friedrich T, Shaxson F, et al. (2009) The spread of conservation agriculture: justification. sustainability and uptake. *Int J Agric Sust* 7: 292-320.
18. González-Sánchez EJ, Veroz-González O, Blanco-Roldan GL, et al. (2015) A renewed view of conservation agriculture and its evolution over the last decade in Spain. *Soil Till Res* 146: 204-212.
19. Cerdà A. Lessons and experience of soil conservation in Spain. 2004. Available from: http://eusoils.jrc.ec.europa.eu/projects/scape/CT_book.pdf
20. González-Sánchez EJ, Ordónez-Fernández R, Carbonell-

Bojollo R, et al. (2012) Meta-analysis on atmospheric carbon capture in Spain through the use of conservation agriculture. *Soil Till Res* 122: 52-60.

21. Friedrich T, Derpsch R, Kassam AH (2012) Global overview of the spread of Conservation Agriculture. *Field Actions Science Reports* 6: 1-7.
22. Kassam AH, Friedrich T, Derpsch R, et al. (2014) Worldwide adoption of Conservation Agriculture. 6th World Congress on Conservation Agriculture. 22-17 June 2014, Winnipeg, Canada.
23. Lal R (1997) Residue management. conservation tillage and soil restoration for mitigating greenhouse effect by CO2-enrichment. *Soil Till Res* 43: 81-107.
24. Lal R (2004) Soil carbon sequestration impacts on global climate change and food security. *Science* 304: 1623-1627.
25. Sorrenson WJ (1997) Financial and economic implications of no-tillage and crop rotations compared to conventional cropping systems. TCI Occasional Paper Series No.9. Rome. FAO.
26. Sorrenson WJ, Lopez-Portillo J, Nuñez M (1997) The economics of no-tillage and crop rotations in Paraguay. *Policy and investment implications. FAO, Final Report to the MAG/GTZ Soil Conservation Project.* 215.
27. Derpsch R (2003) Conservation tillage. no-tillage and related technologies. In: Garcia-Torres L, Benites J, Martinez-Vilela A, Holgado-Cabrera A (Eds.). *Conservation Agriculture: Environment, Farmers Experiences, Innovations, Socio-Economy, Policy*. Springer. Netherlands. 181-191.
28. Hobbs P (2007) Conservation agriculture: what is it and why is it important for future sustainable food production. *J Agric Sci* 145: 127-137.
29. Friedrich T, Kassam AH (2009) Adoption of Conservation Technologies: constraints and opportunities. In: Proceedings of the 4th World Congress on Conservation Agriculture: Innovations for Improving Efficiency, Equity and Environment. 4-7 February 2009. New Delhi. India.
30. Kassam AH, Friedrich T, Derpsch R: (2010) Conservation Agriculture in the 21st Century: a paradigm of sustainable agriculture. In; Proceedings of the European Congress on Conservation Agriculture. 6-8 October 2010. Madrid. Spain.
31. FAO (2011) Save and Grow. FAO, Rome, 99.
32. The Montpellier Panel (2013) Sustainable Intensification: A New Paradigm for African Agriculture. London.
33. Larson WE (1979) Crop residues: Energy production or erosion control. *J Soil Water Conserv* 34: 74-76.
34. HargroveWL(1991) Crop residue management in the Southeast. Crop Residue Management for Conservation. Lexington. KY. Soil and Water Conservation Society.
35. Derpsche R (1999) May 28 Frontiers in conservation agriculture and advances in conservation practice. West Lafayette. USA.
36. Hagen LJ (1996) Crop residue effects on Aerodynamic processes and wind erosion. *Theor Appl Climatol* 54: 39-46.
37. Hernanz JL, Lopez R, Navarrete L, et al (2012) Long-term effects of tillage systems and rotations on soil structual stability and organic carbon stratification in semiarid central Spain. *Soil Till Res* 66: 129-141.
38. Egamberdiev OJ (2007) Dynamics of irrigated alluvial meadow soil properties under the influence of resource saving and soil protective technologies in the Khorezm region. PhD dissertation. National University of Uzbekistan. Tashkent (in Uzbek).
39. Pulatov A, Egamberdiev O, Karimov A, et al. (2012) Introducing conservation agriculture on irrigated meadow alluvial soils (Arenosols) in Khorezm. Uzbekistan. In: Martius C, Rudenko I, Lamers JPA, Vlek PLG (Eds.) *Cotton. Water. Salts and Soums – Economic and Ecological Restructuring in Khorezm. Uzbekistan.* Springer. Dordrecht Heidelberg London New York.195-217.
40. Nurbekov A, Suleymenov M, Friedrich T, et al. (2012) Effect of tillage methods on productivity of winter wheat in the Aral Sea Basin of Uzbekistan. . *J Arid Land Stud* 22: 255-258.
41. Nurbekov A (2007) Final report: Sustainable agricultural practices in the drought affected region of Karakalpakstan (Phase II). FAO/TCP/3102 (A). ICARDA-CAC.
42. Ibragimov N, Evett S, Essenbekov Y, et al (2011) Permanent beds versus conventional tillage in irrigated Central Asia. *Agron J* 103: 1002-1011.
43. Akramkhanov A, Kuziev R, Sommer R, et al. (2012) Soils and soil ecology in Khorezm. In: Martius C, Rudenko I, Lamers JPA, Vlek PLG (eds.) *Cotton. Water. Salts and Soums – Economic and Ecological Restructuring in Khorezm. Uzbekistan.* Springer. Dordrecht Heidelberg London New York, 37-58.
44. Tischbein B, Awan UK, Abdullaev I, et al. (2012) Water management in Khorezm: current situation and options for improvement (hydrological perspective). In: Martius C, Rudenko I, Lamers JPA, Vlek PLG (eds.) *Cotton. Water. Salts and Soums – Economic and Ecological Restructuring in Khorezm. Uzbekistan.* Springer. Dordrecht Heidelberg London New York. 69-92.
45. Bezborodov GA, Shadmanov DK, Mirhashimov RT, et al. (2010) Mulching and water quality effects on soil salinity and sodicity dynamics and cotton productivity in Central Asia. *Agric Ecosyst Environ* 138: 95-102.
46. Brady NC, Well RR (2008) The Nature and Properties of Soils. 14th (ed.). Pearson-Prentice Hall Upper Saddle River. NJ. 960 pp.
47. Skidmore EL, Kumar M, Larson WE (1979) Crop residue management for wind erosion control in the Great Plains. *J Soil Water Conserv* 34: 90-94.
48. Reicosky DC (2007) Conservation agriculture: Environmental benefits of reduced tillage and soil carbon management in water-limited areas of Central Asia. In: Lal R, et al. (eds) *Climate change and terrestrial carbon sequestration in Central Asia*. New York, Taylor & Francis.199-211.
49. Darby GM (1985) Conservation tillage: An importantadaptable tool for soil and water conservation. In: ElSwaify SA, Moldenhauer WC, Lo A (eds.) *Soil*

Erosion and Conservation. Soil Conservation Society of America. Ankeny. IA,649-653.

50. Karabayev M, Wall P, Sayre K, et al. (2012) Conservation agriculture adoption in Kazakhstan: History, Status and Outlooks. CIMMYT Report.

51. Angas P, Lampurlanés J, Cantero-Martínez C (2006) Tillage and N fertilization Effects on N dynamics and Barley yield under semiarid Mediterranean conditions. *Soil Till Res* 87: 59-71.

52. Tursunov M (2009) Potential of conservation agriculture for irrigated cotton and winter wheat production in Khorezm. Aral Sea Basin. PhD dissertation. ZEF/ Rheinische Friedrich-Wilhelms-Universität Bonn. Germany. de/diss_online elektronisch publiziert.

53. Richard WH (1999) Lessons learned in developing and implementing minimum tillage technology with farmers in Northern Kazakhstan.In: Karabayev M, et al. (eds), *Conservation Tillage: A Viable Option/or Sustainable Agriculture in Central Asia.*59-62.

54. Pozharskiy VM, Akimaliev J (2002) Bed-planting winter wheat in Chu valley of the Kyrgyz Republic. International Workshop on Conservation Agriculture for Sustainable Wheat Production in Rotation with Cotton in Limited Water Resource Areas, 14 -18 October., Tashkent. -Uzbekistan.123-126 (in Russian).

55. Ospanbaev J, Karabayev MK, (2009) Outlook for not-till technologies of crop growing in South and Southeast Kazakhstan. In: Suleymenov M, Kaskarbayev JA, Skoblikov VF, et al. (Eds.) No-till With Soil Cover and Crop Rotation: A Basis for Policy Support to Conservation Agriculture for Sustainable Production Intensification. Astana-Shortandy. Kazakhstan.195-199 (in Russian).

56. Rashidov KA (2006) Produktivnost' perspektivnykh sortov pshenitsy pri grebnevom poseve v usloviyakh Tadzhikistana. In: Rashidov KA, Mumindzhanov KHA., Makhmadorov UM (Eds.) Materialy vtoroy Tsentral'no-Aziatskoy konferentsii po zernovym kul'turam. Bishkek P. 240 (in Russian).

57. SydykDS, JarasovSh, SydykovMA, et al. (2008) Rekomendatsii po Resursosberegauyshey Technologii Vozdelyvaniya Zernovykh Kolosovykh Kultur v Usloviyakh Bogarnogo i Oroshaemogo Zemledeliya Yujnogo Kazakhstana.Jebe-Disayin. Shymkent. Kazakhstan. 76. (in Russian).

58. Sanginov S, Khakimov A (2003) Planting winter wheat to growing cotton. In: *Proceedings of the First Central Asian Wheat Conference.*Almaty. Kazakhstan.

59. Qilichev AH, Khalilov N (2008) Go'za qator oralariga ekilgan kuzgi bugdoy hosidorligi va don sifati. Journal AGRO-ILM 2, 163-168 (in Uzbek).

60. Kienzler KM, Lamers JPA, McDonald A, et al (2012) Conservation agriculture in Central Asia – What do we know and where do we go from here. *Field Crops Res* 132: 95-105.

CHAPTER 3

EFFECT OF DIFFERENT MULCHING MATERIALS ON MAIZE GROWTH AND YIELD IN CONSERVATION AGRICULTURE SYSTEMS OF SUB-HUMID ZIMBABWE

W. Mupangwa, I. Nyagumbo, E. Mutsamba

CIMMYT Southern Africa Regional Office, Mount Pleasant, Harare, Zimbabwe

ABSTRACT

The introduction of conservation agriculture (CA) for smallholders increased the competition for crop residues between crop and livestock enterprises of the mixed smallholder farming system. Smallholders practicing CA have resorted to using grass and leaf litter in addition to available crop residues. The effect of these different mulching materials on maize (*Zea mays* L.) growth and yield is not well documented in smallholder CA systems of southern Africa. A two-year experiment was run in 2012/13 and 2013/14 seasons to evaluate the effect of maize residues, grass (*Hyparrhenia filipendula* (L.) Stapf.) and leaf litter that farmers are currently using and residues from leguminous species, sunhemp (*Crotolaria juncea* L.) and *Tephrosia* (*Tephrosia vogelii* ((Hook) *f.*)) on maize nitrogen (N) uptake, growth and yield.

Significant differences in soil water content across treatments were only observed during March in 2012/13 season. Maize residues retained more soil water and *Tephrosia* had the lowest soil water content when seasonal rainfall pattern was erratic. Grass and *Tephrosia* treatments had the lowest chlorophyll content. Conventional ploughing, maize residues and leaf litter had similar chlorophyll content which was significantly higher than grass and *Tephrosia* treatments. At a site with higher initial soil fertility conventional ploughing treatment out yielded the other treatments by 727–1265 kg ha^{-1}. With more degraded sandy soil conventional practice had 119–430 kg ha^{-1} more maize grain than the CA treatments. With adequate fertilization, the mulching materials have a similar effect on maize growth in basins and direct seeding. Further studies on different application rates of mulching materials and mineral N fertilizer, and nutrient release patterns of these residues are critical in order to better understand soil fertility management under smallholder CA systems.

INTRODUCTION

Mixed crop-livestock production characterizes smallholder farming systems in some agro-ecological regions of southern Africa [1,2]. Crops are multi-purpose as they are a source of food and income for the farming families, and livestock derive dry season feed and bedding materials for kraals from the cropping enterprise [3]. The livestock enterprise, dominated by cattle (*Bos indicus* L.) and small stocks such as goats (*Capra hircus* L.), generate manure for crop production, and income sourced through livestock sales is sometimes used for purchasing farming inputs[1,3]. Cattle are also a source of draught power for land preparation and transport on smallholder farms

Citation: W. Mupangwa, I. Nyagumbo, E. Mutsamba. Effect of different mulching materials on maize growth and yield in conservation agriculture systems of sub-humid Zimbabwe. AIMS Agriculture and Food, 2016, 1(2): 239-253. doi: 10.3934/agrfood.2016.2.239

[4]. Livestock also serve social roles in southern Africa, are a status symbol in the rural communities and buffer most farming households against the risk of adverse weather that impede crop production in some years [4,5]. Major crops grown on smallholder farms include maize (*Zea mays* L.), cowpea (*Vigna unguiculata*L. (Walp)), groundnuts (*Arachis hypogaea* L.), bambaranuts (*Voandzeia subterranea* L. (*thouars*)) and these are produced under conventional agriculture. Since early 2000s maize is now grown in conservation agriculture (CA) systems where it is intercropped or rotated with legumes [6].

Low soil productivity in the smallholder farming systems of southern Africa triggered research on different strategies for restoring and improving soil fertility [7,8]. Various combinations of mineral fertilizer and organic nutrient sources have been tested under different socio-economic and biophysical conditions of the smallholder sector of southern Africa [9,10,11,12]. Conservation agriculture (CA) is being promoted as a potential solution to declining soil productivity on smallholder farms [5,13]. Improvements in soil properties due to CA practices were observed in some studies [14,15,16,17] while little changes in critical soil physico-chemical properties were noted in others [18].

On smallholder farms where CA has been introduced, competition for crop residues between crop and livestock limits full use of the practice because both enterprises are critical for farming families in southern Africa [19]. The competition is further compounded by deteriorating pasture quality in most rural communities [20,21] and decreasing arable land sizes due to the increasing human population [4]. In most instances, smallholders are not producing biomass quantities that can meet the CA mulching and livestock feed requirements in the mixed farming systems [19]. Smallholder CA farmers have resorted to applying plant materials that are available in their localities as mulch in combination with any remaining crop residues at the onset of the cropping season [22].

Tree litter from *Uapaca kirkiana* (Benth), *Brachystegia spiciformis* (Benth) and *Julbernardia globiflora*) (Benth) and grass (*Hyparrhenia filipendula* (L.) Stapf.) are available in the Savanna landscape of southern Africa [9,23] and CA farmers have resorted to using them for mulching[22]. *Tephrosia vogelii* ((Hook) *f.*), *Crotolaria juncea* (L.) and *Crotolaria grahamiana* (Wright & Arn) are leguminous species available for soil fertility improvement [8] but can also provide residues for mulching. Stems/litter of woody leguminous species such as *Acacia angustissima* ((Mill.) Kuntze) and *Sesbania sesban* ((L.) Merr.) decompose slowly [24] and this can be capitalized on to provide soil cover in CA systems where crop residues are scarce.

However, the use of such plant material could introduce N immobilization leading to reduced crop yields [24,25,26]. Mineral N fertilizer remains a critical requirement whenever CA is practiced in order to reduce the crop yield penalty [27]. Relay intercropping of maize with annual or perennial legume species can offer alternative soil cover in CA systems that are being promoted in mixed crop-livestock systems of southern Africa [28]. Annual legume species such as velvet bean (*Mucuna pruriens* L.), lablab (*Lablab purpureus L.*) and cowpea (*Vigna unguiculata* L.(Walp)) can be relay cropped with the main cereal crop in CA systems during the growing season [28,29]. Legume species such as *T. vogelii*, *C. grahamiana*, *C. juncea* and *S. sesban* grow and produce lignified plant tissue and their residues can be used for mulching in CA systems. These legumes species have been promoted widely before through agro-forestry and green manuring initiatives in southern Africa [8].

The effect of different mulching materials used in CA systems on maize growth and yield is not well documented under the highly degraded sandy soils of sub-humid areas of Zimbabwe. This study was established to assess the effect of the different mulching materials on maize productivity in CA systems that are being promoted in North-Eastern Zimbabwe. We hypothesized that the different plant materials had similar influence on N uptake and maize yield in the sandy soils of sub-humid Zimbabwe. The objectives were to determine: (1) soil water dynamics, (2) maize leaf chlorophyll content and, (3) maize grain and stover yields under the conventional mouldboard ploughing practice and different mulching materials applied in smallholder CA systems. Maize leaf chlorophyll content was used to assess nitrogen (N) uptake by maize plants. Chlorophyll content is related to N supply to a growing plant [30,31].

MATERIALS AND METHODS

Description of Experimental Sites

A two-year experiment was established at three on-farm and one on-station sites. On-farm and on-station sites were used in the study in order to increase replication of the experiment. Logistically the on-station experimental site enabled researchers to collect detailed measurements more frequently during each cropping season. The experiment was established in Goromonzi and Murehwa districts which lie in agro-ecological region II (800-1000 mm annual rainfall) of Zimbabwe [32] and was run in 2012/13 and 2013/14 cropping seasons. Rainfall season commenced in November and ended in April/May with peak rainfall period stretching from December to February. Granitic sand (*Arenosols, FAO classification*) with low levels of macro and micro-nutrients (Table 1), is the dominant soil type in the two districts [33,34].

Table 1. Chemical and physical characteristics of soil at the on-farm and on-station experimental sites

Site	Soil pH	Avail. P (mg kg^{-1})	Ca	Mg	K	Clay	Silt	Sand
			Cmol$_c$kg^{-1}			(%)	(%)	(%)
DTC	4.9	19	0.30	0.10	0.06	21	5	74
Gono	3.0	16	0.61	0.26	0.05	9	1	90
Murwira	4.1	30	0.88	0.39	0.24	nd	nd	nd
Unyoro	4.0	14	0.90	0.29	0.09	5	7	88
DTC: Domboshawa Training Centre; nd: not determined. pH = 0.01 M $CaCl_2$; available P = Olsen method; Ca, Mg and K = atomic absorption spectroscopy; clay, silt and sand content = hydrometer method.								

Experimental Layout and Description of Treatments

The experiment was laid in a randomized complete block design with three replicates at the on-farm sites and four replicates at the on-station site. The treatments at the on-farm sites were:

- Conventional mouldboard (V100®) ploughing (CP) and seeding after receiving effective rains of more than 20 mm over 2 consecutive days. No crop residues were applied to this treatment as it served as a control;
- Animal traction direct seeding (*Irmãos Fitarelli, Brazil, model #12*) at the on-set of the cropping season with maize residues applied annually as mulch at 3 t ha^{-1} on a dry weight basis;
- Animal traction direct seeding at the on-set of the cropping season with thatching grass applied annually as mulch at 3 t ha^{-1};
- Animal traction direct seeding at the on-set of the cropping season with leaf litter applied annually as mulch at 3 t ha^{-1}. The leaf litter was derived from *Uapaca kirkiana*, *Brachystegia spiciformis* and *Julbernardia globiflora* trees;
- Animal traction direct seeding at the on-set of the cropping season with *C. juncea)* residues applied annually as mulch at 3 t ha^{-1};
- Animal traction direct seeding at the on-set of the cropping season with *T. vogelii* residues applied annually as mulch at 3 t ha^{-1}.

Planting basins are dug in the same position every season using a hand hoe. The planting basins are reopened before the onset of the rains. Spacing of planting basins varies between countries and available options include 75 cm × 60 cm, 90 cm × 50 cm, 90 cm × 60 cm, and 75 cm × 75 cm [35]. In our experiment each planting basin measured 15 cm × 15 cm × 15 cm and 90 cm × 50 cm spacing was used. The on-station (DTC) site had six treatments which are summarized below:

- Conventional mouldboard (V100®) ploughing (CP) at seeding after receiving effective rains. No crop residues were applied to this treatment as it served as a control;
- Planting basins dug in October-November each year with maize residues applied annually as mulch at 3 t ha^{-1} on a weight basis;
- Planting basins dug in October-November each year with thatching grass applied annually as mulch at 3 t ha^{-1};
- Planting basins dug in October-November each year with leaf litter applied annually as mulch at 3 t ha^{-1} and the leaf litter was derived from *Uapaca kirkiana* trees;
- Planting basins dug in October-November each year with sunnhemp (*C. grahamiana*) residues applied annually as mulch at 3 t ha^{-1};
- Planting basins dug in October-November each year with Tephrosia (*T. vogelii*) residues applied annually as mulch at 3 t ha^{-1}.

The CA systems and mulching materials used are summarized in Table 2. Conventional ploughing and animal traction direct seeding at the on-farm sites were conducted after receiving effective planting rain (at least 20 mm over two days). All mulching material was applied at seeding in each season and any mulching materials from the previous season were left on the plots when fresh mulch was applied. At the on-station sites *C. grahamiana* was used because *C. juncea* seed was not enough to cover all experimental sites. Chemical analysis of the different plant materials used as mulch in our experiment was not conducted because available laboratories had no capacity to carry out the tests.

Table 2.CA system and the mulching materials used at the on-farm and on-station experimental sites.

Experimental site	CA system	Mulching material used
On-farm	Direct seeding	Maize residues
		Grass
		Leaf litter
		Sunnhemp residues
		Tephrosia residues
On-station	Planting basins	Maize residues
		Grass
		Leaf litter
		Grahamiana residues
		Tephrosia residues

Experimental Management

Plots measuring 10 m × 10 m were used at each experimental site and spacing of 90 cm × 25 cm was used in the direct seeding system and 90 cm × 50 cm for planting basins. The target maize population at each site was 44,444 plants ha^{-1} giving 1 plant per station for the direct seeding system and 2 plants per station for the planting basins. Maize was grown at all experimental sites and a medium duration variety Pioneer 30G19 (Pioneer Hi-Bred International Inc.) was used in both seasons. Maize was seeded on 12 December 2012 in the 2012/13 season and 6 November 2013 in the 2013/14 season. At seeding each treatment received basal compound D fertilizer (8 N:14 P_2O_5:7 K_2O) at 200 kg ha^{-1}, supplying 16, 12 and 11.6 kg ha^{-1} N, P and K respectively. The maize was top dressed with ammonium nitrate (34.5% N) which was applied at 200 kg ha^{-1} 5-6 weeks after seeding as a single dose. In 2012/13 season AN was applied on 17 January 2013 while in 2013/14 season topdressing was done on 16 December 2013. Initial weed control in the CA treatments was achieved by applying 2.5 L ha^{-1} Glyphosate (480 g L^{-1} active ingredient) soon after seeding maize. Further weed control was done manually whenever weeds had reached 10 cm in height or radius for crawling species.

Data Collection and Harvesting Procedures

Soil Water Measurements at the on-Station Site

Soil water was determined using a Field Scout TM TDR 300 Soil Moisture Meter (Spectrum® Technologies, Inc.). Sampling was done randomly on three spots in the net plot and readings were taken at 20 cm depths. Soil water measurements were taken 14 and 9 times during the 2012/13 and 2013/14 seasons respectively.

Maize Leaf Chlorophyll and Plant Height Measurements at the on-Station Site

In vivo chlorophyll content of maize plant leaves was estimated using a portable chlorophyll meter (SPAD-502, Minolta, Tokyo, Japan). Measurements were taken weekly 7 weeks after sowing on the upper most extended leaf from 5 randomly selected plants per plot until early reproductive stage of the maize crop. Plant height was measured from five randomly selected and permanently tagged maize plants. Height was measured from the ground to the tip of the upper most leaf.

Maize Yield

At harvest, maize grain and stover yields were measured from a net plot consisting of 2 central rows that were 5 m long. Field weights of maize grain and stover yields were recorded before taking 10 maize cobs for moisture correction and to determine the shelling percentage of maize. Grain moisture content was recorded after shelling using a grain moisture meter, a mini GAC® moisture tester (DICKEY-John, USA). Maize final grain (12.5% moisture content) and stover yields were calculated to a hectare basis. Daily rainfall was recorded manually using a standard rain gauge installed at each experimental site.

Statistical Analysis

Soil water, chlorophyll content, plant height and maize yield data were tested for normality using the Shapiro-Wilk test using Statistics 9 program[36]. The data for each parameter was then subjected to analysis of variance with mulching material as the treatment factor in a randomized complete block design. Regression analysis using the linear model was conducted to assess the relationship between soil water and chlorophyll content. The relationship between plant height and leaf chlorophyll content was also assessed in the six treatments.

RESULTS

Seasonal Rainfall and Soil Water Patterns

Seasonal rainfall varied from 555-668 mm in Goromonzi and 615-655 mm in Murehwa during experimentation (Figure 1). The longest dry spells were experienced in 2012/13 cropping season with 12 and 14 continuous days without rain being recorded in Goromonzi and Murehwa during the peak rainfall period. Rainfall was well distributed during the November-February period but dry spells occurred between February and April 2013. Rainfall tailed off in March in both seasons and most rainfall events were recorded during the December-January period.

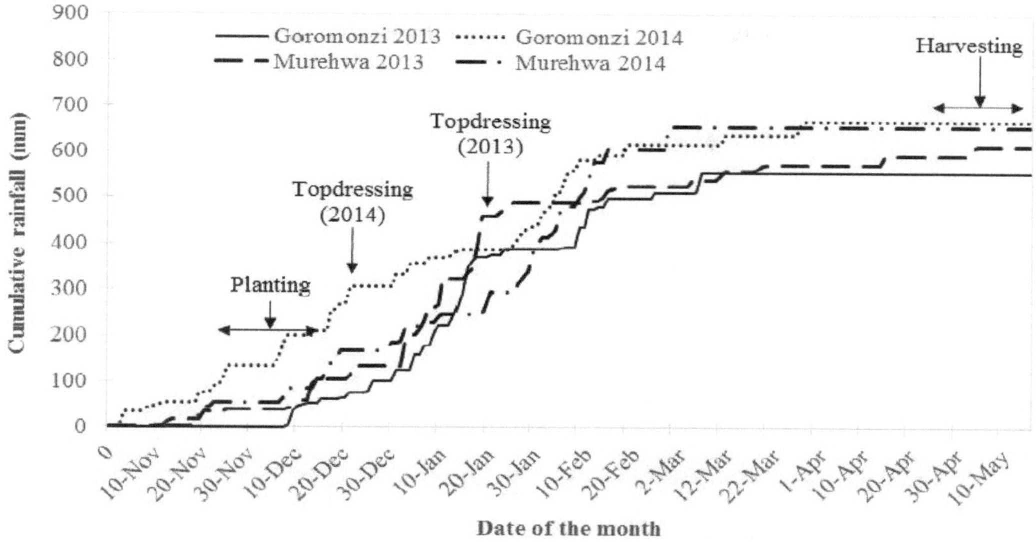

Figure 1. Seasonal rainfall distribution at on-farm and on-station sites in Goromonzi and Murehwa districts during the period of experimentation.

Soil water patterns in the 0-20 cm top soil layer responded to the rainfall received during the cropping period in all treatments. In 2012/13 season maize residue treatment had higher ($p < 0.05$) soil water content than other treatments on 1 February, 8 March, 29 March and 12 April 2013 (Figure 2). *Tephrosia* treatment retained the lowest soil water content during 2012/13 cropping season. In 2013/14 season there were no significant differences in soil water content across the six treatments (Figure 2).

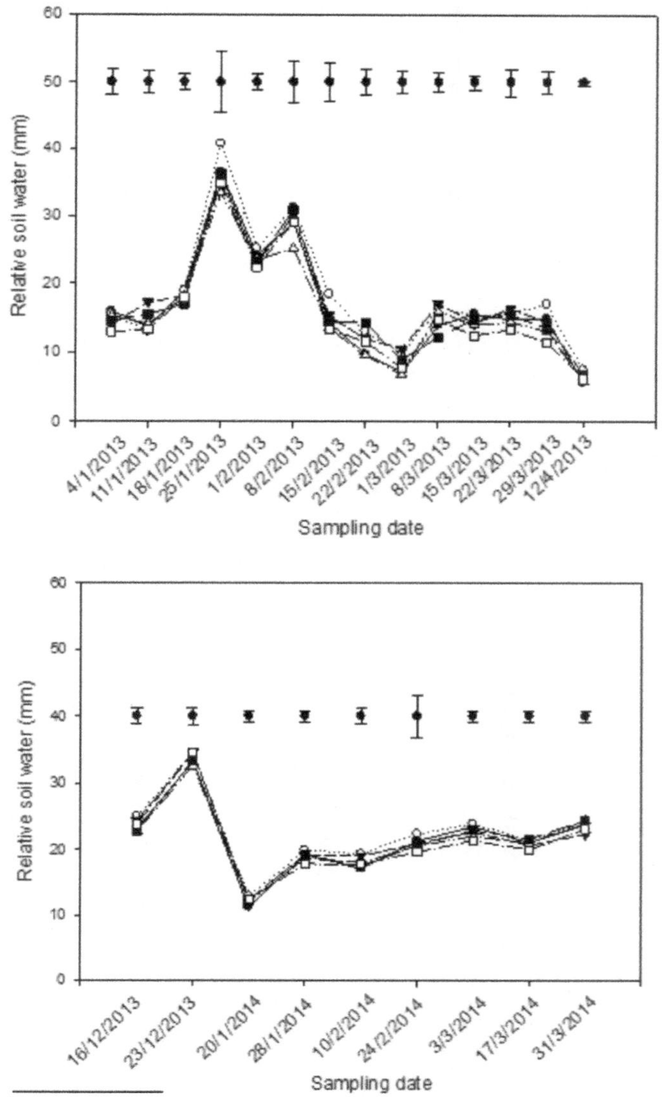

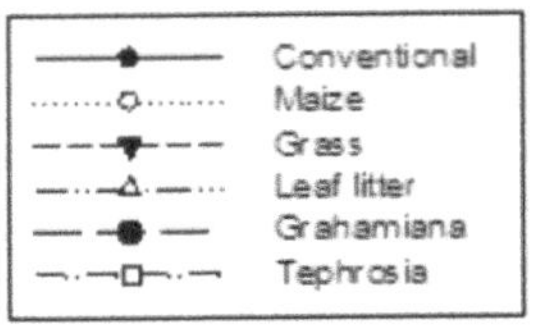

Figure 2. Soil water distributions patterns in the top 0-30 cm soil layer under five different mulching materials and conventional ploughing during 2012/13 and 2013/14 cropping seasons at the on-station experimental site. Vertical bar at each sampling date represents the standard error of the difference between treatment means.

Maize Chlorophyll Content and Plant Height at DTC Site

Leaf chlorophyll content was significantly different on 2 occasions only across the 6 treatments. On 6 January (p = 0.039) and 24 February (p = 0.036) maize plants under the conventional treatment had the higher chlorophyll content than the other treatments (Figure 3). Of the three commonly used mulching materials on smallholder farms, the grass treatment had the lowest chlorophyll content in the maize ear leaf. However, *Tephrosia* mulched treatment had the lowest chlorophyll content during the cropping season. *C. grahamiana* residue treatment had higher chlorophyll content compared with *Tephrosia* mulched plots throughout the season. Generally mulching materials derived from non-leguminous plant species had higher chlorophyll content compared with treatments under *Grahamiana* and *Tephrosia* residue cover. The relationship between soil water and leaf chlorophyll content was not significant in 2013/14 cropping season. The shortest maize plants were observed in the grass and *Tephrosia* mulched treatments during the January-February period (Figure 4). The tallest maize plants were observed in the maize residue mulched treatment. Maize plant growth under *C. grahamiana* and *Tephrosia* residues was slower than CP, maize residues and leaf litter treatments.

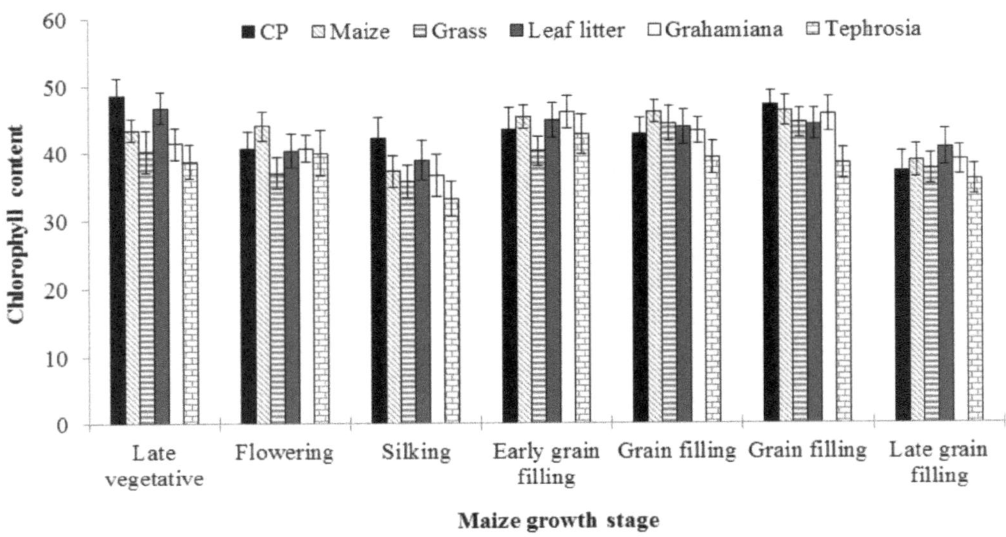

Figure 3. Chlorophyll content (*unitless*) patterns in the maize ear leaf under six treatments tested during the 2013/14 cropping season at the on-station experimental site. Vertical bar represents the standard error of means at each maize growth stage (n = 12).

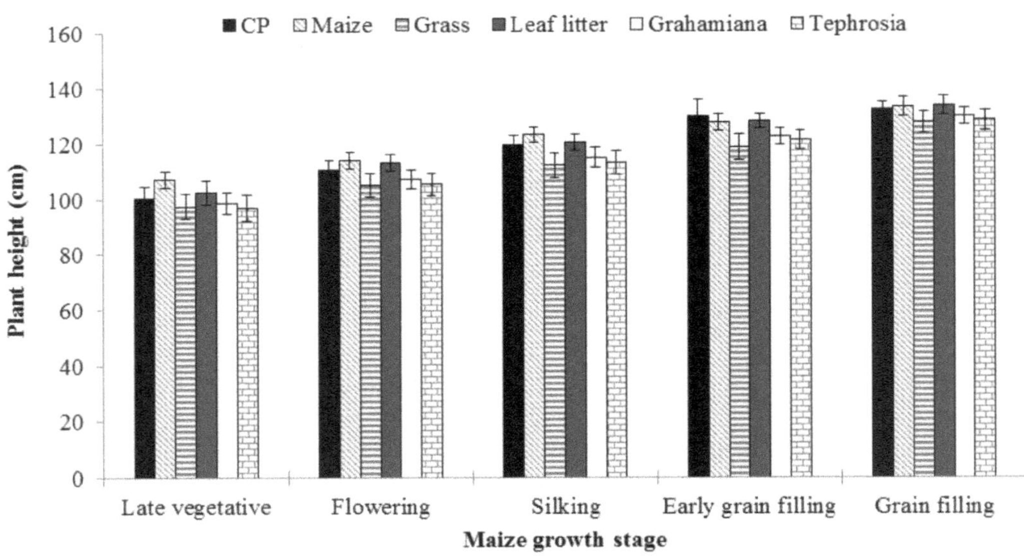

Figure 4. Maize plant height under different mulching materials in CA during the 2013/14 cropping season at DTC. Vertical bar represents the standard error of means at each maize growth stage (n = 12).

Maize Responses to Different Treatments at the DTC On-Station Site

There were no significant differences in plant population across the six treatments in 2012/13 and 2013/14 seasons. Maize grain yield differed significantly across the six treatments in 2012/13 and 2013/14 seasons (Table 3). In 2012/13 season CP treatment had 1045, 965 and 727 kg ha^{-1} more (p = 0.033) grain than maize residues, grass and leaf litter treatments, respectively. The CP treatment significantly out-yielded *C. grahamiana* and *Tephrosia* treatments by 1041 and 1265 kg ha^{-1}. In the same season the six treatments had no significant effect on maize stover yield. In 2013/14 cropping season leaf litter treatment outperformed the *Tephrosia* treatment only. As observed with grain yield, the leaf litter treatment also gave the highest stover yield compared with the other treatments. Regression analysis indicated no significant relationships between maize plant height and chlorophyll content under each treatment.

Table 3. Maize plant population (plants ha^{-1}) and, grain and stover yield responses (kg ha^{-1} on dry matter basis) to different treatments applied at the on-station experimental site in 2012/13 and 2013/14 cropping seasons

Harvest year	Treatment	Plant population	Grain yield	Stover yield
2013	Conventional practice	42,274	2416^{a}	1363
	Maize residues	42,635	1371^{b}	1585
	Grass	43,669	1451^{b}	1318
	Leaf litter	44,205	1689^{b}	1483
	Grahamiana residues	43,083	1375^{b}	1448
	Tephrosia residues	39,842	1151^{b}	944
	P-value	ns	0.033	ns
	SED	1597	348	214
2014	Conventional practice	39,176	4007^{a}	4249
	Maize residues	38,049	4177^{a}	4205
	Grass	37,160	3414^{ab}	3099
	Leaf litter	36,404	4288^{a}	4621
	Grahamiana residues	37,185	3433^{ab}	3692
	Tephrosia residues	36,434	2447^{b}	2890
	P-value	ns	0.010	ns
	SED	1743	666	643
ns: not significant; SED: standard error of the difference between treatment means				

Maize Responses to Different Mulching Materials at the on-Farm Sites

The treatments had no significant influence on maize population across the on-farm experimental sites. In the first season the six treatments had no significant effect on maize grain and stover yields. However, in 2013/14 season significant maize yield differences were detected across the six treatments tested across the three smallholder farms (Table 4). Only the grass mulched treatment significantly outperformed CP by 1230 kg ha^{-1} and *Tephrosia* by 1490 kg ha^{-1} maize grain. Maize stover yield was also influenced (p = 0.005) by the six treatments. Leaf litter and*Tephrosia* treatments had the highest and lowest maize stover yield respectively in 2013/14 season.

Table 4. Maize plant population (plants ha^{-1}) and, grain and stover yield (kg ha^{-1} on dry matter basis) responses to different treatments applied at five farms in 2012/13 and three farms in 2013/14 cropping seasons in Goromonzi and Murehwa districts of Zimbabwe

Harvest year	Treatment	Plant population	Grain yield	Stover yield
2013	Conventional practice	nd	2423	3235
	Maize residues	nd	1993	3017
	Grass	nd	2248	3615
	Leaf litter	nd	2304	3444
	Grahamiana **residues**	nd	2020	3544
	Tephrosia **residues**	nd	2273	3318
	P-value	nd	ns	ns
	SED	nd	332	729

2014	Conventional practice	28,277	2061^{cd}	3443^{abc}
	Maize residues	29,701	3067^{abc}	4212^{ab}
	Grass	28,108	3291^{a}	4343^{a}
	Leaf litter	29,083	3133^{ab}	4414^{a}
	Grahamiana **residues**	29,039	2432^{abcd}	3312^{bc}
	Tephrosia **residues**	29,770	1801^{d}	2693^{c}
	P-value	ns	0.015	0.005
	SED	1504	510	518

nd: not determined in 2012/13 season; ns: not significant; SED: standard error of the difference between treatment means

DISCUSSION

Soil Water Patterns in the Six Treatments

Soil water responses to the seasonal rainfall pattern and the amount of soil water retained under each treatment varied throughout the season. At the time most measurements were taken, the maize residue treatment had more soil water compared with the other treatments. The maize residues used for mulching was a mixture of stems and leaves, and these could have captured more of the rainwater allowing it to infiltrate into the soil. High soil cover substantially reduces surface runoff allowing more time for rainwater to infiltrate into the soil, and can reduce soil evaporation by 30-50% under some agro-ecological conditions [37,38]. *Tephrosia* residues, which comprised mostly twigs, retained the least amount of soil water. The *Tephrosia* twigs/stems could not cover much of the surface area, and hence more soil water could have been lost through evaporation from the *Tephrosia* treatment.

Maize Leaf Chlorophyll Content and Plant Heights in the Different Treatments

Maize plants in the CP treatment had the highest chlorophyll content. This could be attributed to suppressed N uptake by maize plants wherever plant residues were applied as mulch. Plant residues used had high C:N ratio and this could have promoted N immobilization during the cropping season. Plant residues with high C:N ratio (for example >42:1 for maize stover) often stimulate immobilization of soil N by micro-organisms [26] and more N from other sources is required to offset the negative effects of limited N supply to the growing crops. Plant residues with C:N ratios of <20:1 often decompose fast, release N and nutrients to the soil [39] and these would not be ideal for incorporating into smallholder CA systems for mulching purpose. Ideal mulching material could be plant materials that decompose slowly and provide soil cover for much longer during the cropping season. Leaf litter might not be a source of N during the first 60-90 days of its decomposition [23], hence mineral fertilizer becomes a critical input for smallholders using this natural resource in their CA farming system. Of the three commonly used mulching materials on smallholder farms, grass had the least chlorophyll content. Grass species common in miombo dominated forests of Zimbabwe can have C:N ratios >42:1 [23] and their use as surface cover in CA systems can promote N immobilization. The relationship between soil water and leaf chlorophyll content was not significant in the different treatments. The lack of significant relationship between these two parameters suggests that soil water did not limit the uptake of N by the growing maize plants. Soil water was similar across the six treatments at most occasions when measurements were taken. Maize leaf chlorophyll content observed in our study are consistent with findings from previous studies under conventional agriculture in rainfed systems [31,40].

Maize plant height was not significantly influenced by the different treatments despite the differences observed with leaf chlorophyll content across the six treatments at DTC site. These results suggest that the basal and topdressing fertilizer supplied adequate nutrients that offset the negative effects of the different mulching materials on N availability to the maize crop. Basal fertilization at seeding supplied 16, 12 and 11.6 kg ha^{-1} N, P and K respectively, and a further 69 kg N ha^{-1} was applied as topdressing fertilizer. Nitrogen is an important yield determining factor in no-till systems worldwide [27].

Maize Yield Responses to Different Treatments

Maize responded positively to the six treatments in the first season at DTC. Soil at DTC had higher pH and clay content, and maize often grows well at soil pH greater than 5.5 [41]. The CP treatment had higher grain yield than the five mulch treatments in the first season, a trend consistent with observations made at the on-farm sites and leaf chlorophyll content results. These results suggest that in the CP treatment N uptake was higher than the mulched treatments in mulched CA treatments. Residues of legumes such as *Cajanus cajan* (L.), *Leucaena leucocephala* (L.) and *Desmodium* residues cause N immobilization in the first 30-90 days of the cropping season [25]. Similarly residues of species such as *Acacia angustissima* can also immobilize N when they are incorporated in conventional moldboard ploughing system [24].

At the on-farm sites maize yield responses to different treatments varied from site to site and there were no significant yield differences in the first season. The soils at the on-farm sites were highly degraded, with pH below the optimum 5.5 for maize [41]. Decomposing plant materials applied as mulch could not influence soil chemical properties to give a crop response in the first season. However, in the second season maize residue, grass, leaf litter and *Grahamiana* treatments had higher yields than *Tephrosia*. Plant residues from the first season had decomposed and improved soil conditions. For example leaf litter is a source of plant nutrients and the material decomposes easily thereby releasing nutrients to the soil [9,23,42]. *Tephrosia*residue mulching was the least performing treatment across the two cropping seasons. The *Tephrosia* residues consisted of stems and branches, and these decomposed slowly resulting in more N lock up as reflected by low leaf chlorophyll content and shorter maize plants shown by on-station measurements. Tree litter consisting of predominantly twigs immobilizes soil N [23], and this could have happened in the *Tephrosia* residue treatment. The woody plant material has higher C:N ratio and this induces an increased demand for N by decomposers in the soil thereby depriving growing maize plants of N. Our results showed that *C. grahamiana* residues did not suppress maize yield as much as*Tephrosia*, suggesting the former has lower ratios of carbon, N, lignin and polyphenols that influence decomposition of plant material. At the on-station site, maize yield differences between the 2 seasons can be attributed to differences in rainfall distribution during the cropping season. In the first season, rainfall was lower than 2014, with 2 long dry spells experienced between the end of December 2012 and February 2013. Rainfall also tailed off earlier in 2013 where the last effective rains were recorded in the middle of March 2013.

CONCLUSIONS AND RECOMMENDATIONS

The study assessed the effect of leguminous and non-leguminous plant materials used for mulching on maize crop growth under the sub-humid conditions of Zimbabwe. In a season with dry spells during maize grain filling stage, illustrated by 2012/13, maize residue cover conserved more soil water during the cropping season. Maize N uptake fluctuated during the growing season under the conventional and different mulching treatments. Our results show that with adequate basal and topdressing fertilization, the different mulching materials under CA have a similar effect on maize growth and the resultant grain and biomass yields. Adequate fertilization is therefore critical in CA systems where smallholder farmers use different plant materials as mulch. Decomposing residues of *C. grahamiana* and *Tephrosia* can promote N immobilization and are therefore not a direct source of N when the two leguminous species are grown to full maturity. Adequate mineral N fertilization is therefore critical if farmers apply these residues as mulch in CA systems that are being promoted in the smallholder sector.

Our study was conducted over two seasons and would be worth repeating for a longer period in order to get full insights on how the different mulch types influence maize growth and yield under different season qualities. The experiment will generate more valuable information for smallholder CA farmers if different application rates of different mulching materials and mineral fertilizer are tested together. Future studies could also focus on nutrient release patterns of the different plant materials that are used as mulch in smallholder CA systems in southern Africa. The leguminous species used in the study were allowed to grow to maturity before their residue was used for mulching. It would be worthwhile to explore the effect of mulching materials of leguminous residues grown to different growth stages on soil and crop productivity in CA systems.

ACKNOWLEDGEMENTS

This study is embedded in the MAIZE CGIAR Research Program and was funded by the Australian Centre for International Agricultural Research (ACIAR) through project number CSE/2010/022 (*Integrating crops and livestock for improved food security and livelihoods in rural Zimbabwe*). The authors acknowledge the support received from farmers and AGRITEX officers in Goromonzi and Murehwa districts of Zimbabwe. We acknowledge the efforts of Jefias Mataruse, Wadzanayi Mvundura and Miriam Zenda in data collection.

REFERENCES

1. Romney DL, Thorne P, Lukuyu B, et al. (2003) Maize as food and feed in intensive smallholder systems: management options for improved integration in mixed farming systems of east and southern Africa. *Field Crops Res* 84: 159-168.
2. Valbuena D, Erenstein O, Homann-Kee Tui S, et al. (2012) Conservation Agriculture in mixed crop–livestock systems: Scoping crop residue trade-offs in Sub-Saharan Africa and South Asia. *Field Crops Res* 132: 175-184.
3. Rufino MC, Rowe EC, Delve RJ, et al. (2006) Nitrogen cycling efficiencies through resource-poor African crop-livestock systems. *Agric Ecosyst Environ* 112: 261-282.
4. Twomlow SJ, Steyn JT, du Preez CC (2006) Dryland farming in southern Africa. Chapter 19. In: Petersen GA, Unger WP, Payne WA (eds.).*Dryland Agriculture* 2nd Ed. Agronomy Monograph N°. 23. American Society of Agronomy, Madison, Wisconsin. 769-836.
5. Homann Kee-Tui S, Bandason E, Maute F, et al. (2013) *Optimizing livelihood and environmental benefits from crop residues in smallholder crop-livestock systems in southern Africa: Crop residue uses and trade-offs, exploring options for sustainable intensification with stakeholders*. International Crops Research Institute for Semi-Arid Tropics (ICRISAT), Patancheru 502 324, Andhra Pradesh, India. 60.
6. Wall PC, Thierfelder C, Ngwira A, et al. (2013) Conservation Agriculture in Eastern and Southern Africa. *In: Jat, R.A., Graziano de Silva, J., (Eds.), Conservation Agriculture: Global Prospects and Challenges. CABI, Cambridge USA,* ISBN-13: 9781780642598.
7. Waddington SR, Murwira HK, Kumwenda JDT, et al. (eds.) (1998) *Soil Fertility Research for Maize-Based Farming Systems in Malawi and Zimbabwe.* Proceedings of the Soil Fert Net Results and Planning Workshop held from 7 to 11 July 1997 at Africa University, Mutare, Zimbabwe. Soil Fert Net and CIMMYT-Zimbabwe, Harare, Zimbabwe. 312.
8. Waddington SR (2003) Grain legumes and green manures for soil fertility in southern Africa-Taking stock of progress. Proceedings of a Conference held 8-11 October 2002 at Leopard Rock Hotel, Vumba, Zimbabwe. Soil Fert Net and CIMMYT-Zimbabwe. 246.
9. Nyathi P, Campbell B (1995) Interaction effect of tree leaf litter, manure and inorganic fertilizer on performance of maize in Zimbabwe. *Afr Crop Sci J* 3: 451-456.
10. Mtambanengwe F, Mapfumo P (2005) Organic matter management as an underlying cause of soil fertility gradients on smallholder farms in Zimbabwe. *Nutr Cycl Agroecosys* 73: 227-243.
11. Zingore S, Murwira HK, Delve RJ, et al. (2007) Influence of nutrient management strategies on variability of

soil fertility, crop yields and nutrients balances on smallholder farms in Zimbabwe. *Agric Ecosyst Environ* 119: 112-126.

12. Rusinamhodzi L, Corbeels M, Zingore S, et al. (2013) Pushing the envelope. Maize production intensification and the role of cattle manure in recovery of degraded soils in the smallholder farming areas of Zimbabwe. *Field Crops Res* 147: 40-53.
13. Thierfelder C, Rusinamhodzi L, Ngwira AR, et al. (2014) Conservation agriculture in Southern Africa: Advances in knowledge. *Renew Agric Food Syst* 30: 1-21.
14. Govaerts B, Sayre, KD, Goudeseune B, et al. (2009) Conservation agriculture as a sustainable option for central Mexican highlands. *Soil Till Res* 103: 222-230.
15. Thierfelder C, Wall PC (2009) Effects of conservation agriculture techniques on infiltration and soil water content in Zambia and Zimbabwe. *Soil Till Res* 105: 217-227.
16. Mupangwa W, Twomlow S, Walker S (2013) Cumulative effects of reduced tillage and mulching on soil properties under semi-arid conditions. *J Arid Environ* 91: 45-52.
17. Nyamangara J, Marondedze A, Masvaya EN, et al. (2014) Influence of basin-based conservation agriculture on selected soil quality parameters under smallholder farming in Zimbabwe. *Soil Use Manag* 30: 550-559.
18. Cheesman S, Thierfelder C, Eash NS, et al. (2016) Soil carbon stocks in conservation agriculture systems of southern Africa. *Soil Till Res* 156: 99-109.
19. Duncan AJ, Tarawali SA, Thorne PJ, et al. (2013) Integrated crop/livestock systems – a key to sustainable intensification in Africa. *Trop Grassl* 1: 202-206.
20. Delve RJ, Cadisch G, Tanner JC, et al. (2001) Implications of livestock feeding management on soil fertility in the smallholder farming systems of sub-Saharan Africa. *Agric Ecosyst Environ* 84: 227-243.
21. Rufino MC, Dury J, Tittonell P, et al. (2011) Competing use of organic resources, village-level interactions between farm types and climate variability in a communal area of NE Zimbabwe. *Agric Syst* 104: 175-190.
22. Mashango G (2015) Conservation agriculture challenges in developing countries and possible solutions – the case of Gokwe South district, Zimbabwe. *J Techn Sci Special Issue* 1: 67-80.
23. Mtambanengwe F, Kirchmann H (1995) Litter from a Tropical Savanna woodland (miombo): Chemical composition and C and N mineralization. *Soil Biol Biochem* 27: 1639-1651.
24. Chikowo R, Mapfumo P, Leffelaar PA, et al. (2006) Integrating legumes to improve N cycling on smallholder farms in sub-humid Zimbabwe: resource quality, biophysical and environmental limitations. *Nutr Cycl Agroecosys* 76: 219-231.
25. Palm CA, Sanchez PA (1991) Nitrogen release from the leaves of some tropical legumes as affected by their lignin and polyphenolic contents. *Soil Biol Chem* 23: 83-88.
26. Gentile R, Vanlauwe B, van Kessel C, et al. (2009) Managing N availability and losses by combining fertilizer N with different quality residues in Kenya. *Agric Ecosyst Environ* 131: 308-314.
27. Lundy ME, Pittelkow CM, Linquist BA, et al. (2015) Nitrogen fertilization reduces yield declines following no-till adoption. *Field Crops Res* 183: 204-210.
28. Mhlanga B, Cheesman S, Maasdorp B, et al. (2015) Contribution of cover crops to the productivity of maize-based conservation agriculture systems in Zimbabwe. *Crop Sci* 55: 1791-1805.
29. Whitbread AM, Jiri O, Maasdorp B (2004) The effect of managing improved fallows of *mucuna pruriens* on maize production and soil carbon and nitrogen dynamics in sub-humid Zimbabwe. *Nutr Cycl Agroecosys* 69: 59-71.
30. Pandey RK, Maranville JW, Chetima MM (2000) Deficit irrigation and nitrogen effects on maize in a Sahelian environment II. Shoot growth, nitrogen uptake and water extraction. *Agric Water Manag* 46: 15-27.
31. Liu K, Wiatrak P (2011) Corn (*Zea mays* L.) plant characteristics and grain yield response to N fertilization programs in no-till system. *Am J Agric Biol Sci* 6: 279-286.
32. Vincent V, Thomas RG (1961) An agro-ecological survey of Southern Rhodesia: Part I agro-ecological survey. Salisbury, Government Printers.
33. Nyamapfene K (1991) Soils of Zimbabwe. Nehanda Publishers (Pvt) Ltd, Harare, Zimbabwe.
34. Burt R, Wilson MA, Kanyanda CW, et al. (2001) Properties and effects of management on selected granitic soils in Zimbabwe. *Geoderma* 101: 119-141.
35. Twomlow S, Urolov JC, Jenrich M, et al. (2008) Lessons from the field – Zimbabwe's conservation agriculture task force. *J Semi-Arid Trop Agric Res* 6: 1-11.
36. Statistix. Statistix 9: Analytical Software. 2008. Available from: www.statistix.com
37. Hatfield JL, Sauer TJ, Prueger JH (2001) Managing soils to achieve greater water use efficiency: A review. *Agron J* 93: 271-280.
38. Monzon JP, Sadras VO, Andrade FH (2006) Fallow soil evaporation and water storage as affected by stubble in sub-humid (Argentina) and semi-arid (Australia) environments. *Field Crops Res* 98: 83-90.
39. Muchecheti F, Madakadze IC, Soundy P (2012) Production of rape (*Brassica napus* L.) on soils amended with leguminous tree prunings: Yield responses in relation to the chemical composition of the tree prunings. *Afr J Agric Res* 7: 3541-3549.
40. Cairns JE, Sanchez C, Vargas M, et al. (2012) Dissecting maize productivity: Ideotypes associated with grain yield under drought stress and well-watered conditions. *J Int Plant Biol* 54: 1007-1020.
41. Lafitte HR (1994) Identifying production problems in tropical maize: a field guide. Mexico, D.F. CIMMYT.
42. Nezomba H, Mtambanengwe F, Chikowo R, et al. (2014) Sequencing integrated soil fertility management options for sustainable crop intensification by different categories of smallholder farmers in Zimbabwe. *Exp Agric* 51: 17-41.

CHAPTER 4

PLANT CO-EXISTENCE PATTERNS AND HIGH-ARCTIC VEGETATION COMPOSITION IN THREE COMMON PLANT COMMUNITIES IN NORTH-EAST GREENLAND

Oriol Grau,[1] Josep M. Ninot,[1] Aaron Pérez-Haase[1] and Terry V. Callaghan[2,3]

[1] Department of Plant Biology and Institute for Research on Biodiversity, University of Barcelona, Av. Diagonal 643, ES-08028 Barcelona, Catalonia, Spain

[2] Royal Swedish Academy of Sciences, Lilla Frescativägen 4A, SE-114 18, Stockholm, Sweden

[3] Department of Animal and Plant Sciences, University of Sheffield, S10 2TN Sheffield, UK

ABSTRACT

Arctic regions are expected to experience substantial changes in climate in the coming decades. In order to predict potential changes of Arctic vegetation, it is important to understand the distinct role of life forms of plants and of individual species in relation to plant co-existence patterns. Our aim is to investigate if three common Arctic plant patch types dominated by contrasting life forms (by the dwarf shrubs Salix arctica or Dryas octopetala×intermedia or by mosses) are related (a) to the co-existence of vascular plants and species richness at patch scale and (b) to the floristic composition in three distinct plant communities (Salix snowbed, Dryas heath and fell-field) associated with contrasting abiotic regimes. The study was conducted at Zackenberg, in north-east Greenland. Dryas patches showed a clear negative effect on small-scale plant richness and co-existence in the fell-field. Salix and moss patches showed a similar pattern in all the plant communities, although the number of individuals growing in Salix patches was lower than in moss patches. Salix and mosses in the fell-fields hosted a high number of species in spite of the much less vegetated aspect of this harsh, upper zone. The floristic composition varied between plant communities, but it did not change substantially between patch types within each community. This study provides novel background knowledge of plant co-existence patterns at patch scale and of the structure of contrasting Arctic plant communities, which will help to better assess the potential effects of varying abiotic stress regimes on Arctic vegetation.

To access the supplementary material for this article, please see the supplementary files under Article Tools online.

Arctic regions are expected to continue to experience substantial changes in climate in the coming decades according to most recent models and assessments (Symon et al.2005; Solomon et al. 2007; AMAP 2011). There is growing confidence that there will be a marked increase in mean annual temperatures and changes in precipitation regimes, particularly snow (Brown & Mote 2009). The regional models project much larger local temperature increases in winter, compared with other seasons, particularly along Greenland's east coast. This is related to the retreat of sea ice and increased

Citation: Oriol Grau, Josep M. Ninot, Aaron Pérez-Haase and Terry V. Callaghan, "Plant Co-Existence Patterns and High-Arctic Vegetation Composition in Three Common Plant Communities in North-East Greenland," Polar Research 2014, 33, 19235, http://dx.doi.org/10.3402/polar.v33.19235.

precipitation in northern Greenland (Stendel et al. 2008). The marked effect of climate change on organisms (e.g., Post et al. 2009; Callaghan, Tweedie et al. 2011) and on ecosystem functioning (e.g., Euskirchen et al. 2006; Lara et al. 2012; Myers-Smith et al.2011) in the Arctic and the consequences of these effects at a larger scale have led to an increasing importance of understanding climate–ecosystem interactions in northern latitudes, particularly as not all ecosystems have responded as expected to climate change (Callaghan, Tweedie et al. 2011; Van Bogaert et al. 2011).

Snow cover and length of the growing season are crucial factors determining the characteristics and functioning of plant communities occurring in deglaciated areas in the Arctic (Evans et al. 1989; Elberling et al. 2008; Callaghan, Johansson et al. 2011). For instance, substantial phenological changes have already been detected recently in plants in response to advanced snowmelt in north-eastern Greenland (Høye et al. 2007; Schmidt et al. 2012) and to manipulated and natural extreme thaw events in the sub-Arctic in winter (Bokhorst et al. 2009). Variations of snow cover regimes lead to marked changes of abiotic soil parameters, such as water content, temperature or depth of the active layer; snow acts both as a resource for vegetation (e.g., moisture store) and as an abiotic stressor (e.g., constraining the length of the growing season). The dynamic balance between variations in the resource and abiotic stresses on plants is important because this balance is expected to determine the nature of plant–plant interactions (Lortie et al.2004) and thus control the plant co-existence patterns. The distinct role of contrasting life forms of plants and of individual species in relation to co-existence patterns in Arctic environments remains poorly explored (Pajunen et al. 2011), although this issue is crucial to understand potential changes of vegetation under varying abiotic stress regimes.

Dwarf shrubs are a dominant life form in cold regions and have expanded during the last decades in northern latitudes (Hallinger & Wilmking 2011; Myers-Smith et al. 2011; Rundqvist et al. 2011). They have been described as potential facilitators or as drivers of plant co-existence and plant community composition in cold environments, such as in alpine, sub-Arctic and Arctic ecosystems (Klanderud & Totland 2004; Graae et al. 2011; Grau et al. 2012). On the other hand, in cold regions, mosses are also very abundant and expected to decrease under a warmer climate (Wijk et al. 2003; Walker et al. 2006) or with increasing snow cover (Wahren et al. 2005); mosses have been described as crucial drivers of vascular plant recruitment and plant composition in the sub-Arctic (Soudzilovskaia et al. 2011). Nevertheless, it remains unknown if and how shrubs or mosses affect the plant co-existence patterns at patch scale and the plant community composition under contrasting abiotic regimes in High-Arctic ecosystems.

Our aim is to investigate if distinct common Arctic plant patch types dominated by dwarf shrubs and mosses are related to the co-existence of vascular plants and species richness at patch scale and the floristic similarity in three distinct plant communities. The plant communities are Salix snowbed, Dryas heath and fell-field, which are associated with contrasting abiotic regimes, with regard to snow cover, soil water saturation, active layer depth or soil temperature (see Table 1). The patch types selected within these three plant communities were patches with high dominance of Salix arctica Pall., Dryas octopetala L.×intermedia Vahl.—hereafter Salix and Dryas, respectively—and patches without any dominating dwarf shrub but covered with a thin layer of mosses. These three patch types show obvious structural differences between them (for instance, the density of the covering canopy, litter accumulation, biomass dry weight/ fresh weight; see patch type descriptions below and Table 1), which may result in different plant co-existence patterns. We hypothesize that the structural differences between the distinct patch types (i.e., Salix, Dryas and mosses) will lead to varying species co-existence and richness patterns at patch scale independently of the community type in which they occur.

Table 1 Attributes of the communities sampled and mean values (±standard deviation) of some relevant biotic and abiotic variables

		Fell-field	***Dryas*heath**	***Salix*snowbed**
Altitude (m a.s.l.)[a]		300–600	150–300	<150 m
Slope angle (°)		13	11	0
Cover of main patch types (%)	*Dryas*	14.07±7.38	40.29±5.03	11.59±9.78
	Salix	7.85±3.42	25.17±7.25	50.09±8.76
	Mosses	6.44±4.09	19.27±17.57	24.17±14.13
Leaf dry weight per area (g/m^2) at patch scale	*Dryas*	92±19	71±19	98±13
	Salix	83±21	74±10	70±4
Leaf N*(%)[b]	*Dryas*	1.46±0.25	1.62±0.32	1.65±0.36
	Salix	1.95±0.27	2.38±0.16	2.33±0.28
Leaf C/N	*Dryas*	30.57±4.20	26.56±4.36	28.18±9.23
	Salix	22.15±2.48	18.11±1.24	18.68±2.11
Average snow depth[c]		No snow or irregular snow cover	61 cm	95 cm
% Water saturation (per vol., at 0–5 cm)[c]		<40	40–60	65–90

Soil T°C at 5 cm[c,d]		Not determined	9.1±3.8	7.6±3.8
Soil parameters[c]		Increasing soil water content from fell-field to snowbed		
		Increasing soil element stocks from fell-field to snowbed		
		Decreasing active layer depth from fell-field to snowbed		

[a]According to Bay (1998).

[b] % N is significantly lower in the fell-field in both *Salix* and in *Dryas* patches, compared to the heath and to the snowbed. *Dryas* patches also show significantly lower values of % N compared to those in *Salix* patches. See Supplementary Fig. S4.

[c]Data from Elberling et al. 2008.

[d]Average of hourly readings in July and August.

MATERIAL AND METHODS

Study Area

The study was conducted in July 2011 along the south-western slope of Mount Aucellabjerg, in the Zackenberg valley, near the Zackenberg Research Station (74°30′N, 20°30′W), on the north-eastern coast of Greenland. This area was deglaciated about 10 000 years ago and currently its High-Arctic climate is strongly affected by the wide and dense belt of polar pack ice on the coast, which makes the climate more continental, with very cold winters, little precipitation and sunny summers (Meltofte & Rasch 2008). During the polar night, monthly mean air temperatures are below −20°C, with dominating northerly winds. In the snow-free summer period, the mean monthly air temperatures vary between 3°C and 7°C in July and August and daily temperatures rarely get below zero during this warmest part of the summer season (Hansen et al. 2008). The growing season at Zackenberg starts in late May in early snow-free areas, while extensive snow cover may prevail into early summer in snowdrift areas (Meltofte & Rasch 2008). The average annual accumulated precipitation at Zackenberg was 261 mm for the years 1996–2005, of which 10% was rain and 7% was mixed precipitation. For the period 1958–2005, there has been an increase of 1.9 mm/year in the annual precipitation and a significant annual warming of 2.25°C for the period 1991–2005 (Hansen et al. 2008).

Plant Communities

We selected three main plant communities occurring along the altitudinal gradient (35 m to 450 m a.s.l.) in Aucellabjerg. These were representative of the vegetation commonly found at low, medium and high elevation, respectively. The communities were the *Salix* snow bed, occurring mainly at the bottom of the valley, with high vegetation cover; the *Dryas* heath, which is found at intermediate altitudes, still with substantial plant cover; and the fell-field, dominated also by *Dryas* but with sparse vegetation (Bay 1998; Table 1). These three plant communities are also associated with decreasing snow thickness from *Salix* snow beds in the valley bottom to the fell-fields at higher altitude; wind-blown snow accumulates on valley bottoms, where snowdrifts persist until early summer (Hinkler et al. 2008). At higher altitudes the effect of wind becomes progressively stronger, and snow is frequently blown away throughout the winter season. Therefore, there is intermediate snow accumulation in the *Dryas* heath and less accumulation in the fell-fields. Soil water content and storage of organic matter increases from the fell-field to the *Salix* snowbed, whereas the active layer depth shows the opposite trend (Table 1); the maximum active layer depth (as determined by the 0°C isotherm; Christiansen et al. 2008) is reached by the end of August. It is about 80 cm deep at the *Dryas* heath and about 45 cm in *Salix* snowbeds (Elberling et al. 2008; Meltofte & Rasch 2008).

Patch Types and Study Species

In each plant community, we searched for three distinct patch types, each dominated (>80% cover) by (1) *Salix*, (2) *Dryas* or (3) mosses, as described below.

Salix is a dominating dwarf shrub in the Zackenberg valley (Bay 1998) which thrives in a wide niche range, from sand and almost barren moraine tills, to snowbeds—where it is dominant—and open fell-fields. It shows wide morphological plasticity, forming diffuse mats with long twigs and large leaves in disturbed spots or under low competitive conditions, and smaller structures in fell-fields or in denser tundra. *Salix* comprises a larger component of musk oxen (*Ovibos moschatus*) and collared lemming (*Dicrostonyx groenlandicus*) diet than *Dryas* (Klein & Bay 1994; Berg et al. 2008).

Dryas is also a dominating dwarf shrub in the study area and is an Arctic–alpine species, which has a circumpolar distribution in the Northern Hemisphere. It has also a wide niche, although it is less opportunistic and also less adapted to long lasting snow cover than *Salix* (Bay 1998). In the Zackenberg region, it forms dense cushion-shaped mats which protrude from the bare soil in the fell-fields, or from flat ground or small convexities in moister tundra and snowbeds. *Dryas* mats have this compact, dense structure as they retain a great amount of dead leaves attached to the stem. It shows higher dry weight/area (Table 1) and higher dry weight/fresh weight ratios than *Salix*(Supplementary Fig. S1a). *Dryas* dominates in the heaths but also occurs sparsely in the snowbed and in the fell-fields.

The third patch type is comprised of mosses without any dominant co-occurring shrub; the distinct moss species forming this patch type could not be identified, partly because of their early stage of seasonal development during the sampling period. The mosses showed a rather homogeneous structural aspect and stage of development; this patch type was dominated by poorly developed moss carpets or tiny layers of prothalli, although some small spots with cushion-shaped acrocarpic mosses (such as*Polytrichum* sp.) and sparse macrolichens (such as *Cetraria ricetroum* Opiz, *Peltigera rufescens* (Weiss) Humb. and *Stereocaulon* gr. *alpinum* Laurer ex Funck) were present. In this patch type there was very little biomass production and litter accumulation compared to *Salix* and *Dryas* patches. The plants growing in the mosses were therefore not generally affected by shade from a covering canopy as occurred in the shrub patches.

Data Collection

We selected three study plots at similar altitudes within each plant community (i.e., *Salix* snowbed, 35–43 m a.s.l.; *Dryas* heath, 182–240 m a.s.l; and fell-field, 415–450 m a.s.l; Fig. 1), separated by a few hundred metres from each other, but with similar orientation (mainly south-west). In a radius of approximately 10 m (Fig. 1) within each plot, we searched for the three distinct patch types. To clearly elucidate the effects of each patch type, mixed patches were rejected. Thus only those where *Dryas* or *Salix* or mosses clearly dominated (>80%) were considered. For instance, we avoided those *Dryas* or *Salix* patches which contained high cover of mosses. Within each patch type we sampled four 25×25 cm quadrats, which were treated as subsamples (n=4 subsamples×3 patch types×3 plots×3 plant communities=108 quadrats in total; see Fig. 1). Each quadrat was surveyed with a rigid frame divided into 100 2.5×2.5 cm squares. The quadrats were separated by only a few meters (generally 2–8 m), and they were always put on patches which were bigger than 25×25 cm. In each quadrat we recorded species present and estimated the number of individuals. Clonal species growing in adjacent 2.5 cm squares within the frame were considered as one single individual when counting the total number of individuals, unless they could be clearly identified as separate individuals (i.e., if visibly recently germinated). However, it is acknowledged that determining an individual of a clonal plant is difficult in practice (Callaghan et al. 1999).

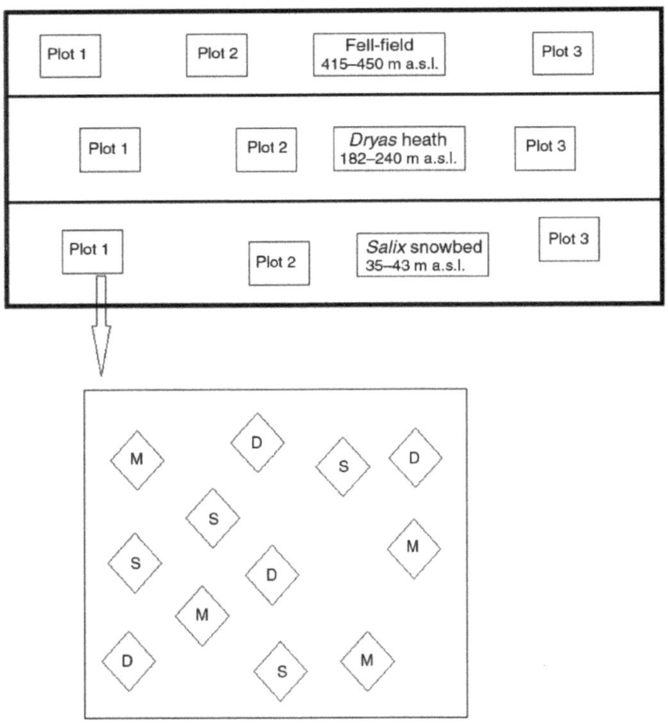

Figure 1: Scheme of the experimental design: in each of the three plant communities—fell-field, *Dryas* heath and *Salix* snowbed—three plots were chosen, which were separated by a few hundred meters from each other; within a radius of ca. 10 m in each plot, three patch types were analyzed: M=mosses; D=*Dryas*; and S=*Salix*. Four 25×25 cm quadrats (subsamples) of each of these three patch types were chosen randomly within each plot.

In order to describe the patch types in terms of aboveground biomass and availability of nitrogen in leaves, we sampled leaves of *Salix* and *Dryas*. We clipped a surface of 10×10 cm for *Salix* and 7×7 cm for *Dryas* (the latter showed a more uniform canopy and this amount was considered sufficient). The fresh leaves were kept moist in small, sealed plastic bags and promptly taken to the laboratory. There green leaves were removed and fresh weights obtained. Leaves were then dried at 70°C for 24 h and the dry weight measured. Leaf dry matter content (dry weight/fresh weight) and the ratio of leaf dry weight to area covered were calculated (Supplementary Fig. S1). Nitrogen and carbon concentrations were determined with an elemental analyzer (EA1108, Series 1; Carlo Erba Instrumentazione, Milan, Italy). To evaluate the% cover of each patch type in each plant community, we recorded the occurrence of the distinct patch types every 5 cm along five parallel lines of 5 m length in each study plot (see Table 1).

Data Analyses

The data were analyzed using R software (R Development Core Team 2012). To investigate whether the total number of species and individuals varied, we used a linear mixed model with Poisson error structure, using the lme4 package (Bates & Maechler2010). The number of species or the number of individuals was the response variable; "plant community" and "patch type" were fixed factors and "patch type" was nested within "plot" as a random factor in the model.

Other parameters were analyzed to investigate the specific composition and structure of the patch types and plant communities studied.

We used species accumulation curves to calculate the β-diversity values, which are a degree measure of uniformity of the species pool if interpreted together with the species fidelity values. To estimate species richness at the patch scale in each community, species accumulation curves were obtained by counting the number of species found when increasing the area sampled for each patch type in each community; all of the quadrats of a given patch type×plant community combination were taken as a whole and were added in a random order for 100 times to obtain the final accumulation curves with the Vegan package (Oksanen et al. 2009). From these curves we calculated the β-diversity (Whittaker 1972), defined here as the ratio relating the final species richness in the accumulation curve (n=12 quadrats) to the mean initial richness value in the accumulation curve (n=1 quadrat). Species fidelity was calculated as implemented in Ginkgo (a vegetation data analyzer developed by de Cáceres 2012) for all species in each patch type×plant community combination. The fidelity measures the degree to which a species is confined to a given group (Legendre & Legendre 2003). We used the phi fidelity statistic (φ) as defined by Chytrý et al. (2002):

$$\phi = \frac{N \cdot np - n \cdot Np}{\sqrt{n \cdot Np \cdot (N - n) \cdot (N - Np)}}$$

This equation takes into account the number of quadrats in the data set (N); the number of relevés in the particular vegetation unit (Np); the number of occurrences of the species in the data set (n); and the number of occurrences of the species in the particular vegetation unit (*np*). The value 1 indicates that the species and the vegetation units are completely faithful to each other; only the more faithful (φ>0.3) and the less faithful (φ<−0.3) species are shown.

To assess the floristic similarity between each patch type×plant community combination, we performed a principal components analysis (PCA) by using the FactoMineR package (Husson et al. 2011). Abundance data was transformed by the Hellinger transformation (Legendre & Gallagher 2001) with the Ginkgo multivariate data analyzer (de Cáceres2012). The Hellinger distance more strongly reduces the highest abundance values compared to low values and avoids the similarities derived from sharing absent species (Legendre & Legendre 2003; Borcard et al. 2011). The results are identical to PCA based on a symmetrical matrix of Hellinger distances between objects.

RESULTS

Plant Species Richness and Co-Existence Patterns

The total number of individual vascular plants counted was 974 in the snowbed community, 994 in the heath and only 556 in the fell-field. Forty eight species were recorded. The number of individuals in *Salix* patches was lower than in moss patches (Fig. 2a; Supplementary Table S1). Species number did not differ between the distinct patch types in snowbed and heath communities, but these numbers were significantly lower in *Dryas* patches in the fell-field community (Fig. 2b; Supplementary Table S1).

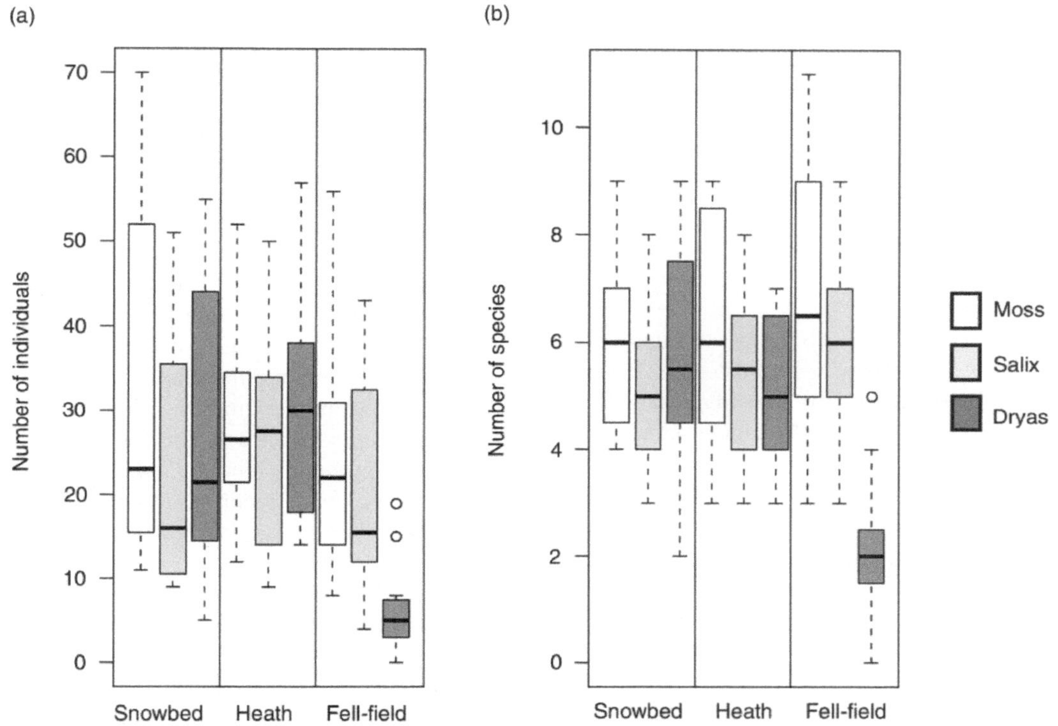

Figure 2: (a) Number of individuals and (b) number of species found in the *Salix* snowbed, the *Dryas* heath and the fell-field in each patch type (moss, *Salix*, *Dryas*).

Species accumulation curves for the distinct patch type×plant community combinations (Fig. 3) indicated that *Dryas* patches generally accumulate (assuming that associations among species are directly caused by the presence of the dominant species) the lowest absolute number of species when considering all the sampled quadrats, irrespective of the plant community. In the fell-field, *Dryas* patches promoted a clearly lower final accumulation of species with increasing sampling area compared to *Salix* and moss patches in this community. *Dryas* patches also showed rather lower species accumulation in the heath, but this difference became less evident in the snowbed community. Species accumulation was especially high in moss patches in all plant communities. The effect of *Dryas* on the β-diversity was highest in the fell-field and lowest in the other communities (Supplementary Table S2).

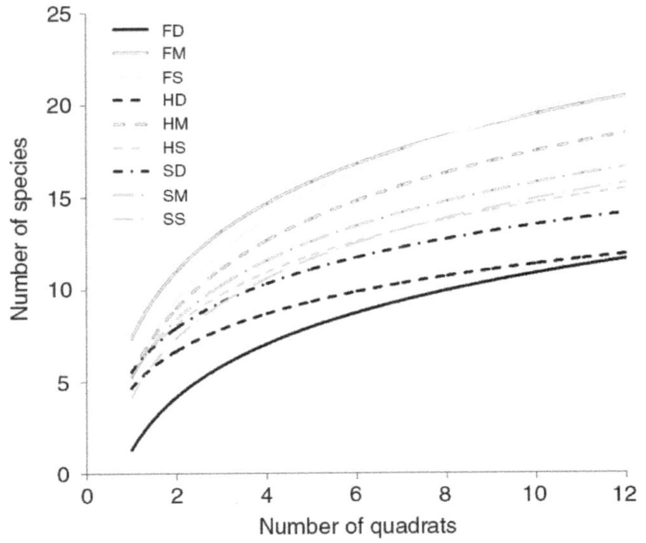

Figure 3: Species accumulation curves for all the distinct patch type×plant community combinations. FD=fell-field *Dryas* patches, FM=moss patches; FS=*Salix* patches; HD=*Dryas* heath *Dryas* patches; HM=moss patches; HS=*Salix* patches; SD=*Salix* snowbed *Dryas* patches; SM=moss patches; and SS=*Salix* patches.

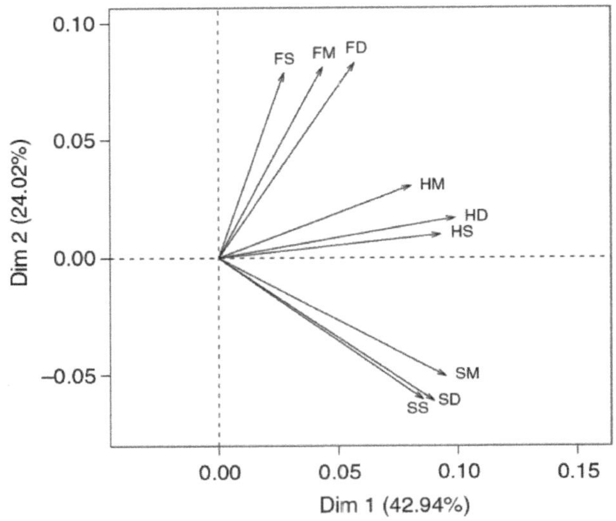

Figure 4: Principal component analysis based on the floristic similarity of the distinct patch type×plant community combinations. FD=fell-field *Dryas* patches; FM=moss patches; FS=*Salix* patches; HD=*Dryas* heath *Dryas* patches; HM=moss patches; HS=*Salix* patches; SD=*Salix* snowbed *Dryas* patches; SM=moss patches; and SS=*Salix* patches.

Plant Communities

The PCA in Fig. 4 shows the ecological distances (i.e., differences in species similarity based on the Hellinger distance) between the distinct patch type×plant community combinations and indicates that the three plant communities studied may be clearly distinguished from each other by their flora. However, there was no clear difference between patch types if plant communities were analyzed separately (results not shown), suggesting that the plots sampled within the plant communities were less variable than plots of the same patch in different communities. When analyzing only presence/absence of species in a PCA ordination, therefore irrespective of their local abundance (Supplementary Fig. S3), we observed a very similar pattern and we did not detect a clear segregation related to patch types either.

Species Fidelity

In agreement with the species ordination shown in Fig. 4, the pool of species showing high or low fidelity to the distinct patch type×plant community combinations (Table 2) varied more between communities than it did between patch types within each plant community. The highest fidelities were reached in moss patches in the fell-field and the lowest in this same community but in *Salix* and *Dryas* patches. No species showed high fidelity to *Dryas* patches in the fell-field. When the fidelity was analyzed for the whole species pool, irrespective of the patch type, several species showed high fidelity to each plant community. In the *Salix* snowbeds the species with highest fidelity (φ value higher than 0.3) were *Hierochloe alpina*, *Arctagrostis latifolia*, *Luzula confusa* and *Alopecurus borealis*; in the heath the high fidelity species were *Poa arctica*, *Kobresia myosuroides*and *Festuca brachyphylla*; in the fell-field the species showing high fidelity were mostly of the genus *Saxifraga* (*S. cernua*, *S. integrifolia*, *S. nivalis*, *S. oppositifolia* and *S. platysepala*) but also the grass *Poa glauca*.

Table 2 List of species with highest (positive values) and lowest (negative values) fidelity to the distinct patch type×plant community combinations; −1 indicates no fidelity to a plant community, whereas +1 indicates maximum fidelity. Only the species with values of fidelity equal to or higher than 0.3 and those with values equal to or lower than −0.3 are shown

	Dryas		**Mosses**		***Salix***	
Fell-field	*Luzula confusa*	−0.41	*Saxifraga platysepala*	0.62		
			Poa glauca	0.49	*Poa glauca*	0.41
			Saxifraga cernua	0.45		
			Sagina intermedia	0.43	*Sagina intermedia*	0.32
			Festuca rubra ssp. richardsonii	0.39	*Luzula confusa*	−0.35
			Saxifraga nivalis	0.39		
Dryas heath	*Kobresia myosuroides*	0.52	*Poa arctica*	0.41	*Poa arctica*	0.34

	Festuca brachyphylla	0.43	*Potentilla hyparctica*	0.3	*Carex bigelowii*	0.3
*Salix*snowbed	*Hierochloe alpina*	0.52	*Alopecurus borealis*	0.44	*Hierochloe alpina*	0.52
	Arctagrostis latifolia	0.41	*Salix arctica*	0.34	*Arctagrostis latifolia*	0.33
	Salix arctica	0.34	*Arctagrostis latifolia*	0.33	*Luzula confusa*	0.3
	*Stellaria longipes*s.l.	0.3				

DISCUSSION

Dominant species most likely determine the species associations at a site. In our study, *Dryas* patches had a clear negative effect on small-scale plant richness and co-existence in the fell-field. We found that the negative impact of *Dryas* on plant co-existence can occur in High-Arctic ecosystems depending on the environmental constraints, even though it has been commonly described as a facilitator low shrub in northern latitudes (Chapin et al. 1994; Cooper et al. 2004). Klanderud & Totland (2004) predicted *Dryas* would have a negative impact on alpine, but not on High-Arctic, community diversity. *Salix* and moss patches supported a high number of species in the less vegetated upper zone. In fact, *Salix* and mosses showed a similar pattern across all the plant communities analyzed.

Plant co-existence and eventual establishment within the dominant vegetation are governed by many factors, including both external factors—e.g., availability of airborne seeds (Klanderud & Totland 2007)—and factors modulated by the species already present (Pajunen et al. 2011), which may also affect microhabitat characteristics—e.g., the accumulation of allelopathic substances or the structure of the canopy of the dominant species, which has important effects on the availability of light to plants growing below (Shevtsova et al. 1995; Totland & Esaete 2002), or on the provision of moisture, shelter and nutrients. Viable seed rain would be expected to decrease with increasing altitude (Thompson 1978). However, the equally high number of individuals and species observed in *Salix* and in moss patches in the fell-field compared to the other two communities suggests that the differing pattern of plant co-existence (and expected eventual establishment) observed in *Dryas* patches is not primarily limited by seed availability. Structural differences between mosses and *Salix* may explain the lower number of individuals occurring in *Salix* patches; however, rather similar growth forms like the dwarf shrubs *Salix* and *Dryas* did not promote a similar pattern of plant co-existence. Our results therefore suggest that structural differences between patch types did not clearly explain the co-existence patterns observed between patches in terms of number of individuals and species. Alternative explanations should be considered instead.

A possible explanation for the pattern observed in *Dryas* compared to the other patch types, irrespective of the physical structure of the dominant species, could be that *Dryas* lacks energy-demanding N-fixing nodulation at higher elevations, whereas at lower elevations it becomes nodulated (Kohls et al. 1994). Therefore, we speculate that *Dryas* growing under the severe conditions found in the fell-fields may act as a stronger competitor for soil N and may thus reduce its recruiting capacity for potential neighbor competitors (Tilman et al. 2001). This would contrast with the theory that predicts that species richness is higher in relatively harsh environmental conditions compared to more favorable sites due to the relatively minor role of competition in these systems (e.g., Grime 1979). Our study suggests that this would not apply to such very extreme conditions in the High-Arctic as those found in the fell-field. On the other hand, *Dryas* at the fell-field might actively inhibit recruitment to a higher degree than *Salix* and mosses by producing allelopathic substances, as plants may respond to varying stress factors by changing their content of secondary plant metabolites (Berg et al. 2008). In any case, further research is needed in relation to possible N-fixation and allelopathy to better understand the observed patterns of plant co-existence in these poorly studied Arctic plant communities.

The high β-diversity observed in *Dryas* patches in the fell-field (Supplementary Table S2) seems to be related to the low fidelity of the few species occurring at the patch scale (Figs. 2b, 3; Supplementary Table S3), which favours a high species turnover between the distinct sampled quadrats. The species found in *Dryas* patches in the fell-field differed greatly between quadrats but generally co-occurred with *Polygonum viviparum*, which was highly abundant in all quadrats (Supplementary Table S3). All the other patch type×plant community combinations showed higher species richness and higher fidelity values, but lower β-diversity, as the species pool was more uniform between quadrats and more similar to the species pool at the community scale.

Floristic similarity based on the Hellinger distance (Fig. 4) varied between plant communities, but it did not change substantially between patch types within each community. Even though plant co-existence patterns and species richness varied at patch scale, the floristic similarity was determined at community scale, as indicated by the similarity between patch types within each community and by the marked differences between communities (Fig. 4; Supplementary Fig. S3).

The fell-field is the most environmentally harsh community, as low plant cover, aboveground primary production and foliar N content are all low (Table 1). Moss and the *Salix* patches act as small biodiversity "hot spots" in this highly stressed plant community.

The floristic similarity of *Dryas* heath is intermediate between the fell-field and the snowbed (Fig. 4). However, the similar species richness and co-existence patterns of *Dryas* heath and the *Salix* snowbed indicate that these two plant communities may be influenced by a similar overall stress regime despite the differences in the characteristics of the abiotic stress occurring in each community (Table 1). These two communities have a lower overall stress balance than that found at the fell-field.

Our results suggest that the richness and the plant co-existence patterns vary between patch types if influenced by strong abiotic stress. It is predicted that the study area will undergo changes in precipitation and snow cover in the future

(Stendel et al. 2008; Brown & Mote 2009; AMAP 2011); altering the abiotic stress regimes will affect the nature of plant–plant interactions (Lortie et al. 2004) and eventually the plant co-existence patterns. If these predicted changes promote stronger abiotic stress, the Arctic plant communities dominated by *Dryas* may decrease and this will show a decrease in species richness through increased negative interactions of this dwarf shrub on the co-occurring plant species. Further, any changes on the stress regimes altering moss and *Salix* cover occurring in the fell-fields may promote changes in species richness and diversity at the community scale.

CONCLUSIONS

Under a strong abiotic stress regime such as that found at the fell-field, the distinct dominant patch types (i.e., mosses, *Salix* and *Dryas*) did not show uniform plant species richness and co-existence patterns; in the comparatively more benign communities (i.e., *Salix* snowbed and *Dryas* heath) the distinct patch types showed similar patterns. We therefore predict that any future environmental changes altering the occurrence of the dominant patch types in fell-fields in the High-Arctic may greatly alter plant species richness and co-existence patterns. Further, the contrasting abiotic conditions in the three plant communities strongly determined the floristic composition based on ecological distances at community scale.

This study provides novel insight on plant co-existence patterns and on the structure of contrasting Arctic plant communities in this remote region in north-east Greenland. Our research will help to better assess the potential effects of varying abiotic stress regimes on Arctic vegetation.

SUPPLEMENTARY INFORMATION

Table S1. Results of the generalized linear mixed models with Poisson error structure on the number of individuals (above) and number of species (below)

ralized linear mixed model: Fixed effects: number of individuals ~ Community * Patch type Random effect: Plot (patch type nested within plot)				
	Estimate	Std. error	z-value	Pr(>|z|)
(Intercept)	3.29759	0.25925	12.720	< 2e-16 ***
Heath community	-0.01204	0.36637	-0.033	0.9738
Fell-field community	-0.14048	0.36697	-0.383	0.7019
Salix patch	-0.35914	0.16283	-2.206	0.0274 *
Dryas patch	-0.18979	0.16068	-1.181	0.2375
Heath community:*Salix* patch	0.26353	0.22739	1.159	0.2465
Fell-field community:*Salix* patch	0.15904	0.23037	0.690	0.4900
Heath community:*Dryas* patch	0.25541	0.22468	1.137	0.2556
Fell-field:*Dryas* patch	-1.18432	0.24854	-4.765	1.89e-06 ***
Number of observations: 108				
Number of groups:				
Plot 9	Patch type within plot 27			

Generalized linear mixed model: Fixed effects: number of species ~ Community * Patch type Random effect: Plot (patch type nested within plot)				
	Estimate	Std. error	z-value	Pr(>|z|)
(Intercept)	1.783e+00	1.388e-01	12.853	< 2e-16 ***
Heath community	4.109e-02	1.948e-01	0.211	0.833
Fell-field community	1.551e-01	1.911e-01	0.811	0.417
Salix patch	-1.823e-01	1.750e-01	-1.042	0.297
Dryas patch	-4.256e-02	1.686e-01	-0.252	0.801
Heath community:*Salix* patch	3.922e-02	2.437e-01	0.161	0.872
Fell-field community:*Salix* patch	-1.633e-07	2.384e-01	0.000	1.000
Heath community:*Dryas* patch	-1.806e-01	2.418e-01	-0.747	0.455
Fell-field:*Dryas* patch	-1.130e+00	2.809e-01	-4.023	5.74e-05 ***

Number of observations: 108	
Number of groups:	
Plot 9	Patch type within plot 27

Table S2. Multiplicative β-diversity: final species richness in the accumulation curve/mean initial number of species from species accumulation curves

Patch type × plant community combination	β-diversity
Dryas patches in the *Salix* snowbed	2.446
Dryas patches in the *Dryas* heath	2.478
Moss patches in the fell-field	2.733
Moss patches in the *Salix* snowbed	2.761
Salix patches in the *Dryas* heath	2.814
Moss patches in the *Dryas* heath	2.992
Salix patches in the *Salix* snowbed	3.267
Salix patches in the fell-field	3.589
Dryas patches in the fell-field	5.678

Table S3. Models describing the best fit to species abundance distributions and the three most abundant species found in each patch type and plant community. The differences estimated by these fitted models depend basically on how abundant the most abundant species are compared to the rest of species. We firstly classified the species from most to least abundant (i.e., rank-abundance plots, not shown) using the Biodiversity R package (Kindt 2004) in each patch type × plant community combination. In this classification we considered all the individuals of each species from all quadrats together. Then, we selected the model which best fitted their species abundance distribution, that is, with the lowest Akaike Information Criterion (AIC) value compared to all the other potential models (Kindt & Coe 2005). Log-normal and Zipf models were fitted as a generalized linear model with a logarithmic link function. A preemption model was fitted as a purely nonlinear model (see Supplementary Fig. S2 for a graphic description of the fitted models). The model which best fitted the species abundance distribution differed between patch types in the fell-field and in the heath but it became similar for all patch types in the snowbed, where species abundances (i.e., number of individuals for each species) in all patch types showed a similar decreasing pattern from the most to the least abundant species. Moss and Salix patches showed a similar pattern in the fell-field and in the heath, whereas Dryas patches showed a varying distribution in each plant community

		Dryas	Mosses	*Salix*
Fell-field	Best model	Zipf	Log-normal	Zipf
	Most abundant species	*Polygonum viviparum* *Salix arctica* *Festuca ovina*	*Polygonum viviparum* *Dryas octopetala* *Saxifraga cernua*	*Polygonum viviparum* *Dryas octopetala* *Festuca ovina*
Dryas heath	Best model	Log-normal	Log-normal	Zipf
	Most abundant species	*Polygonum viviparum* *Luzula confusa* *Salix arctica*	*Polygonum viviparum* *Poa arctica* *Salix arctica*	*Polygonum viviparum* *Luzula confusa* *Dryas octopetala*
Salix snowbed	Best model	Pre-emption	Pre-emption	Pre-emption
	Most abundant species	*Equisetum arvense* *Polygonum viviparum* *Salix arctica*	*Equisetum arvense* *Polygonum viviparum* *Luzula confusa*	*Equisetum arvense* *Hierochloe alpine* *Polygonum viviparum*

(a)

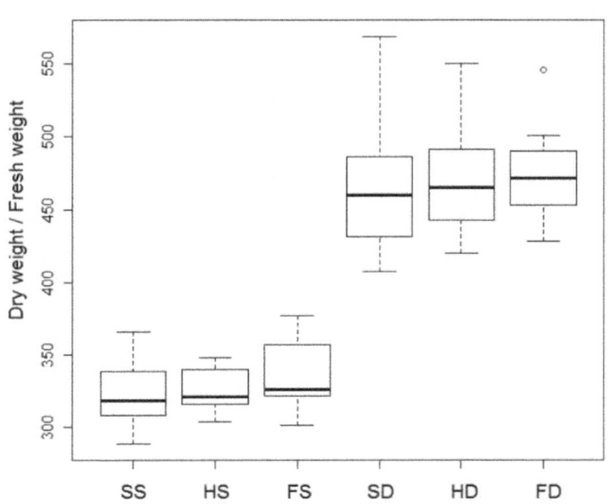

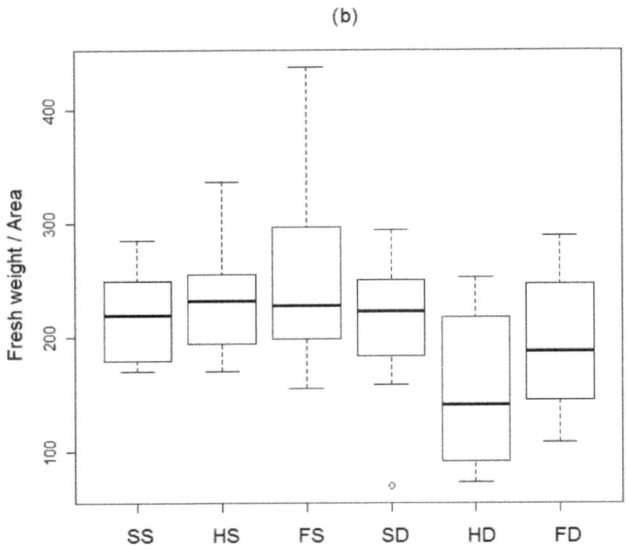

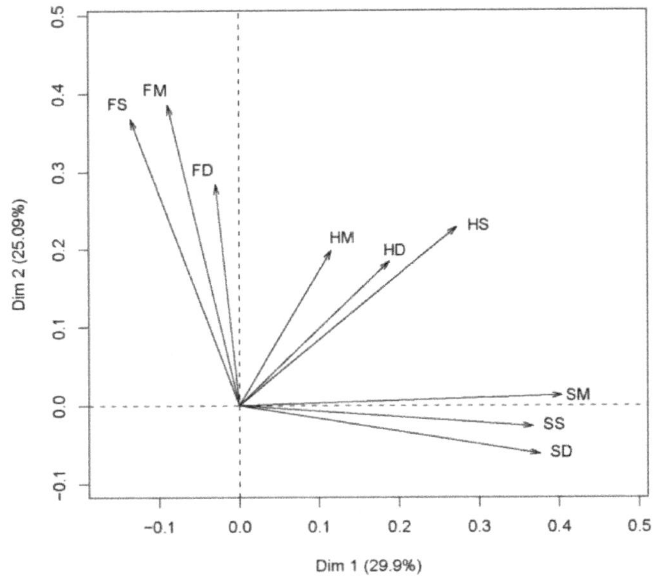

Figure S1. (a) Leaf dry weight/leaf fresh weight ratio (mg/g) and (b) leaf fresh weight/soil area covered (g/m2) in Salix and Dryas patches in the Salix snowbed, the Dryas heath and the fell-field. Terms are abbreviated as follows: fell-field Dryas patches (FD), moss patches (FM) and Salix patches (FS); Dryas heath Dryas patches (HD), moss patches (HM) and Salix patches (HS); Salix snowbed Dryas patches (SD), moss patches (SM) and Salix patches (SS).

Figure S3. Principal component analysis based on the presence and absence of species in the distinct patch type × plant community combinations. Terms are abbreviated as follows: fell-field Dryas patches (FD), moss patches (FM) and Salix patches (FS); Dryas heath Dryas patches (HD), moss patches (HM) and Salix patches (HS); Salix snowbed Dryas patches (SD), moss patches (SM) and Salix patches (SS).

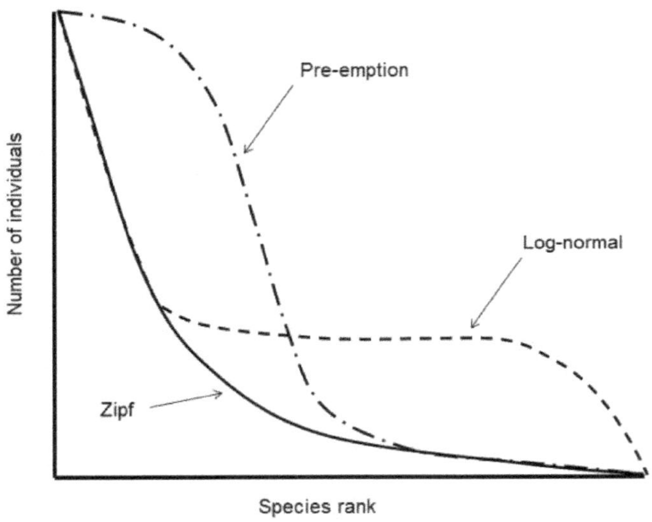

Figure S2. Pre-emption, log-normal and Zipf species abundance distribution models.

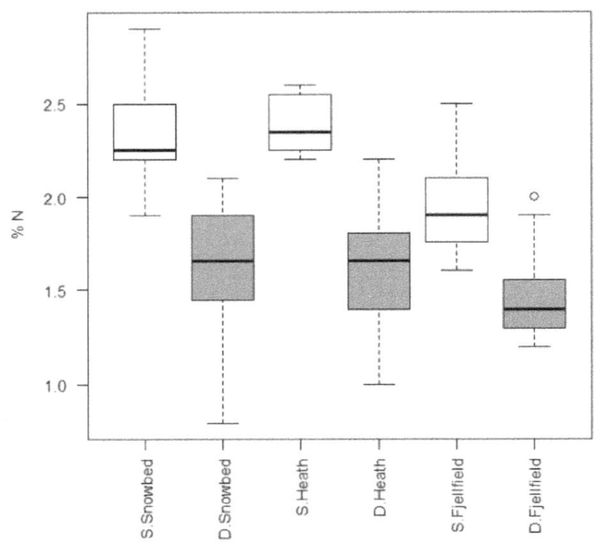

Figure S4. Leaf N (%) in Salix (S) and in Dryas (D) growing in the Salix snowbed, in the Dryas heath and in the fell-field.

ACKNOWLEDGEMENTS

The research leading to these results has received funding from INTERACT (grant agreement no. 262693) under the European Community's Seventh Framework Program, and from the research project no. 2009SGR451 from the AGAUR Agency of the Catalan Government. We thank Niels M. Schmidt for his useful advices and the staff at the Zackenberg Research Station for logistical support. We also thank two anonymous reviewers and Bente J. Graae for providing helpful comments on earlier versions of the manuscript.

REFERENCES

1. AMAP (Arctic Monitoring and Assessment Programme) 2011. *Snow, water, ice and permafrost in the Arctic (SWIPA)*. Oslo: Arctic Monitoring and Assessment Programme.
2. Bates D. & Maechler M. 2010. *lme4: linear mixed-effects models using S4 classes. R package version 0.999375-33*. Accessed on the internet at http://CRAN.R-project.org/package=lme4 on 1 June 2010.
3. Bay C. 1998. *Vegetation mapping of Zackenberg valley, northeast Greenland*. Copenhagen: Danish Polar Center and Botanical Museum, University of Copenhagen.
4. Berg T.B., Schmidt N.M., Høye T.T., Aastrup P.J., Hendrichsen D.K., Forchhammer M.C. & Klein D.R. 2008. High-Arctic plant–herbivore interactions under climate influence.*Advances in Ecological Research 40*, 275–298.
5. Bokhorst S., Bjerke J.W., Tømmervik H., Callaghan T.V. & Phoenix G.K. 2009. Winter warming events damage sub-Arctic vegetation: consistent evidence from an experimental manipulation and a natural event. *Journal of Ecology 97*, 1408–1415.
6. Borcard D., Gillet F. & Legendre P. 2011. *Numerical ecology with R*. Dordrecht: Springer.
7. Brown R.D. & Mote P.W. 2009. The response of Northern Hemisphere snow cover to a changing climate. *Journal of Climate 22*, 2124–2145.
8. Callaghan T.V., Johansson M., Brown R.D., Groisman P.Y., Labba N., Radionov V., Bradley R.S., Blangy S., Bulygina O.N., Christensen T.R., Colman J.E., Essery R.L.H., Forbes B.C., Forchhammer M.C., Golubev V.N., Honrath R.E., Juday G.P., Meshcherskaya A.V., Phoenix G.K., Pomeroy J., Rautio A., Robinson D.A., Schmidt N.M., Serreze M.C., Shevchenko V.P., Shiklomanov A.I., Shmakin A.B., Sköld P., Sturm M., Woo M. & Wood E.F. 2011. Multiple effects of changes in Arctic snow cover. *Ambio 40*, 32–45.
9. Callaghan T.V., Press M.C., Lee J.A., Robinson D. & Anderson C. 1999. Spatial and temporal variability in the responses of Arctic terrestrial ecosystems to environmental change. *Polar Research 18*, 1–7.
10. Callaghan T.V., Tweedie C.E., Åkerman J., Andrews C., Bergstedt J., Butler M.G., Christensen T.R., Cooley D., Dahlberg U., Danby R.K., Daniëls F.J., de Molenaar J.G., Dick J., Mortensen C.E., Ebert-May D., Emanuelsson U., Eriksson H., Hedenås H., Henry H.R.G, Hik D.S., Hobbie J.E., Jantze E.J., Jaspers C., Johansson C., Johansson M., Johnson D.R., Johnstone J.F., Jonasson C., Kennedy C., Kenney A.J., Keuper F., Koh S., Krebs C.J., Lantuit H., Lara M.J., Lin D., Lougheed V.L., Madsen J., Matveyeva N., Mcewen D.C., Myers-Smith I.H., Narozhniy Y.K., Olsson H., Pohjola V.A., Price L.W., Rigét F., Rundqvist S., Sandström A., Tamstorf M., Van Bogaert R., Villarreal S., Webber P.J. & Zemtsov V.A. 2011. Multi-decadal changes in tundra environments and ecosystems: synthesis of the International Polar Year Back to the Future Project. *Ambio 40*, 705–716.
11. Chapin F.S. III, Walker L.R., Fastie C.L. & Sharman L.C. 1994. Mechanisms of primary succession following deglaciation at Glacier Bay, Alaska. *Ecological Monographs 64*, 149–175.
12. Christiansen H.H., Sigsgaard C., Humlum O., Rasch M. & Hansen B.U. 2008. Permafrost and periglacial geomorphology at Zackenberg. *Advances in Ecological Research 40*, 151–174.
13. Chytrý M., Lubomír T., Jason H. & Zoltán B.D. 2002. Determination of diagnostic species with statistical fidelity measures. *Journal of Vegetation Science 13*, 79–90.
14. Cooper E.J., Alsos I.G., Hagen D., Smith F.M., Coulson S.J. & Hodkinson I.D. 2004. Plant recruitment in the High Arctic: seed bank and seedling emergence on Svalbard. *Journal of Vegetation Science 15*, 115–124.
15. de Cáceres M. 2012. Ginkgo user's manual. Version 1.4. Accessed on the internet athttp://biodiver.bio.ub.es/veganaweb/resources/manual-ginkgo/manual.pdf on 1 October 2012.
16. Elberling B., Tamstorf M.P., Michelsen A., Arndal M.F., Sigsgaard C, Illeris L., Bay C., Hansen B.U., Christensen T.R., Hansen E.S., Jakobsen B.H. & Beyens L. 2008. Soil and plant community-characteristics and dynamics at Zackenberg. *Advances in Ecological Research 40*, 223–248.
17. Euskirchen E.S., McGuire A.D., Kicklighter D.W., Zhuang Q., Clein J.S., Dargaville R.J., Dye D.G., Kimball J.S., McDonald K.C., Melillo J.M., Romanovsky V.E. & Smith N.V. 2006. Importance of recent shifts in soil thermal dynamics on growing season length, productivity, and carbon sequestration in terrestrial high-latitude ecosystems. *Global Change Biology 12*, 731–750.
18. Evans B.M., Walker D.A., Benson C.S., Nordstrand E.A. & Petersen G.W. 1989. Spatial interrelationships between terrain, snow distribution and vegetation patterns at an Arctic foothills site in Alaska. *Holarctic Ecology 12*, 270–278.
19. Graae B.J., Ejrnæs R., Lang S.I., Meineri E., Ibarra P.T. & Bruun H.H. 2011. Strong microsite control of seedling recruitment in tundra. *Oecologia 166*, 565–576.
20. Grau O., Ninot J., Blanco-Moreno J.M., van Logtestijn R.S.P, Cornelissen J.H.C. & Callaghan T.V. 2012. Shrub–tree interactions and environmental changes drive treeline dynamics in the Subarctic. *Oikos 121*, 1680–1690.
21. Grime J.P. 1979. *Plant strategies and vegetation processes*. Chichester, UK: John Wiley & Sons.
22. Hallinger M. & Wilmking M. 2011. No change without a cause—why climate change remains the most plausible reason for shrub growth in Scandinavia. *New Phytologist 189*, 902–908.
23. Hansen B.U., Sigsgaard C., Rasmussen L., Cappelen J., Hinkler J., Mernild S.H., Petersen D., Tamstorf M.P., Rasch M. & Hasholt B. 2008. Present-day climate at Zackenberg.*Advances in Ecological Research 40*, 111–149.
24. Hinkler J., Hansen B.U., Tamstorf M.P., Sigsgaard C. & Petersen D. 2008. Snow and snow-cover in central

northeast Greenland. *Advances in Ecological Research 40*, 175–195.

25. Høye T.T., Post E., Meltofte H., Schmidt N.M. & Forchhammer M.C. 2007. Rapid advancement of spring in the High Arctic. *Current Biology 17*, 449–450.
26. Husson F., Julie J & Lê S. 2011. *FactomineR*. Accessed on the internet athttp://factominer.free.fr/ on 1 December 2011.
27. Klanderud K. & Totland Ø. 2004. Habitat dependent nurse-effects of the dwarf shrub*Dryas octopetala* on alpine and Arctic plant community structure. *Ecoscience 11*, 410–420.
28. Klanderud K. & Totland Ø. 2007. The relative role of dispersal and local interactions for alpine plant community diversity under simulated climate warming. *Oikos 116*, 1279–1288.
29. Klein D.R. & Bay C. 1994. Resource partitioning by mammalian herbivores in the High Arctic. *Oecologia 97*, 439–450.
30. Kohls S.J., van Kessel C., Baker D.D., Grigal D.F. & Lawrence D.B. 1994. Assessment of N_2 fixation and N cycling by *Dryas* along a chronosequence within the forelands of the Athabasca Glacier, Canada. *Soil Biology and Biochemistry 26*, 623–632.
31. Lara M.J., Villarreal S., Johnson D.R., Hollister R.D., Webber P.J. & Tweedie C.E. 2012. Estimated change in tundra ecosystem function near Barrow, Alaska between 1972 and 2010. *Environmental Research Letters 7*, 015507, doi: 10.1088/1748–9326/7/1/015507.
32. Legendre P. & Gallagher E.D. 2001. Ecologically meaningful transformations of ordinations of species data. *Oecologia 129*, 271–280.
33. Legendre P. & Legendre L. 2003. *Numerical ecology*. 3rd edn. Amsterdam: Elsevier.
34. Lortie C., Brooker R.W., Choler P., Kikvidze Z., Michalet R., Pugnaire F.I. & Callaway R.M. 2004. Rethinking plant community theory. *Oikos 107*, 433–438.
35. Meltofte H. & Rasch M. 2008. The study area at Zackenberg. *Advances in Ecological Research 40*, 101–110.
36. Myers-Smith I., Forbes B., Wilmking M., Hallinger M., Lantz T., Blok D., Tape K., Macias-Fauria M., Sass-Klaassen U., Lévesque E., Boudreau S., Ropars P., Hermanutz L., Trant A., Siegwart Collier L., Weijers S., Rozema J., Martin Schmidt N., Schaepman-Strub G., Wipf S., Rixen C., Ménard C., Venn S., Goetz S., Andreu-Hayles L., Elmendorf S., Ravolainen V., Welker J., Grogan Pz., Epstein H. & Hik D. 2011. Shrub expansion in tundra ecosystems: dynamics, impacts and research priorities. *Environmental Research Letter6*, 045509, doi: 10.1088/1748-9326/6/4/045509.
37. Oksanen J., Kindt R., Legendre P., O'Hara P., Simpson G.L., Solymos P., Steven M.H.H. & Wagner H. 2009. Vegan: Community Ecology Package. R package, version 1, 15–3. Accessed on the internet at http://cran.r-project.org/web/packages/vegan/index.html on 1 June 2010.
38. Pajunen A.M., Oksanen J. & Virtanen R. 2011. Impact of shrub canopies on understory vegetation in western Eurasian tundra. *Journal of Vegetation 22*, 837–846.
39. Post E., Forchhammer M.C., Bret-Harte S., Callaghan T.V., Christensen T.R., Elberling B., Fox A.D., Gilg O., Hik D.S., Ims R.A., Jeppesen E., Klein D.R., Madsen J., McGuire A.D., Rysgaard S., Schindler D.E., Stirling I., Tamstorf M.P., Tyler N.J.C., van der Wal R., Welker J., Wookey P.A., Schmidt N.M. & Aastrup P. 2009. Ecological dynamics across the Arctic associated with recent climate change. *Science 325*, 1355–1358.
40. R Development Core Team 2012. *R: a language and environment for statistical computing*. Vienna: R Foundation for Statistical Computing.
41. Rundqvist S., Hedenas H., Sandstrom A., Emanuelsson U., Eriksson H., Jonasson C. & Callaghan T.V. 2011. Tree and shrub expansion over the past 34 years at the tree-line near Abisko, Sweden. *Ambio 40*, 683–692.
42. Schmidt N.M., Kristensen D.K., Michelsen A. & Bay C. 2012. High Arctic plant community responses to a decade of ambient warming. *Biodiversity 13*, 191–199.
43. Shevtsova A., Ojala A., Neuvonen S., Vieno M. & Haukioja E. 1995. Growth and reproduction of dwarf shrubs in a Subarctic plant community: annual variation and above-ground interactions with neighbours. *Journal of Ecology 83*, 263–275.
44. Solomon S., Qin D., Manning M., Chen Z., Marquis M., Averyt K.B., Tignor M. & Miller H.L. Jr. (eds.) 2007. *Climate change 2007. The physical science basis: contribution of Working Group I to the fourth assessment report of the Intergovernmental Panel on Climate Change*. Cambridge: Cambridge University Press.
45. Soudzilovskaia N.A., Graae B.J., Douma J.C., Grau O., Milbau A., Shevtsova A., Wolters L. & Cornelissen J.H.C. 2011. How do bryophytes govern generative recruitment of vascular plants? *New Phytologist 190*, 1019–1031.
46. Stendel M., Christensen J.H. & Petersen D. 2008. Arctic climate and climate change with a focus on Greenland. *Advances in Ecological Research 40*, 13–43.
47. Symon C., Arris L. & Heal B. (eds.) 2005. *Arctic climate impact assessment*. Cambridge: Cambridge University Press.
48. Thompson K. 1978. The occurrence of buried viable seeds in relation to environmental gradients. *Journal of Biogeography 5*, 425–430.
49. Tilman D., Reich P.B., Knops J., Wedin D., Mielke T. & Lehman C. 2001. Diversity and productivity in a long-term grassland experiment. *Science 294*, 843–845.
50. Totland Ø. & Esaete J. 2002. Effects of willow canopies on plant species performance in a low-alpine community. *Plant Ecology 161*, 157–166.
51. Van Bogaert R., Haneca K., Hoogesteger J., Jonasson C., De Dapper M. & Callaghan T.V. 2011. A century of tree line changes in sub-Arctic Sweden show local and regional variability and only a minor role of 20th century climate warming. *Journal of Biogeography 38*, 907–921.
52. Van Wijk M., Clemmensen K.E., Shaver G.R., Williams M., Callaghan T.V., Chapin F.S. III, Cornelissen J.H.C.,

Gough L., Hobbie S.E., Jonasson S., Lee J.A., Michelsen A., Press M.C., Richardson S.J. & Rueth H. 2003. Long-term ecosystem level experiments at Toolik Lake, Alaska, and at Abisko, Northern Sweden: generalisations and differences in ecosystem and plant type responses to global change. *Global Change Biology 10*, 105–123.

53. Wahren C.H.A., Walker M.D. & Bret-Harte M.S. 2005. Vegetation responses in Alaskan Arctic tundra after 8 years of a summer warming and winter snow manipulation experiment. *Global Change Biology 11*, 537–552.

54. Walker M.D., Wahren C.H., Hollister R.D., Henry G.H.R, Ahlquist L.E., Alatalo J.M., Bret-Harte M.S., Calef M.P., Callaghan T.V., Carroll A.B., Epstein H.E., Jónsdóttir I.S., Klein J.A., Magnússon B., Molau U., Oberbauer S.F., Rewa S.P., Robinson C.H., Shaver G.R., Suding K.N., Thompson C.C., Tolvanen A., Totlandt Ø., Turner P.L., Tweedie C.E., Webber P.J. & Wookey P.A. 2006. Plant community responses to experimental warming across the tundra biome. *Proceedings of the National Academy of Sciences of the United States of America 103*, 1342–1346.

55. Whittaker R.H. 1972. Evolution and measurement of species diversity. *Taxon 21*, 213–251.

CHAPTER 5

TOWARDS A STANDARD PLANT SPECIES SPECTRAL LIBRARY PROTOCOL FOR VEGETATION MAPPING: A CASE STUDY IN THE SHRUBLAND OF DOÑANA NATIONAL PARK

Marcos Jiménez[1] and Ricardo Díaz-Delgado[2]

[1] Remote Sensing Area, National Institute of Aerospace Technologies (INTA), Ctra. Ajalvir s/n, Torrejón de Ardoz, 28850 Madrid, Spain

[2] Remote Sensing and GIS Lab., Doñana Biological Station, CSIC, Avda. Americo Vespucio, 41092 Sevilla, Spain

ABSTRACT

One of the main applications of field spectroscopy is the generation of spectral libraries of Earth's surfaces or materials to support mapping activities using imaging spectroscopy. To enhance the reliability of these libraries, spectral signature acquisition should be carried out following standard procedures and controlled experimental approaches. This paper presents a standard protocol for the creation of a spectral library for plant species. The protocol is based on characterizing the reflectance spectral response of different species in the spatiotemporal domain, by accounting for intra-species variation and inter-species similarity. A practical case study was conducted on the shrubland located in Doñana National Park (SW Spain). Spectral libraries of the five dominant shrub species were built (*Erica scoparia*, *Halimium halimifolium*, *Ulex australis*, *Rosmarinus officinalis*, and *Stauracanthus genistoides*). An estimation was made of the separability between species: on one hand, the Student's t-test evaluates significant intra-species variability ($p < 0.05$) and on the other hand, spectral similarity value (SSV) and spectral angle mapper (SAM) algorithms obtain significant separability values for dominant species, although it was not possible to discriminate the legume species *Ulex australis* and *Stauracanthus genistoides*.

INTRODUCTION

In order to develop understanding of the effects of ecological processes and perturbations on the spatiotemporal distribution of plant communities, organizations responsible for managing protected natural areas require vegetation mapping at species level [1]. Furthermore, the ability to map and monitor changes in species composition over time optimizes management plans in these areas and is useful in tracking the spatiotemporal trends of invasive and keystone species [2].

Hyperspectral remote sensing has the potential to provide quantitative information on spatial cover, species composition, and the physicochemical status of vegetation, a capability that has already been demonstrated [3]. Among hyperspectral techniques, imaging spectroscopy is well developed on airborne platforms and has recently gained ground with the deployment of Unmanned Aerial Vehicles (UAV), although the process is still in its initial stages for spaceborne instruments [4]. In this respect, forthcoming space missions, such as EnMAP [5] or PRISMA [6], will bolster to move forward in the consolidation of hyperspectral techniques. Field spectroscopy predates the

Citation: Jiménez, M.; Díaz-Delgado, R. Towards a Standard Plant Species Spectral Library Protocol for Vegetation Mapping: A Case Study in the Shrubland of Doñana National Park. ISPRS Int. J. Geo-Inf. 2015, 4, 2472-2495. doi:10.3390/ijgi4042472.

development of airborne or spaceborne imaging spectroscopy by many years [7], and has experienced remarkable growth over the past two decades in terms of its use and its application in different scientific disciplines.

Although imaging spectroscopy data looks promising for plant species mapping, few operational approaches exist because of our limited biophysical understanding of when remotely sensed signatures indicate the presence of unique species within and across plant communities [8]. Two of the main drawbacks in this respect are: high spectral similarity among species with similar ecological adaptations [9]; and, in contrast, high spectral variability response within species due to the variation ranges in plant constituents (such as tissue chemistry and structure) across environmental gradients [10]. Besides these drawbacks, vegetation phenology is layered on the top, a factor reducing inter-species similarity while at the same time increasing intra-species variation ranges [11].

In order to improve mapping efficiency using imaging spectroscopy, the analysis techniques applied to imagery could provide more satisfactory results by introducing ground-truth data to characterize the spectral response of every plant species studied [12]. However, field spectroscopy is the primary method used to collect ground-truth data for the development of reference spectral libraries [13]. In addition, a vast amount of literature documenting the potential of using plant spectral libraries for vegetation mapping is available (see Section 2). Nevertheless, in order to create consistently unique and detectable spectral signatures among species, spectral libraries should take into account the spatiotemporal variability of plant species across plant communities and throughout the seasons [14].

To ensure improvements in the reliability of these spectral libraries using imaging spectroscopy, reflectance spectral signature acquisition on field campaigns should be based on standard procedures and follow controlled experimental approaches [15]. Furthermore, the important constraints imposed by the complexity of natural illumination and huge variability in the components of a field spectroscopy campaign (*i.e.*, measurement sites, selected targets, sampling protocol), demand a complete metadata system to properly report on what has been measured and the conditions in which the measurements were carried out [16].

At present, there is no standard harmonized protocol for the measurement of field reflectance in plants [17]. In this paper, we propose a complete methodology for the creation of a spectral library suitable for plant species mapping. This protocol seeks to establish standard procedures for the acquisition of plant spectra in the field, based on a number of solid bases for any plant type, but also including flexible procedures depending on the type of plant considered (*i.e.*, forest, shrub, pasture). One of these solid bases for the protocol is to estimate inter-species spectral similarity and intra-species variability by accounting for environmental gradients and phenological changes. To demonstrate the usability of the protocol, a practical case study was conducted in the shrub communities located in the stabilized sand dune ecosystem in Doñana National Park (SW, Spain).

BACKGROUND: CURRENT STATUS OF FIELD SPECTROSCOPY FOR PLANT SPECTRAL LIBRARY GENERATION

Before outlining the proposed protocol, this section notes several general issues in generating plant spectral libraries: it summarizes the current status of field spectroscopy in the solar spectrum; it highlights the most specific aspects of the acquisition of plant spectra in the field, and finally, it lists the most widely used spectral separability metrics.

Field Spectroscopy

Field spectroscopy is the measurement of high-resolution spectral radiance or irradiance in the field and is applied to retrieve the reflectance or emissivity spectral signatures of terrestrial surface targets. Compared with airborne or spaceborne imaging spectroscopy, the sensing instrument in the field can remain fixed over the subject of interest for much longer, thereby reducing the path length between the instrument and the object being measured [7].

The rugged and portable spectroradiometers currently available have evolved from the non-imaging spectrometers commonly used in the laboratory. Field spectroradiometer manufacturers fundamentally offer two types of instruments: (1) small, light devices that are designed to work only in the visible near infrared (VNIR: 350–1000 nm), with a signal-to-noise ratio (SNR) of around 250:1; (2) larger, heavier devices that are sensitive to the entire solar spectrum, with refrigerated shortwave infrared (SWIR: 1000–2500 nm) detectors and an SNR of around 1000:1. Depending on the application, these spectroradiometers can be configured for the sampling interval and spectral resolution. A typical configuration sets a full width at half maximum (FWHM) of nearly 3 nm in the VNIR spectral region, and about 10 nm in the SWIR region, although for more specific applications (*i.e.*, fluorescence measurements) a FWHM of 1 nm is required in the VNIR.

Spectral radiance measurement using the supplied fiber optic bundle with the option to attach different optics for Field Of View (FOV) variation has become widespread in recent years. Currently, field spectroscopy using hyperspectral imaging sensors or cameras is becoming increasingly relevant and is promising, although problems of scale and non-linear effects in this near-object imaging observation still need to be addressed [18]. The most widely used acquisition methodology to obtain near-ground reflectance is the *single-beam*, where the same instrument is used to measure both the target and the calibration panel spectral radiance. For these cases, the Spectralon® (Labsphere, North Sutton, NH, USA) has become the standard material for the panels. Measurements are typically taken with field spectroradiometers that are hand-held, usually with the sensor head mounted on a pole or yoke to keep it away from the operator's body. For more automated remote observations UAVs are also being tested. Although challenging to use, they offer a high degree of automation and fast throughput [19]. In this regard, it is worth highlighting the novel carrier-lift system MUFSPEM@MED (Mobile Unit for Field SPEctral Measurements at the MEDiterranean) [13], which provides extremely controlled measurements and is automatically operated from the ground. Spectral libraries are collections of spectra that characterize the reflectance or emissivity spectral response of terrestrial surfaces and materials. Because spectral acquisition is largely variable, the simplest way to build a spectral library is by computing mean spectra and applying the standard deviation of the measured target to collect endmembers [16]. However, more complex cases can also include the spatiotemporal variation of a surface or material in order to characterize different properties or conditions (*i.e.*, different phenological stages in a plant spectral library [20]).

The existence of extensively documented metadata on spectral libraries enhances the suitability, long-term usability, and quality assurance of data from other researchers [21]. In Rashaid *et al.* [22] the most important field spectroscopy metadata was highlighted by advanced users, and Jimenez *et al.* [23] introduced an approach to establish a standard metadata system for field spectroscopy based on ISO and OGC standards.

Spectral Signature Acquisition

The spectral response of plants is a function of the optical properties of their constituents and structural attributes [10]. The chemical properties of pigments, water, and dry-matter content create distinct absorption features across the reflectance spectra. Because canopies have an intricate architecture, with gaps in the leaves and branches, the scattering and directional response may vary according to structural properties such as the Leaf Area Index (LAI) and Leaf Angle Distribution (LAD) [10]. The theoretical function describing the relationship between the incident flux and the distribution of what is reflected is referred to as the Bidirectional Reflectance Directional Function (BRDF) [24]. It is therefore, necessary to account for the geometry of observation when obtaining plant spectra in order to rigorously compare the acquired measurements under different illumination and observation conditions.

Spectral characterization of plant species is always problematic due to variations in biophysical properties, physiology, environmental parameters, and phenology [25]. Even so, it is important to highlight that a large number of remote-sensing vegetation-mapping studies have focused on the acquisition of field spectra over different vegetation types: grasslands [26,27]; shrublands [28,29], marshlands [30], forest [31,32], and sub-aquatic [33].

Recent works have addressed the generation of a spectral library of plant species using field spectroscopy. Zomer *et al.*[14] described field reflectance measurements in wetland vegetation in California, Texas, and Mississippi and the corresponding spectral library generation. Manakos *et al.* [13] presented the first steps on how to organize a spectral database for common Mediterranean land cover types. More recently, Nidamauri *et al.* [20] built spectral libraries for different crops taking into account their phenological stages. As an extremely helpful starting point, the SPECCHIO [34] project provides an online spectral database where research groups within the remote sensing community can interchange plant spectral libraries.

Spectral Separability Metrics

By comparing plant spectral signatures between and within species and detecting differences and distances in their spectral shape and reflectance, we can estimate the uniqueness of the spectral response of individual plant species. The most widespread spectral analytical procedures (*i.e.*, spectral matching, discriminant analysis) could be applied to reflectance, by looking for subtle differences in the spectra, to the first derivative spectra, or to the continuum applied. Continuum-removal is a normalization technique, resulting in a curve with values from 0 to 1, which emphasizes the location and depth of individual absorption features [35].

An extremely high variation value in the spectral reflectance response between plant individuals of the same species has been reported [36]. Several physical and empirical-based methods have been proposed to quantify this intra-specific variability. Physical methods use radiative transfer codes, which calculate the spectral response of plants as a function of their constituents' content. Among the many different codes published, the canopy bidirectional reflectance model, known as SAIL (Scattering by Arbitrary Inclined Leaves), the more recent version SAILH, and the PROSPECT leaf optical model are the most widely used to study spectral and directional reflectance response of a plant canopy in relation to vegetation biophysical characteristics [37]. However, numerical simulations of these codes are based on actual values of measured biophysical parameters (*i.e.*, chlorophyll content, LAI) in the study area [10]. Meanwhile, empirical methods aim to estimate variation directly from the spectra collected in the field or laboratory for plant individuals measured with replicated levels of every relevant plant parameter [25].

Spectral similarity measurements between two species estimate the similarity of both the shape and amplitude reflectance curves in individual canopies. There are two categories of similarity measurement: stochastic and deterministic [38]. Depending on the considered spectrum region, we can calculate a separate, punctual measurement for every wavelength range, or a global measurement when the entire spectrum is used in the calculation. Deterministic measurements include the Spectral Angle Mapper (SAM) [39], the Euclidean Distance, and Cross-Correlation. Stochastic measurements make use of the inherent properties of sampled data as self-information and define certain spectral information criteria, such as divergence, probability, and entropy. Univariate and multivariate statistical techniques including parametric and non-parametric analysis of variance, canonical and discriminant analysis are widely used spectral separability metrics [11].

PLANT SPECTRAL LIBRARY PROTOCOL

Figure 1 schematizes the proposed protocol for generating a plant spectral library and displays the two different parts. On one hand, a sampling protocol [40], which is a combination of a sampling strategy and observation procedure, includes every single aspect of field campaign preparation and spectral acquisition. On the other hand, the data processing protocol caters for the preparation of spectral files, spectral library creation, and spectral separability quantification between and within the plant species presented. In a parallel effort, the time invested in metadata collection is surpassed by its benefits in reducing system bias and variability [17]. In this sense, the protocol emphasizes the most important metadata needed to describe the spectral library according to Rashaid *et al.* [22]

and Jimenez *et al.* [23]. The spectral library sampling protocol has to be constrained in accordance with the ultimate goal of spectral measurement. In this work, the spectral library collection aims to support imaging spectroscopy analysis in plant mapping. Our protocol is based on an empirical approach in which field spectroscopy campaigns are designed to acquire the field spectra of plant species, by covering representative variations related to phenology and different environmental gradients. The collected spectral libraries will therefore characterize specific plant spectral signatures by identifying their uniqueness among species. Moreover, the sampling protocol proposes different technical aspects according to the type of vegetation considered: grass, shrub, or trees (underwater vegetation is not considered).

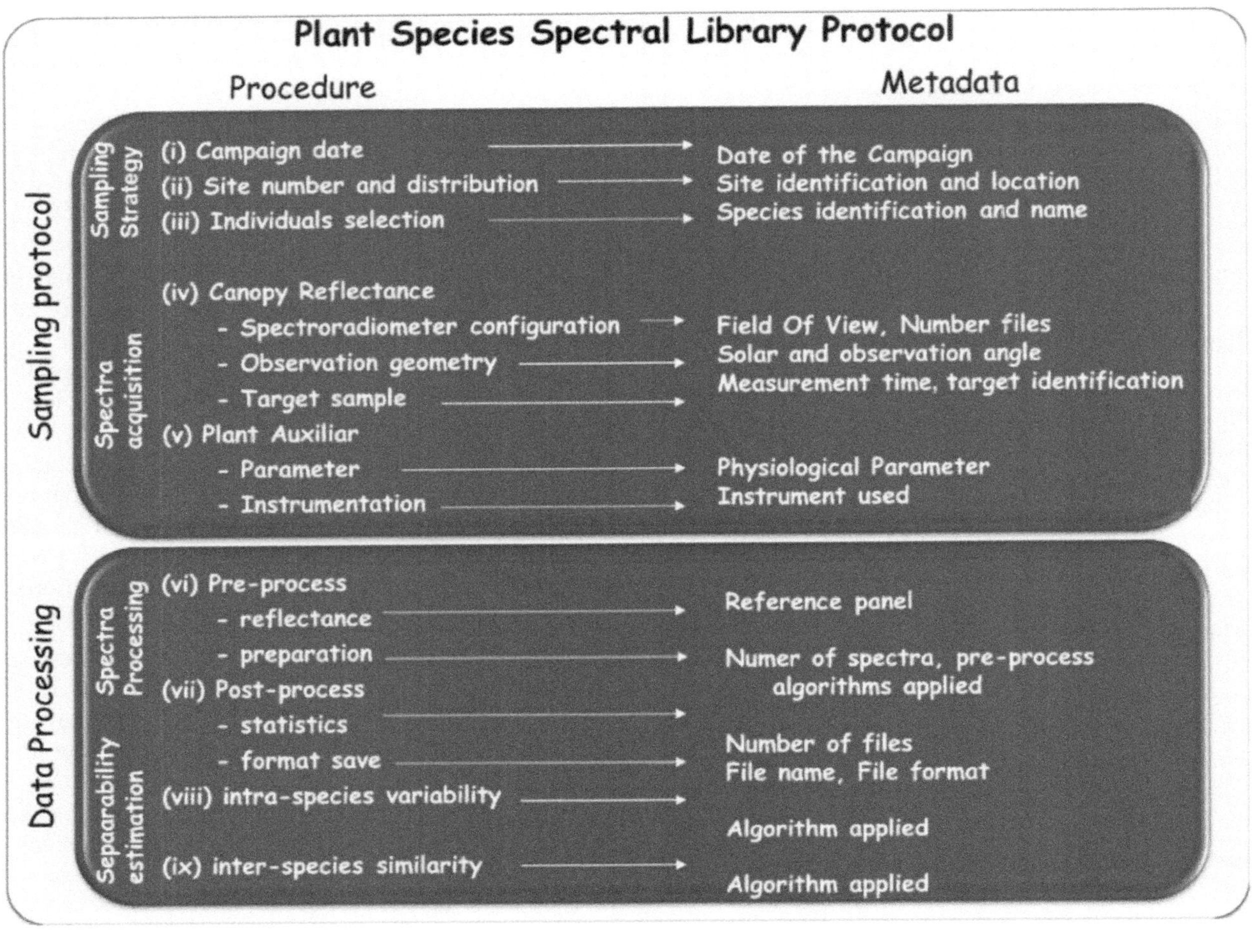

Figure 1. Field spectroscopy protocol for plant spectral library collection. The complete procedure and corresponding metadata are indicated for the sampling protocol and data processing.

Sampling Protocol

The sampling strategy designed should cover spatiotemporal variations for every species, taking into account the choice of the measurement sites and appropriate dates, and the selected sample individuals. The essential metadata for this section are location and identification of the measurement sites, as well as field campaign dates:

(i) Field campaign dates should cover the phenological stages of the different species and identify seasonal differences in leaf aging, leaf drop, flowering and fruiting, and processes that affect relative proportions and the structural arrangement of chemicals exposed to the sensor over time. Depending on the plant community considered, two to four different dates throughout the year may be required to characterize spectral variation.

(ii) The number of measurement sites and their spatial distribution has to cover the different plant communities presented, and should be stratified to account for environmental gradients (*i.e.*, soil type, topography). Stratified areas might be easily identified by using ancillary vegetation and topography maps of the study area, taking into account the extent of the area and its accessibility (*i.e.*, marshlands or very dense vegetation). The size of the measurement site is not relevant because the aim of the spectral library is the characterization of the species signature and endmember generation.

(iii) The individuals to be measured may be selected by random sampling, but selected individuals should represent the "normal" state of the plant species and avoid damaged, sick, or juvenile individuals. The total number of individuals will be a function of the number of the definitive sampling sites, after the stratification process is completed. Several more individuals may be selected for plant communities with marked heterogeneity related to the environmental gradients observed. As an overall recommendation, at least 30 individuals are required for statistical reasons [41].

Acquisition of reflectance signatures in the field includes both the canopy reflectance and ancillary plant parameter measurements. The measurement technique for plant reflectance is based on spectral collection with a field spectroradiometer covering the whole plant canopy at the time of measurement. Reflectance measurements will represent the spatial assemblage of canopy constituents plus soil and understory plant elements, meaning that "contact probe" or laboratory measurements are not representative. Ancillary plant parameters may range from very simple information about plant status (*i.e.*, only record phenological stage) to measuring vegetation parameters with proper instrumentation (*i.e.*, pigment concentration or LAI). Metadata key elements for this section are: the spectroradiometer's FOV applied, solar, and observation angles, measurement time, and ancillary plant parameters:

(iv) In selecting and configuring the field spectroradiometer, the vegetation type must be considered. For instance, to accomplish lignin and cellulose absorption bands presented in woody species the field spectroradiometer should have sensing capabilities for the SWIR region [10]. For the appropriate determination of observation geometry, the instrument's FOV should be broad enough to include every constituent of the canopy, capturing the entire size of tree canopies, shrubs or forbs by using poles, ladders, or cranes. However, the FOV should not be too large as to prevent signals from external elements around the measured plant. To select observation angles, nadiral sampling is suggested in order to avoid hotspot effects. The whole canopy has to be sampled moving the spectroradiometer's head within the canopy, recording various files per measurement to account for canopy variability. At this point, depending on the spectroradiometer used, attention must be paid to the actual pixel size with the fore-optic or fiber optic applied [42]. As recommended in general field spectroscopy protocols [43], the measurement time must be as close as possible to solar noon. For plant spectra acquisition, a field spectroradiometer could be configured to acquire radiance or DN mode. After field spectroradiometer optimization, the radiance of the reference panel and plant canopy are registered. In order to minimize the variation in the output plant reflectance due to changes in solar illumination, the canopy and calibration panel radiance and sensor dark current should be measured nearly simultaneously.

(v) The measurement of physicochemical parameters for the sampled individuals during spectral acquisition is strongly recommended. A vast amount of literature for this physiological parameter acquisition on site is available. For example, Jonckheere *et al.* [44] review several direct and indirect methods for LAI estimation. Methods for chlorophyll content or leaf water content measurements can be found in Gitelson [45] and Colombo *et al.* [46], respectively.

Data Processing and Spectral Separability Calculations

Spectra processing consists of pre-processing treatments to prepare reflectance readings for each canopy and post-processing procedures to record the data and metadata in standard format. Key metadata elements for this section are: file name, file size, applied algorithms, and selected output format:

(vi) Field spectroradiometers usually have the option to register target and reference panel radiance measurements either jointly in the same file or in separate files. It is important to be aware that spectral data is usually saved in a proprietary format, which can only be read by specific software distributed by the manufacturer. Depending on the spectral noise level, polishing techniques, such as the Savitzky-Golay filter, might be applied but it is advisable to be conservative for the sake of preserving information [47]. In addition, it is also possible to apply spectral transformation (*i.e.*, continuum-removed, derivative). Similarly, all metadata on spectral acquisition should be addressed, by identifying the metadata source and data type: for example, recording information about the configuration of the spectroradiometer in the file, or saving a picture of the sampled plant species.

(vii) During post-processing of spectra, the average reflectance and standard deviation are calculated for every measurement of the same target. To make data available for the scientific community, a common file format can be used: ASCII, Hierarchical Data File (HDF), jcamp-dx, or commercial imaging software format like the ENVI spectral library (Exelis Visual Information Solutions, Inc., Boulder, CO, USA). Similarly, metadata files should be generated and appended to the spectral library in a widely used and standard format. In this respect, Jimenez *et al.* [23] have suggested writing metadata files in Extensible Markup Language (XML) following international standards.

Once the canopy spectra for every individual has been measured, we can easily quantify spectral separability among species taking into account within-species variability due to physiochemical differences measured on site, and the similarity between species as well.

(viii) Intra-species spectra variability may be checked based on the different phenological states and physiochemical parameters. The coefficient of variation (CV) or simple least square regression are statistical techniques that help to identify sensitive wavelength. In addition, univariate and multivariate statistical techniques, including both parametric and non-parametric analysis of variance, can aid in the comparison of the spectral response over the different measurement dates in order to identify in which season the species easily discriminate from each other.

(ix) The inter-species similarity index [11] can provide an estimation of differences in the spectral signature of plant species. In order not to losepossible subtle differences between species spectra, the protocol proposes the use of two different separability metrics for comparison, both applied either to the original and to the continuum-removed spectra.

CASE STUDY: SHRUBLANDS OF DOÑANA NATIONAL PARK

Study Area

Doñana National Park (DNP) is located on the southwest coast of Spain. It is one of the most important wetlands in Europe [48] and was designated a UNESCO World Heritage Site in 1995. DNP has a Mediterranean-type climate with some oceanic influences and mean annual precipitation of around 550 mm. Rainfall has a very well-defined seasonality, mostly concentrated between October and March (wet season) and almost absent between June and August (dry season). Three main land units have been traditionally distinguished in DNP: inland marshes, mobile sand dunes, and stabilized sand dunes. Our study was carried out in the stabilized sand dunes of the Doñana Biological Reserve (DBR), the core area of the DNP (Figure 2).

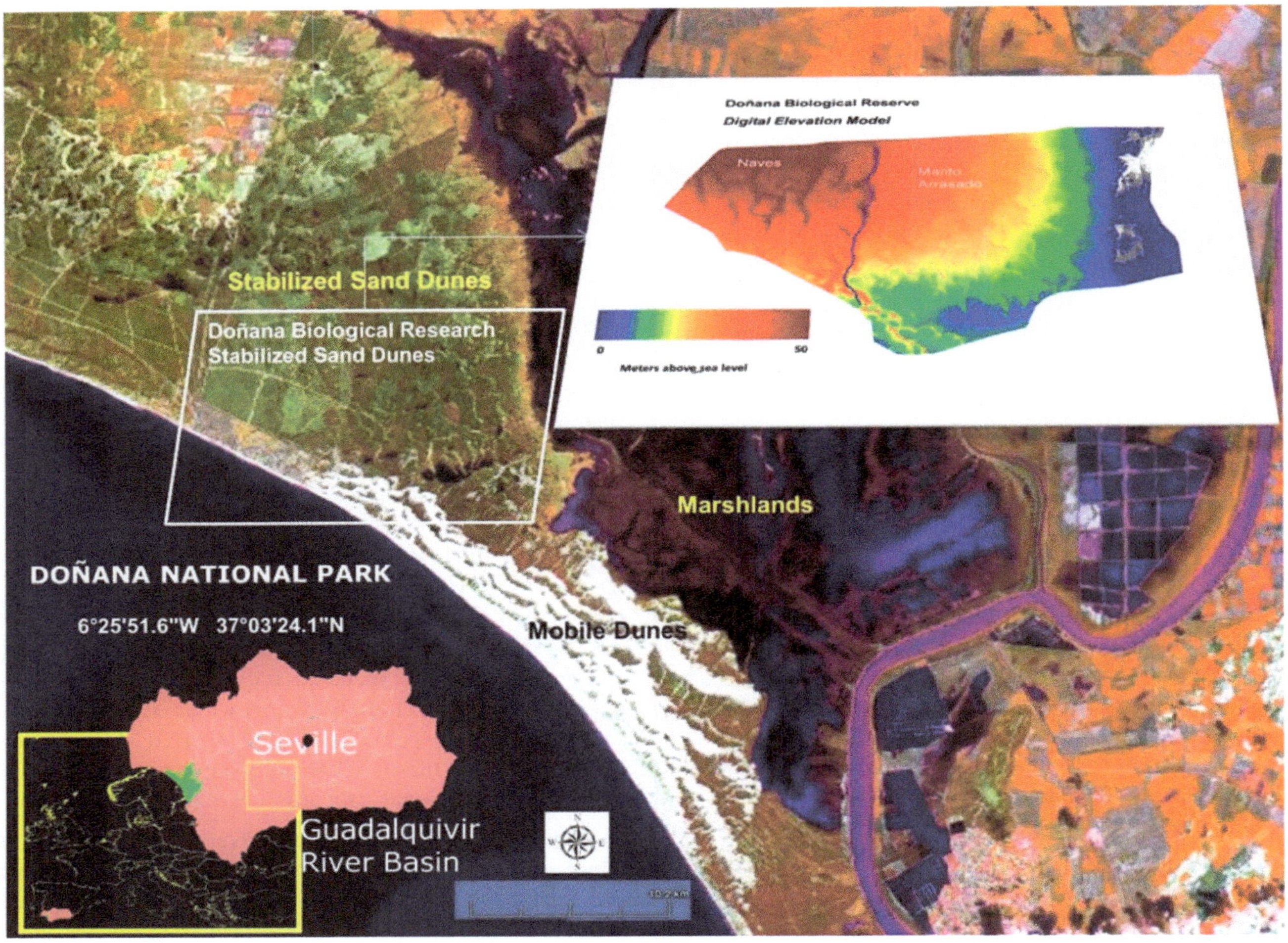

Figure 2. Location of Doñana National Park (SW, Spain). Square indicates the stabilized sand dunes ecosystem of Doñana Biological Reserve. The sub-image shows a digital elevation model of the study area.

Stabilized sand dunes exhibit a rolling topography (see Figure 2) due to the old dune morphology colonized by vegetation which is more or less dependent on groundwater supply. There are three large vegetation zones in the stabilized sand dunes of DBR: an elevated zone dominated by xerophytic shrubs (Naves), a lower zone dominated by hygrophytic shrub (Manto Arrasado), and the transitional grasslands (Vera). The current vegetation is composed of remnants of Mediterranean plant communities such as cork oak woodlands, juniper woodlands (*Juniperus phoenicea subs turbinata*), umbrella pine plantations (*Pinus pinea*), grasslands, and Mediterranean shrubland communities.

Three main types of shrub communities can be found on the stabilized sand dunes depending on the water table depth: Monte Blanco (Xerophytic sites), located on crests of former dunes where water table in summer is usually deeper than 4 m, and even over 3 m below the soil surface. This community is dominated by *Rosmarinus officinalis, Halimium commutatum, Halimium halimifolium, Juniperus phoenicea,* and *Cistus libanotis*. The Monte Negro (Hygrophytic sites) community is located at low depressions, where the water table in summer rarely lies 1 m below the soil surface and temporary ground flooding occurs in winter. This plant community is dominated by *Erica scoparia, Erica ciliaris, Calluna vulgaris, Ulex minor, Myrtus communis and Cistus salvifolius*. Finally, the Monte Intermedio community is located on the slopes of the dune ridges where the water table depth is in transition and there is no surface flooding. This community is dominated by *Halimium halimifolium* and *Ulex australis*. Spatial distribution of shrubland is determined, at all scales, by the rolling topography that modulates the groundwater level [49].

As for the structural characteristics of the studied shrub

species, which may influence the canopy spectral response, three different leaf types were found: sclerophyll, semi-sclerophyll and spiny leaves [50]. Evergreen sclerophylls like, *E. scoparia,*are characterized by small, thick, leathery leaves, presenting a high surface-to-volume ratio. High leaf consistency and density is a trait that improves drought resistance in the Mediterranean summer climate [51]. These plants are isomorphic and their LAI values are constant throughout the year. Semi-sclerophyll species, also called xerophytic malacophylls, such as *H. halimifolium* and *R. officinalis*, have leaves with less density. Although they do not present foliar dimorphism, they may show important variations in LAI due to leaf substitution during summer. Spiny leaves developed by legume species *S. genistoides* and *U. australis* occupy the lower levels of LAI in Doñana's shrublands [52].

Shrub Species Spectral Library

Shrubland communities play an important role in DNP as a feeding and refuge habitat for local fauna. According to field studies conducted and available literature [49,50], five species are the most abundant and dominant across the stabilized sand dunes ecosystem: *Erica scoparia*, *Halimium halimifolium*, *Ulex australis*, *Rosmarinus officinalis*, *Stauracantus genistoides*. Table 1 shows several physiochemical characteristics of these dominant shrub species, the spectral library protocol was specially adapted to collect the reflectance signature of these shrubland species, taking into account their characteristics, such us plant height and canopy size. Furthermore, due to the LAI variations is a very important parameter in Mediterranean shrublands [50,51], *in situ* LAI measurements are acquired as ancillary information for the spectral library generation.

Table 1. Characteristics of the shrub dominant species in the stabilized sand dunes ecosystem of Doñana National Park

Plant Physiochemical Parameters	***Erica scoparia***	***Halimium halimifolium***	***Rosmarinus officinalis***	***Stauracanthus genistoides***	***Ulex australis***

Family	Ericaceae	Cistaceae	Lamiaceae	Leguminosae	Leguminosae
Common name	Brezo de escobas	Jaguarzo blanco	Romero	Tojo Morisco	Tojo
Leaf type	Schlerophyll	Semi-Schlerophyll	Semi-Schlerophyll	spines	spines
Average Height	100 cm–3 m	60–150 cm	100–150 m	60–100 cm	60–100 cm
Canopy Diameter	1–2 m	50–100 cm	50–100 cm	50–100 cm	50–100 cm
Regeneration	roots	seeds	Seeds	roots	roots
Root morphology	Horizontal	Horizontal / vertical	Horizontal	Horizontal/vertical	Horizontal/vertical
Leaf water potential	−3.0 Mpa	−4.0 Mpa	−11.0 Mpa	−1.7 Mpa	−2.7 Mpa

Below we follow the guidelines of the protocol (Section 3) to describe the procedures for generating the spectral library of the dominant shrub species in DNP:

(i) To determine field campaign dates, we took into account the very sharp differences in water availability between the wet and dry seasons in DNP. Therefore, in 2006 and 2007, several field spectroscopy campaigns were carried out with measurements collected during the end of the dry season (late March and early April) and the end of the wet season (late August and early September). In the wet season campaigns, no measurements were taken during the flowering period of the Doñana shrubland, which takes place mainly between May and June.

(ii) In the stabilized sand dune ecosystem, micro-topography is the major factor conditioning the sampling stratification. A 10 m pixel digital elevation model and the DBR Ecological Map were used for the stratification of the study area. To select sites along the stratified areas, locations for plant spectral acquisition and LAI measurements were determined using maps, but the choice was constrained by proximity to pathways due to the great difficulty of accessing areas with dense vegetation. A total of 16 measurement sites were selected, 12 of them intended for the acquisition of spectral signatures, while in the four other sites spectral signatures were acquired together with LAI measurements. Figure 3 shows sampling sites distribution, eight of those were located in the Naves, and the other eight sites in Manto Arrasado.

(iii) Two types of canopy spectral measurements with two different aims were recorded: type (I) in order to generate spectral libraries for the five dominant species, 30 individuals for every dominant species were non-randomly selected covering the LAI ranges presented in the stabilized sand dune ecosystems. These measurements were carried out in spectral signature and LAI measurement sites (red tacks in Figure 3) and the individuals were marked and measured in wet and dry seasons; type (II) in order to calculate the spectral variability among the most abundant shrub species, canopy spectral reflectance was also measured for the five dominant species and other abundant shrub species (*Lavandula stoechas*, *Thymus mastichina*, *Cistus libanotis*, *Halimium commutatum, Phylirrea angustifilia*). These measurements were carried out in spectral signature sites (yellow tacks in Figure 3) only in the dry season and for 30 randomly selected individuals in every species; damaged or individuals in early growing stages were avoided.

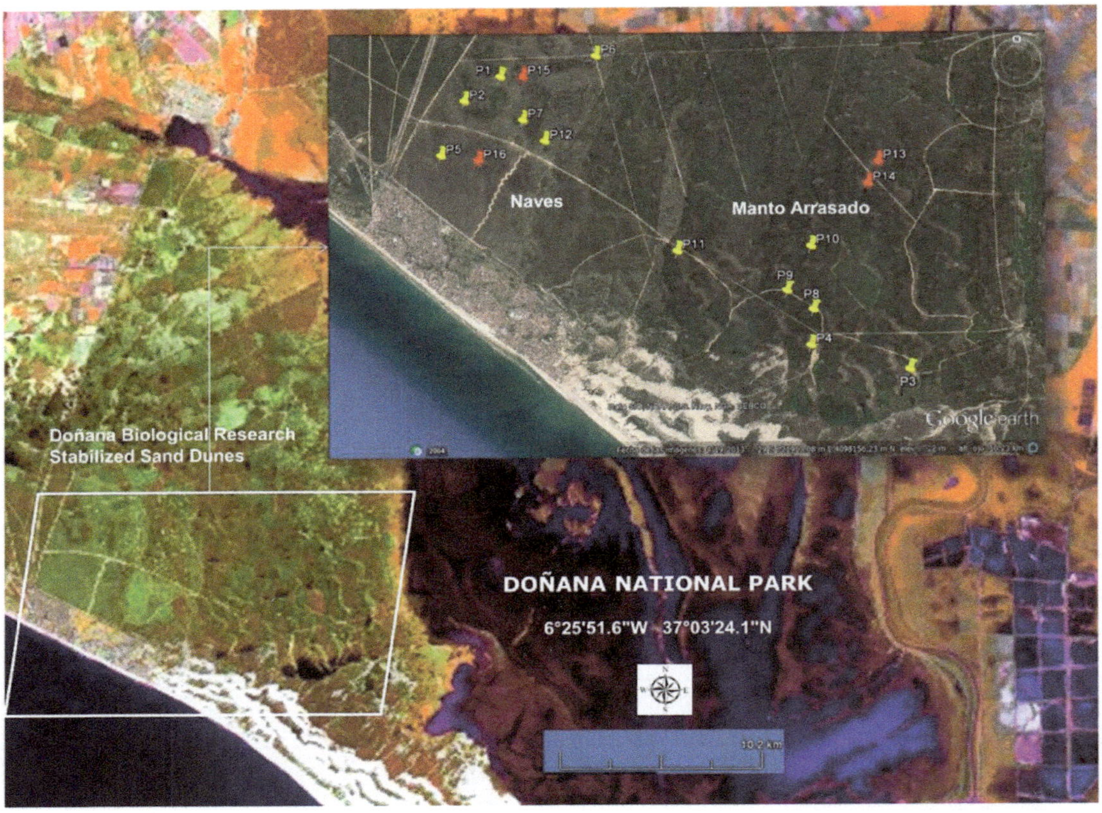

Figure 3. Site location for plant spectral library measurements. Yellow tacks identify spectral signatures acquisition sites and red tacks identify spectral signatures and LAI measurements sites.

(iv) Shrub species have a considerable woody part, which recommended the use of a spectroradiometer with a SWIR detector. The field spectroradiometer chosen was the ASD FieldSpec3 (Analytical Spectral Devices, Boulder, CO, USA), which measures incoming radiance using a fiber optic which is adaptable with a fore optic lens. It has a spectral range from 350 to 2500 nm with a 3 nm spectral resolution and a sampling interval of 1.4 nm in the VNIR spectral regions and 10 nm and 2 nm in the SWIR. Measurements were acquired directly with the fiber optic (FOV 25°) with no fore-optics mounted. The fiber was nadir-oriented and held 1 m above the canopy (GIFOV = 44 cm). For tall individuals a leader was used to maintain a constant distance between the fiber and canopy. Target and reference panel radiance were alternatively acquired and the instrument's dark current was measured before each sampling. At least five files per plant, averaging five spectra per file, were recorded on a whole sweep of the canopy. During sweeping the pistol grip was continuously turned clockwise and counter-clockwise to minimize spectral misalignments of the ASD fiber [42]. Measurements were concentrated in two hours around noon. The field campaigns were always carried out under cloudless skies.

(v) LAI measurements were carried out in some individuals at the same time as the reflectance spectral acquisitions. We used an indirect method and took hemispherical canopy pictures for each plant. These 180° pictures were always taken from beneath the canopy looking upwards with a NIKON FC-E8 fisheye lens adapted to a Nikon 4000 Coolpix digital camera. The pictures were always gathered at nearly sunset [53]. In this process, the Plant Area Index (PAI) which also includes non-photosynthetic elements, such as branches, was estimated instead of the LAI. The PAI was calculated by processing the corresponding hemispherical canopy picture with Hemiview® 2.1 (Delta-T Devices, Ltd.,

Cambridge, UK). This software estimates gap fraction over the entire picture and calculates the PAI. In Figure 3, the sites for the PAI measurements are indicated with red tacks.

(vi) Spectral files acquired using ASD Fieldspec3 were imported to ASCII format using ViewSpec® software. Canopy reflectance per plant (five measurements) was calculated by dividing target and reference panel radiance. All collected spectra were parabolic-corrected for the spectroradiometer sensitivity drift at the junctions of the spectral regions of the three different detectors. No spectral polishing algorithm was applied.

(vii) For each plant individual, an average reflectance spectral signature and standard deviation (SD) file was generated. Spectral reflectance at wavelengths of around 1400 nm, 1940 nm, and 2400 nm were excluded due to the presence of excessive noise caused by atmospheric water absorption. The output file was saved in the ENVI spectral library format (Exelis Visual Information Solutions, Inc., Boulder, CO, USA).

(viii) Once the canopy reflectance spectra for every plant individual was calculated, the spectral library for the five dominant species were generated using the spectra from Type-I measurements. The species coefficient of variation (CV) in dry season was calculated using the spectra for which PAI was measured. CV was calculated for every wavelength and an averaged value was calculated for the entire spectral region. To determine the season when dominant species show greatest variability and most differences between each other, parametrical statistical tests were applied to compare species of Type-I measurement per season. A standard series of two-group t-tests were performed for every wavelength to test the null hypothesis that there were no significant differences between means. This test allows for the identification of the wavelength region where the species show higher spectral separability. Normality was previously explored and confirmed with the Kolmogorov–Smirnov normality test.

(ix) Similarity indexes between species were calculated. In this case, global similarity methodologies were applied on Type-II measurements. Two deterministic algorithms were calculated using the original and continuum-removal algorithm applied to entire spectra. These two algorithms were the Spectral Similarity Value (SSV) [54] and SAM. SSV takes into account curve shapes by comparing correlation and separation calculated by Euclidean distance. By definition, the spectral similarity scale has a minimum of 0 and a maximum of the square root of 2. Low values in the similarity scale indicate similar spectra. SAM calculates the angular distance between two reflectance vectors across wavelengths, and is less sensitive to the amplitude of both spectra. Range values vary from 0 to $\pi/2$, where a lower SAM angle means a higher similarity between two spectra.

RESULTS

The main outcome of our work are the spectral libraries for the five dominant shrub species in DNP; they provide significant support to vegetation mapping using airborne imaging spectroscopy (reported in Jiménez *et al.* [55]). However, intra-species variability between dominant species is also conveyed which contributes to determining the suitable season and spectral region for mapping activities. We also report differences between the five dominant shrub species and other less abundant species present in the study area.

Dominant Species Spectral Library

Figure 4 shows an average spectral signature for the five dominant species of Doñana's shrublands. The spectral library curves for the dry and wet season show the mean and SD across the measured PAI gradient found in the stabilized sand dunes plant communities. Figure 4 also shows the averaged PAI values and corresponding CV calculated for every species during dry season. Although *R. officinalis* exhibits the highest CV, the greater PAI range was found for *H. halimifolium*, meaning that PAI changes have a greater influence on the spectral response of *R. officinalis.* The lowest CV values and the lowest PAI spectral response variations were found in both legume species *S. genistoides* and *U. australis*.

A visual examination of the spectral library graphics (see Figure 4) highlights the similarity between species, as might be expected for a group of species growing in a very similar environment with very similar conditions (poor soils, high radiations levels, and long drought periods). Focusing on the five dominant species for both seasons, distinct absorption features in their spectral signatures may be related to their structural arrangement. For green leaves, chlorophyll a and b and accessory pigments (*i.e.*, carotenoids) dominate absorption features in the visible spectrum (400–700 nm). Water creates absorption features in the near-infrared at 970 nm and 1188 nm, respectively. However, in the shortwave infrared, relatively low reflectance values and strong water absorption regions are found in green leaves.

When comparing the spectral signatures between seasons, subtle differences are revealed, such as lower variability (lower SD) values for all the species in the wet season, and deeper NIR water absorption bands, which almost disappear in legume species in the dry season. The most evident differences are noticeable for lignin and cellulose absorption bands, centered at 2100 and 2310 nm, which are deeper in dry season spectra for all species. This is because those bands are masked in wet season due to the higher water content. Muddle differences are found in the reflectance response between species: on one hand, we found higher reflectance in the VIS part for dry season spectra, mostly due to a lower concentration of leaf pigments; on the other hand, lower reflectance in the NIR part, except for *E. scoparia,* due to a higher structural arrangement of tissues in crowns, measured through leaf and branch density, angular distribution, and clumping [45].

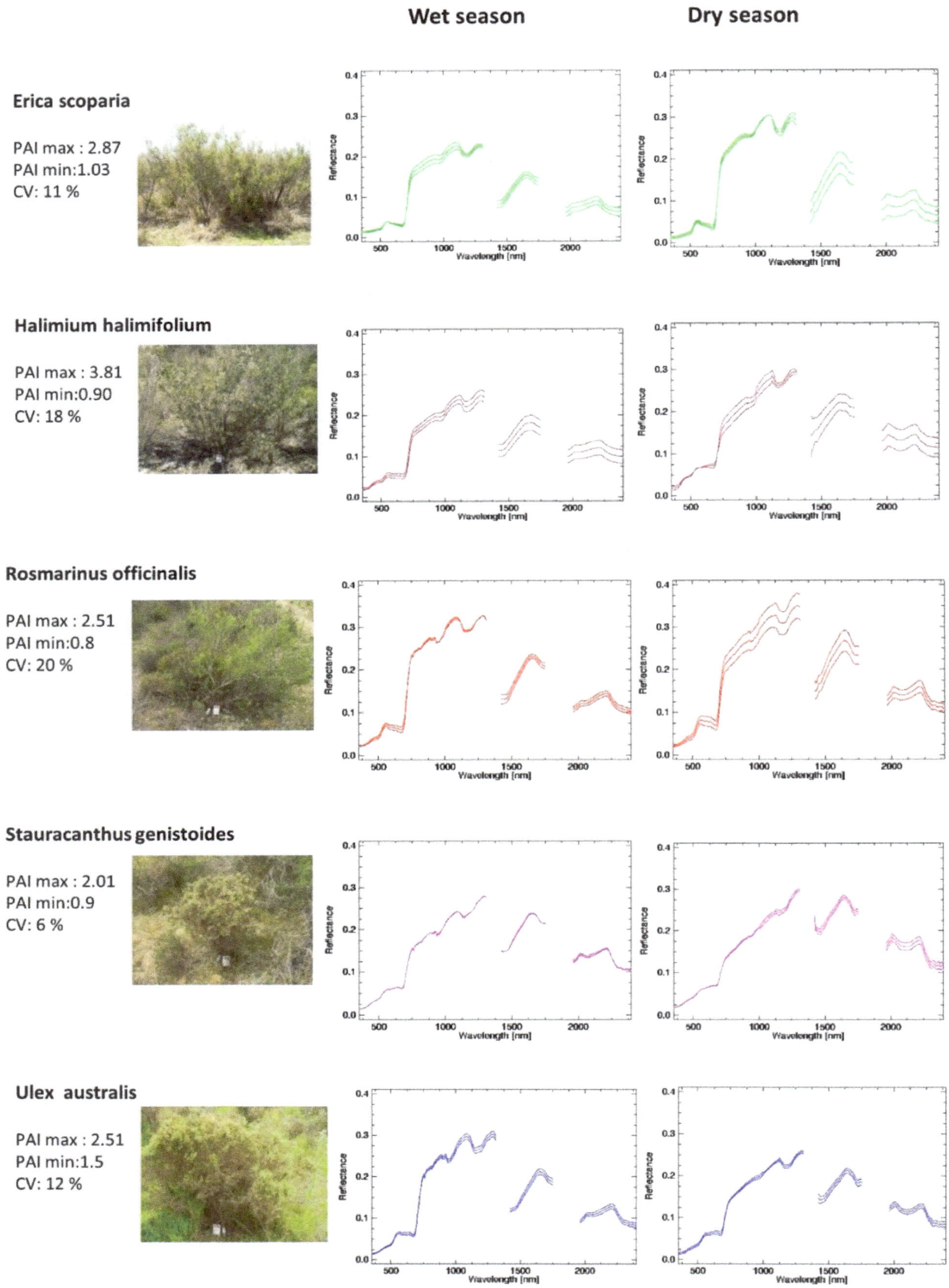

Figure 4. Spectral library of the five dominant shrub species in Doñana National Park. The graphics show the average and standard deviation spectral reflectance for dry and wet season. PAI and CV values are shown for every species spectral response in dry season.

Intra-Species Variability

The standard t-test evaluates if the intra-species variability was significantly different between species. The means of two species were considered significantly different when calculated t-values were greater than the t-critical value ($p < 0.05$). The t-value was calculated assuming pooled variance (rather than separate variances), because the number of spectral samples was different for each species and their variances were unequal/no-paired (F-test). Single graphics in Figure 5 show every two-species comparison, where the curve represents the calculated t-value per wavelength. The red line in the graphics identifies t-critical values corresponding with the degrees of freedom in the sample.

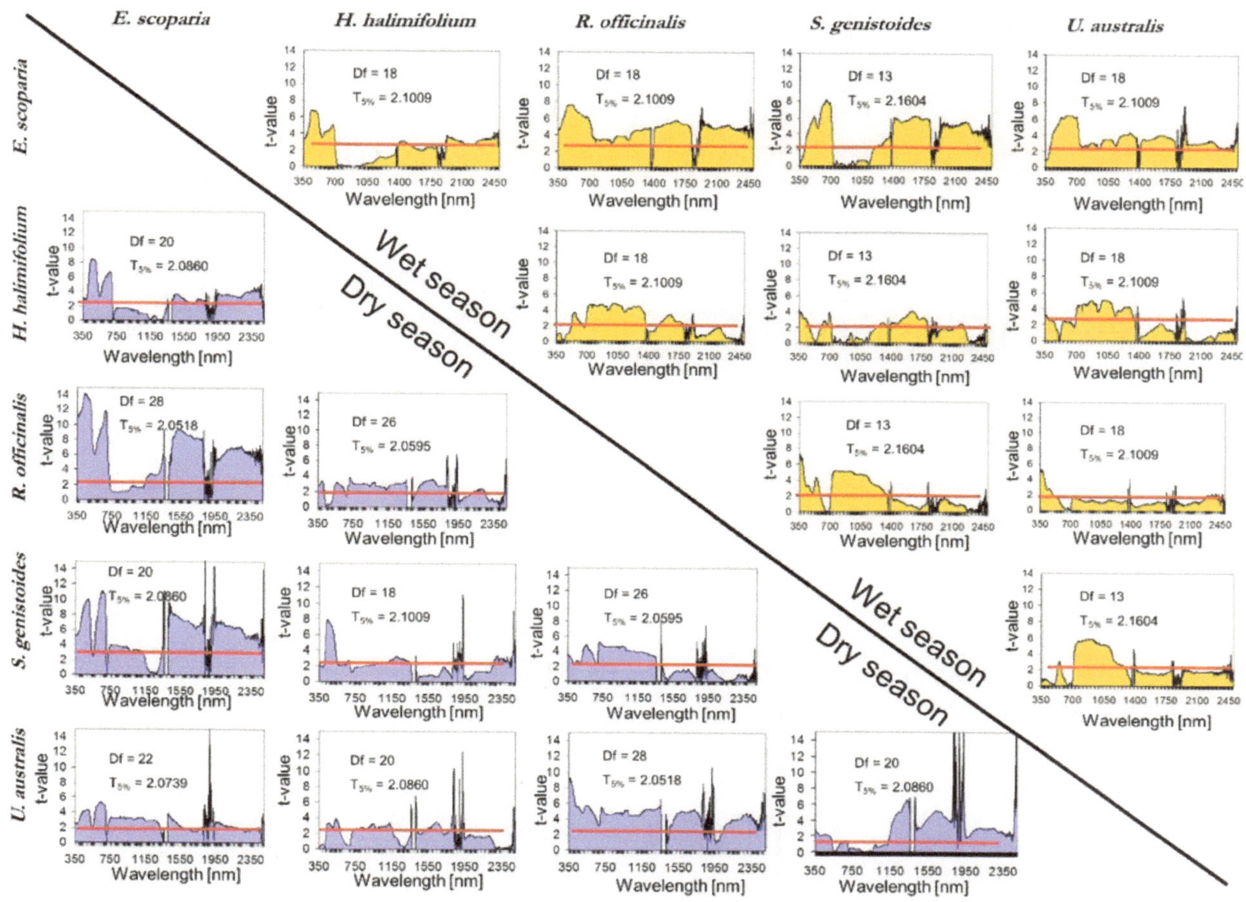

Figure 5. Intra-species variability for the DNP shrubland dominant species. t-test for species-to-species comparison for dry and wet season separately. The red line in the graphics identifies t-critical values according to the corresponding degrees of freedom.

According to Figure 5, dry season measurements show a slightly higher discrimination power among species than those collected during wet season. Although in dry season the species show more variable signatures (see Figure 4), they are also more different among themselves and its variability is less significant. For example, in the VIS part of the spectrum five species during dry season, and *E. scoparia* has higher calculated t-values in the dry season than in the wet season.

It is evident that some bands have more power for discriminating between species, since they have a higher frequency of statistically different mean reflectance. The tests were calculated independently for each wavelength value. We therefore used this local similarity algorithm to locate the spectral regions where the differences were systematically higher. For the dominant shrub species the best spectral regions to discriminate between species ($p < 0.05$), were found at (480, 690, 770, 979, 1600, and 2100 nm) both during the dry and wet seasons.

Inter-Species Similarity

Spectral reflectance similarity indexes were calculated among the dominant shrub species and also for less abundant species present in the Doñana shrubland communities. Figure 6 presents similarity indexes between all the shrub species studied and were calculated using SSV and SAM algorithms, by comparing both the original and the continuum-removed applied spectra. The highest dissimilarity values identified, and represented in white, were 0.4 for SSV and 0.6 for SAM. InFigure 6, shrub species name were underlined according to Doñana's shrubland community. Furthermore, to assess the importance of foliage type in the spectral signature and its discriminability, shrub species names were colored according to foliage type.

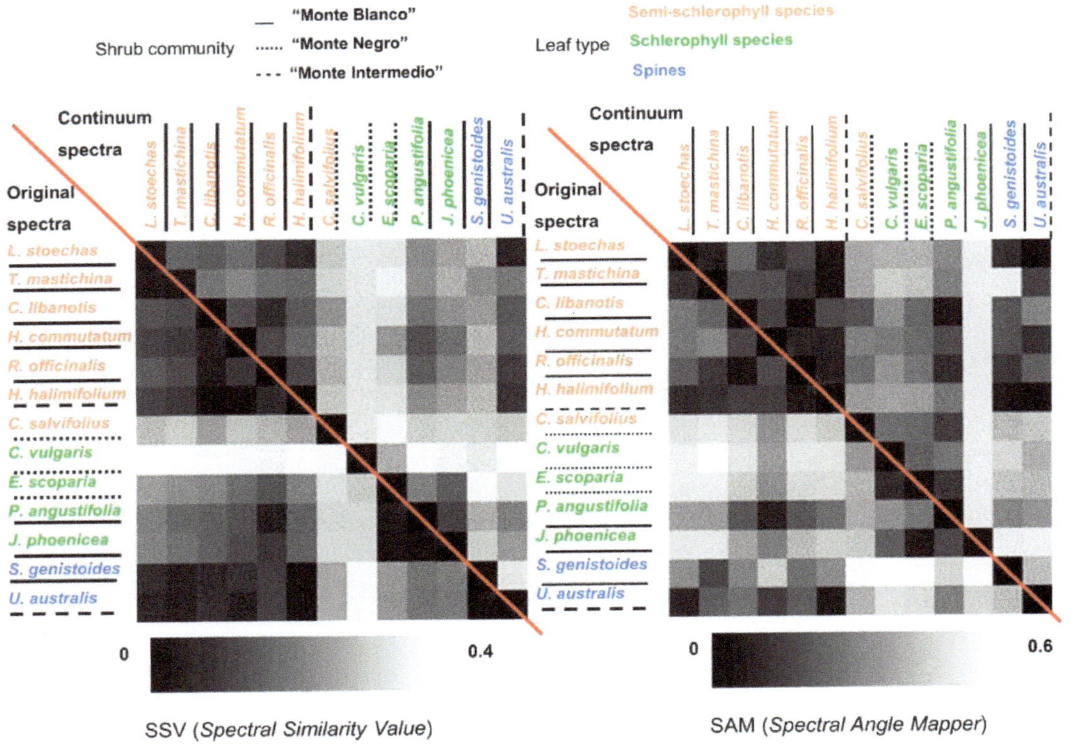

Figure 6. Matrix of spectral similarity indexes in grayscale, where black indicates total similarity and white the highest dissimilarity for the dominant and the less abundant shrub species in Doñana National Park. Foliage type is also shown by coloring the species name. Species names are also underlined according to the shrubland community they belong to.

Leaf Type Comparison: Within and Between Leaf Type Groups

Semi-schlerophyll species showed intermediate similarity values, with values of less than 0.2 for both algorithms and spectral signature process type. For this leaf type, continuum-removed spectra had higher SSV values than the SAM algorithm. Furthermore, *C. salvifolius* reached the higher values of dissimilarity, 0.3 for SSV and 0.4 for SAM, when compared to the other semi-schlerophyll species. Schlerophyll species presented smaller dissimilarity values, for both algorithms and spectral signature process type, than the rest of the leaf types. In this case, the SSV algorithm showed higher values than SAM, meaning a greater separability capacity. Finally, spiny legume species also showed intermediate values of similarity. The SAM algorithm produced the best separation.

When comparing semi-schlerophyll and schlerophyll species, considerable dissimilarity values for both algorithms were found (SSV = 0.3, SAM = 0.5). In this case, the SAM algorithm from the original spectra produced a slightly better separability. Spiny legume species were found to be much more dissimilar to schlerophyll species than to semi-schlerophyll species for both algorithms, reaching values of around 0.25 for SSV and 0.35 for the SAM. In this case, the SAM calculated from the original spectra yielded slightly higher dissimilarity values for both leaf types; *S. genistoides* was the spiny legume species with higher separability values than the schlerophyll species.

Within and between-Communities Species Comparison

Monte Intermedio species showed the lowest dissimilarity values for the species within it (SSV = 0.15, SAM = 0.25). Although *H. halimifolium* is a semi-schlerophyll species and *U. australis* is a spiny legume species, it had lower dissimilarity values than expected for species of different plant communities. In contrast, higher separability values were found within the Monte Negro species. In this case, even the comparison between two schlerophyll species *E. scoparia* and *C. vulgaris*provided high separability values, and showed the highest differences with *C. salvifilolius*. Monte Blanco showed different levels of dissimilarity, with intermediate values (SSV = 0.2 and SAM = 0.3) within semi-schlerophyll species, and higher values between schlerophyll species *P. angustifolia* and *J. phoenicia* in comparison to other semi-schlerophyll species and the spiny legume *S. genistoides*.

In the comparison between species of different communities, all three communities presented very separable values among each other. The highest dissimilarity values were found for Monte Negro compared with the other two communities. SAM produced the highest differences (SAM = 0.5 and SSV = 0.35). Dissimilarity values decreased when comparing Monte Blanco species with Monte Intermedio. Again, SAM was found to provide the highest differences between communities.

Comparison within Dominant Species

Considering dissimilarity values between the dominant species, *E. scoparia* had the highest values compared with the other species. In this comparison, SAM from original spectra showed the best performance. Other high separability values were found for the spiny legume *S. genistoides* when compared to the schlerophyll *E. scoparia*. Furthermore, between the two spiny legume species the separability values were intermediate and acceptable to discriminate them. In contrast, the lowest dissimilarity values were found between the spiny legume *U. australis* and the semi-schlerophyll *H. halimifolium*. Although separability values between these two Monte Negro community species were lower, the discrimination power is still suitable for mapping activities. Finally, the semi-schlerophyll species *H. halimifolium* and *R.officinalis* showed intermediate separability values.

DISCUSSION

In this work, we proposed a protocol for developing a plant spectral library focusing on its application in supporting mapping activities with imaging spectroscopy. The proposal aims to ensure standard and highly repeatable measurements. Nevertheless, it also has to be flexible enough to adapt the procedures (*i.e.*, field spectroradiometer, observation geometry, number of dates, and individuals) to vegetation type, to species diversity, and to the environmental heterogeneity of the study area.

One of the main drawbacks of the protocol is that plant spectral signatures were acquired over the entire canopy, including the spectral response to the background structural arrangements and understory vegetation. The principal argument underpinning this approach is that the plant spectral signature will support future analysis of airborne or satellite imaging spectroscopy, which captures the whole canopy signal plus background spectral response in its pixels. Therefore, this field-level spectral signature is designed to integrate the at-sensor-level signal. For example, in Manevski *et al.* [29], the soil background from the acquired canopy spectra was treated as integrated parts of the vegetation spectra in typical Mediterranean semi-arid environments.

Another tricky point of the protocol is the procedure to estimate plant separability. We believe that the spectral library generation should evaluate the spectral variation throughout space and time, which means that intra-species variation is crucial *a priori* information to decide the optimum image acquisition date and to choose the best bands for species discrimination. Consequently, inter-species similarity quantification is secondary for the protocol, but very important to be able to estimate the degree of uniqueness of the species studied. In this sense, several methods of similarity metrics are widely available, as we pointed out in section 2.3. We recommend applying at least two different algorithms not just to the original spectral signatures, according to the possible, subtle differences between plant species.

In fact, the protocol was designed to be applied to different types of vegetation (*i.e.*, grasslands, shrublands, marshlands, forest), with the exception of sub-aquatic vegetation which implies more specific technical aspects for spectral signature acquisition. For other vegetation types, the sampling protocol should be adapted mainly in terms of the observation geometry.

The application of the protocol to Doñana's shrublands communities allows refinement of the methodology. The case study presented in this work was developed for a plant community with around 20 shrub species, where five are dominant. The majority of these shrub species, including dominant species, have canopy sizes of between two or three meters, and no more than three meters in height. The determinant environmental gradients for spatial patterns in species distribution (*i.e.*, micro-topography and ocean proximity) were addressed with the stratification sampling strategy. In any case, the Doñana shrubland spectral library is a typical mapping activity case, while in other complex cases of species diversity and characteristics, and environmental gradients, the protocol might be more challenging in terms of sampling strategy and spectra acquisition. However, in this case the most relevant constraint was the high plant density found for some areas of the Monte Negro shrubland.

In terms of the Doñana shrubland spectral libraries, the dominant species showed typical shrub spectra with differences enhanced according to the season in which the measurement took place. During the dry season, reduced water content in the leaves is revealed by low reflectance in the NIR and SWIR regions, while lignin and cellulose bands were more pronounced. Groundwater availability due to the ancient dune micro-topography is a major factor in the establishment and growth of plants; as a result, considerable PAI ranges were found for the dominant shrub species, especially for the more widespread *H. halimifolium.* Although significant spectral variation in shrubs in terms of PAI changes was found for the spectral amplitude (SSV) but not in the shape (SAM), larger-amplitude spectral variability reinforces the need for empirical estimation of the spectral changes due to local environmental gradients in the area.

T-test evidenced that during the dry season the analyzed species had better separability than in the wet season. During drought conditions, the stabilized sand dune ecosystem shows a more homogeneous background with most of the underbrush vegetation being in a senescent stage. The spectral regions found with higher separability significance are usually related to plant water content (*i.e*, water absorption bands) and indirectly to pigments, lignin, and cellulose contents.

The application of the protocol to other, less abundant species helped to identify discrimination features suitable to enhance mapping with imaging spectroscopy. Plant species discrimination is a very challenging application, even more so for species of the same vegetation type and living in similar environmental conditions. The use of different separation algorithms on both the original and continuum-removed spectra is strongly recommended in order to increase species separability.

CONCLUSIONS

This paper presents an approach to a standard protocol for spectral library generation to support vegetation mapping using imaging spectroscopy. The proposed protocol is based on field spectroscopy measurements to characterize the reflectance spectral response of the distinct species throughout spatiotemporal changes. To this end, we carried out several field campaigns in different seasons, following a stratified sampling strategy, taking into account the phenological stages and different spectral responses due to environmental gradients present in the study area. In addition, measurement of ancillary vegetation parameters together with spectral reflectance contributed to the empirical identification of spectral variability linked to environmental conditions. After the spectral libraries were collected, the separability between species was quantified by considering intra-species variability and inter-species similarity. Endmembers obtained from the spectral libraries were used as the main input for vegetation mapping with imaging spectroscopy of Doñana shrubland communities [55].

Overall, the analyzed species appeared very similar in spectral terms, as expected for a group of species living in an environment with very similar conditions, poor soils, high insolation, and low water availability. Nevertheless, separability indexes and PAI values calculated for the dominant shrub species were statistically significant, although Monte Negro species were the most difficult to discriminate. The spectral libraries built were used for planning the airborne imaging spectroscopy mapping campaigns in the Doñana National Park shrublands [55].

ACKNOWLEDGMENTS

The authors are extremely grateful to the staff of the INTA Remote Sensing Laboratory for their scientific support and the staff of the Doñana Biological Research Station for their outstanding logistical support received in the course of the work. We appreciate the contribution of Antonio Pou, who largely supervised this work. We also thank Patrick Vaughan and Angela de Santis for field assistance in collecting PAI field measurements.

AUTHOR CONTRIBUTIONS

Both authors are intellectually responsible for the research conducted. Marcos Jiménez designed the work presented and prepared the manuscript. Ricardo Díaz-Delgado was substantially involved in field spectroscopy acquisition and in reviewing the manuscript.

REFERENCES

1. Schmeller, D.S. European species and habitat monitoring: where are we now? *Biodivers. Conserv.* **2008**, *17*, 3321–3326.
2. He, K.S.; Rocchini, D.; Neteler, M.; Nagendra, H. Benefits of hyperspectral remote sensing for tracking plant invasions. *Divers. Distrib.* **2011**, *17*, 381–392.
3. Ustin, S.; Zarco Tejada, P.; Jacquemoud, S.; Asner, G. Remote sensing of environment: State of the science and new directions. In *Remote Sensing for Natural Resources Management and Environmental Monitoring. Manual of Remote Sensing*, 3rd ed.; Ustin, S.L., Ed.; John Wiley & Sons, Inc.: Hoboken, NJ, USA, 2004; Volume 4, pp. 679–729.
4. Schaepman, M.E.; Ustin, S.L.; Plaza, A.J.; Painter, T.H.; Verrelst, J.; Liang, S. Earth system science related imaging spectroscopy—An assessment. *Remote Sens. Environ* **2009**, *113*, 123–137.
5. Kaufmann, H.; Segl, K.; Kuester, T.; Rogass, C.; Foerster, S.; Wulf, H.; Hofer, S.; Sang, B.; Storch, T.; Mueller, A.; *et al.* The Environmental Mapping and Analysis Program (EnMAP)—Present status of preparatory phase. In Proceedings of the International Geoscience and Remote Sensing Symposium (IAGARSS'13), Melbourne, Australia, 21–26 July 2013.
6. Romano, F.; Santini, F.; Simoniello, T.; Ananasso, C.; Corsini, G.; Cuomo, V. The PRISMA hyperspectral mission: Science activities and opportunities for agriculture and land monitoring. In Proceedings of the International Geoscience and Remote Sensing Symposium (IGARSS'13), Melbourne, Australia, 21–26 July 2013.
7. Milton, E.J.; Schaepman, M.E.; Anderson, K.; Kneubühler, M.; Fox, N. Progress in field spectroscopy. *Remote Sens. Environ.* **2009**, *113*, 92–109.
8. Asner, G.P.; Jones, M.O.: Martin, R.E.; Knapp, D.E.; Hughes, R.F. Remote sensing of native and invasive species in Hawaiian rainforests. *Remote Sens. Environ.* **2008**, *112*, 1912–1926.
9. Lewis, M.M. A strategy for mapping arid vegetation associations with hyperspectral imagery. In Proceedings of Eleventh Australian Remote Sensing and Photogrammetry Conference, Brisbane, Australia, 2–6 September 2002; pp. 647–655.
10. Asner, G.P. Biophysical and biochemical sources of variability in canopy reflectance. *Remote Sens. Environ.* **1998**, *64*, 234–253.
11. Clark, M.L.; Roberts, D.A. Species-level differences in hyperspectral metrics among tropical rainforest trees as determined by a tree-based classifier. *Remote Sens.* **2012**, *4*, 1820–1855.
12. Warner, T.A. Remote sensing analysis: From project design to implementation. In *Manual of Geospatial Sciences*, 2nd ed.; Bossler, J.D., McMaster, R.B., Rizos, C., Campbell, J.B., Eds.; Taylor and Francis: London, UK, 2010; Chapter 17; pp. 301–318.
13. Manakos, I.; Manevski, K.; Petropoulos, G.P.; Elhag, M.; Kalaitzidis, C. Development of a spectral library for Mediterranean land cover types. In Proceedings of 30th EARSeL Symp.: Remote Sensing for Science, Education and Natural and Cultural Heritage, Paris, France, 31 May–3 June 2010; pp. 663–668.
14. Zomer, R.J.; Trabucco, A.; Ustin, S.L. Building spectral libraries from wetlands land cover classification and hyperspectral remote sensing. *J. Environ. Manag.* **2008**, *90*, 2170–2177.
15. Ruby, J.G.; Fischer, R.L. Spectral signatures database for remote sensing applications. *Proc. SPIE* **2002**, *4816*, 156–163.
16. Hueni, A.; Malthus, T.; Kneubuehler, M.; Schaepman, M. Data exchange between distributed spectral databases. *Comput. Geosci.* **2011**, *37*, 861–873.
17. Pfitzner, K.; Bollhöfer, A.; Esparon, A.; Bartolo, B.; Staben, G. Standardized spectra (400–2500 nm) and associated metadata: An example from northern tropical Australia. In Proceedings of IEEE International Geoscience and Remote Sensing Symposium, Honolulu, Hawaii, USA, 25–30 July 2010.
18. Buddenbaum, H.; Stern, O.; Stellmes, M.; Stoffels, J.; Pueschel, P.; Hill, J.; Werner, W. Field Imaging spectroscopy of beech seedlings under dryness stress. *Remote Sens.* **2012**, *4*, 3721–3740.
19. Hruska, R.; Mitchell, J.; Anderson, M.; Glenn, N.F. Radiometric and geometric analysis of hyperspectral imagery acquired from an unmanned aerial vehicle.

Remote Sens. **2012**, *4*, 2736–2752.

20. Nidamanuri, R.R.; Zbell, B. Use of field reflectance data for crop mapping using airborne hyperspectral image. *ISPRS J. Photogramm. Remote Sens.* **2011**, *66*, 683–691.
21. Hueni, A.; Nieke, J.; Schopfer, J.; Kneubühler, M.; Itten, K.I. The spectral database SPECCHIO for improved long-term usability and datasharing. *Comput. Geosci.* **2009**, *35*, 557–565.
22. Rasaiah, B.; Jones, S.; Bellman, C.; Malthus, T.J. Critical metadata for spectroscopy field campaigns. *Remote Sens.* **2014**,*6*, 3662–3680.
23. Jiménez, M.; González, M.; Amaro, A.; Fernández-Renau, A. Field spectroscopy metadata system based on ISO and OGC standards. *ISPRS Int. J. Geo-Inf.* **2014**, *3*, 1003–1022.
24. Nicodemus, F.E.; Richmond, J.C.; Hsia, J.J.; Ginsberg, I.W.; Limperis, T. *Geometrical Considerations and Nomenclature for Reflectance*; National Bureau of Standards, US Department of Commerce: Washington, DC, USA, 1977.
25. Thenkabail, P.; Lyon, J.; Huete, A. Advances in hyperspectral remote sensing of vegetation and agricultural croplands. In *Hyperspectral Remote Sensing of Vegetation*; Thenkabail, P.S., Lyon, J.G., Huete, A., Eds.; CRC Press/Taylor and Francis Group: Boca Raton, FL, USA, 2011; pp. 3–36.
26. Miao, X. Estimation of yellow starthistle abundance through CASI-2 hyperspectral imagery linear spectral mixture models. *Remote Sens. Environ.* **2006**, *101*, 329–341.
27. Möckel, T.; Dalmayne, J.; Prentice, H.C.; Eklundh, L.; Purschke, O.; Schmidtlein, S.; Hall, K. Classification of grassland successional stages using airborne hyperspectral imagery. *Remote Sens.* **2014**, *6*, 7732–7761.
28. Roberts, D.A.; Gardner, M.; Church, R.; Ustin, S.; Scheer, G.; Green, R.O. Mapping chaparral in the Santa Monica mountains using multiple endmember spectral mixture models. *Remote Sens. Environ.* **1998**, *65*, 267–279.
29. Manevski, K.; Manakosa, I.; Petropoulosa, P.; Kalaitzidis, C. Discrimination of common Mediterranean plant species using field spectroradiometry. *Int. J. Appl. Earth Obs. Geoinf.* **2011**, *13*, 922–933.
30. Silvestry, S.; Marani, M.; Marani, A. Hyperspectral remote sensing of salt marsh vegetation. *Phys. Chem. Earth.* **2003**,*28*, 15–25.
31. Kalacska, M. Ecological fingerprinting of ecosystem succession: Estimating secondary tropical dry forest structure and diversity using imaging Spectroscopy. *Remote Sens. Environ.* **2007**, *108*, 82–96.
32. Somers, B.; Asner, G.P. Hyperspectral time series analysis of native and invasive species in Hawaiian rainforests.*Remote Sens.* **2012**, *4*, 2510–2529.
33. Fyfe, S.K. Spatial and temporal variation in spectral reflectance: Are seagrass species spectrally distinct? *Limnol. Oceanogr.* **2003**, 464–479.
34. SPECCHIO. Online Spectral Database. Available online: http://www.specchio.ch (accessed on 11 August 2015).
35. Clark, R.N.; Roush, T.L. Reflectance spectroscopy: Quantitative analysis techniques for remote sensing applications. *J. Geophys. Res.* **1984**, *89*, 6329–6340.
36. Price, J.C. How unique are spectral signatures? *Remote Sens. Environ.* **1994**, *49*, 181–186.
37. Jacquemoud, S.; Verhoef, W.; Baret, F.; Bacour, C.; Zarco-Tejada, P.J.; Asner, G.P.; Francois, C.; Ustin, S.L. PROSPECT + SAIL models: A review of use for vegetation characterization. *Remote Sens. Environ.* **2009**, *113*, S56–S66.
38. Homayouni, S.; Roux, M. Hyperspectral Image analysis for material mapping using spectral matching. In Proceedings of ISPRS Congress 2004, Istanbul, Turkey, 12–23 July 2004.
39. Kruse, F.A.; Lefkoff, A.B.; Boardman, J.B.; Heidebrecht, K.B.; Shapiro, A.T.; Barloon, P.J.; Goetz, A.F.H. The Spectral Image Processing System (SIPS)—Interactive visualization and analysis of imaging spectrometer data. *Remote Sens. Environ.* **1993**, *44*, 145–163.
40. International Organization for Standardization (ISO). *Geographic Information—Metadata*; International Organization for Standardization: Geneva, Switzerland, 2003.
41. Mac Arthur, A.; Alonso, L.; Malthus, T.; Moreno, J. Spectroscopy field strategies and their effect on measurements of heterogeneous and homogeneous earth surfaces. In Proceedings of the 2013 Living Planet Symposium, Edinburgh, UK, 9–13 September 2013.
42. Mac Arthur, A.; MacLellan, C.; Malthus, T.J. The fields of view and directional response functions of two field spectroradiometers. *IEEE Trans. Geosci. Remote Sens.* **2012**, *50*, 3892–3907.
43. Salisbury, J.W. *Spectral Measurements Field Guide*; Tech. Rep. ADA362372; Defense Technology Information Centre: Fort Belvoir, VA, USA, 1998.
44. Jonkheere, I.; Fleck, S.; Nackaerts, K.; Muys, B.; Coppin, P.; Weiss, M.; Baret, F. Review of *in-situ* methods of leaf area index determination. Part I. Theories, sensors and hemispherical photography. *Agric. For. Meteorol.* **2004**, *121*, 19–35.
45. Gitelson, A.A. Nondestructive estimation of foliar pigment (chlorophylls, carotenoids, and anthocyanins) contents. In*Hyperspectral Remote Sensing of Vegetation*; Thenkabail, A., Lyon, P.S., Huete, J.G., Eds.; CRC Press: Boca Raton, FL, USA, 2011; pp. 141–166.
46. Colombo, R.; Busetto, L.; Meroni, M.; Rossini, M.; Panigada, C. Optical remote sensing of vegetation water content. In*Hyperspectral Remote Sensing of Vegetation*; Thenkabail, A., Lyon, P.S., Huete, J.G., Eds.; CRC Press: Boca Raton, FL, USA, 2011; pp. 227–244.
47. Schmidt, K.S.; Skidmore, A.K. Smoothing vegetation spectra with wavelets. *Int. J. Remote Sens.* **2004**, *25*, 1167–1184.
48. García Novo, F.; Marín Cabrera, C. *Doñana: Agua y Biosfera*; Ministerio de Medio Ambiente: Sevilla, España, 2005.
49. Muñoz Reinoso, J.C.; García Novo, F. Multiscale control

of vegetation patterns: The case of Doñana (SW Spain). *Lands. Eco.* **2005**, *20*, 51–61.

50. Zunzunegui, M.; Díaz Barradas, M.C.; Ain-Lhout, F.; Clavijo, A.; García Novo, F. To live or to survive in Doñana dunes: Adaptive responses of woody species under a Mediterranean climate. *Plant Soil* **2005**, *273*, 77–89.
51. Gratani, L.; Varone, L. Adaptive photosynthetic strategies of the Mediterranean maquis species according to their origin. *Photosynthetica* **2004**, *42*, 551–558.
52. Ain-Lhout, F. Seasonal differences in photochemical efficiency and chlorophyll and carotenoid contents in six Mediterranean shrub species under field conditions. *Photosynthetica* **2004**, *42*, 309–407.
53. Chen, J.M.; Cihlar, J. Quantifying the effect of canopy architecture on optical measurements of leaf area index using two gap size analysis methods. *IEEE Trans. Geosci. Remote Sens.* **1994**, *33*, 777–787.
54. Sweet, J.; Granahan, J.; Sharp, M. An objective standard for hyperspectral image quality. In Proceedings of AVIRIS Workshop, Jet Propulsion Laboratory, Pasadena, CA, USA, 23–25 February 2000.
55. Jimenez, M.; Pou, A.; Díaz-Delgado, R. Cartografía de especies de matorral de la Reserva Biológica de Doñana mediante el sistema hiperespectral aeroportado INTA-AHS: Implicaciones en el seguimiento y estudio del matorral de Doñana. *Revis. de Teledetec.* **2011**, *36*, 98–102.

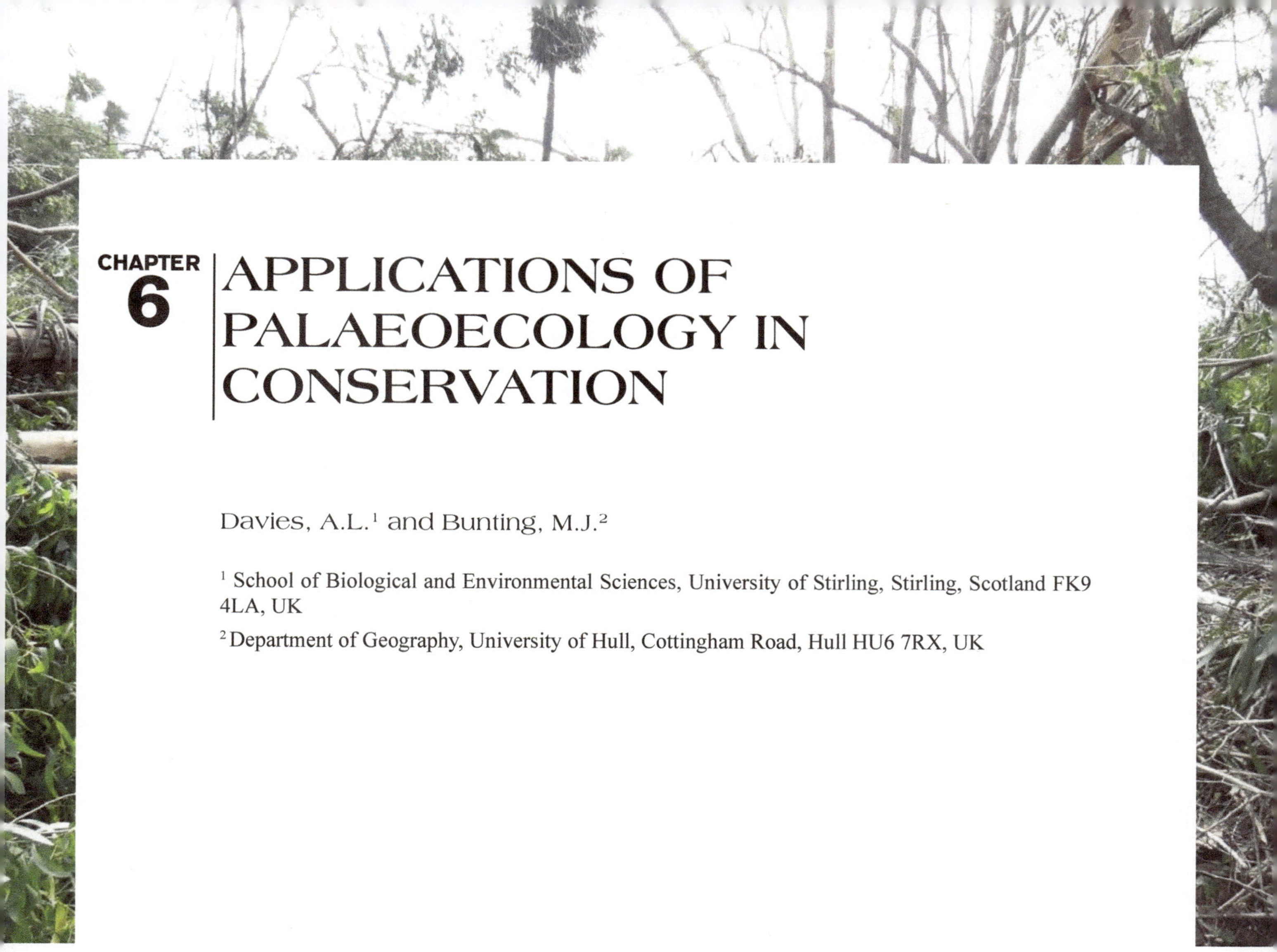

CHAPTER 6

APPLICATIONS OF PALAEOECOLOGY IN CONSERVATION

Davies, A.L.[1] and Bunting, M.J.[2]

[1] School of Biological and Environmental Sciences, University of Stirling, Stirling, Scotland FK9 4LA, UK

[2] Department of Geography, University of Hull, Cottingham Road, Hull HU6 7RX, UK

ABSTRACT

We review several case studies (water quality, upland management, woodland management) where palaeoecological data are able to contribute to current debate in conservation practice and ecology, as part of a wider consideration of the relationships between palaeoecology and conservation. We conclude that, although there are structural and inherent barriers that currently limit flow of information and collective working, the current environment of challenging policy targets and rapid environmental change makes better collaboration particularly urgent.

INTRODUCTION

The prospect of environmental change beyond the range of variability within which conservation practice has developed provides a powerful stimulus to evaluate the effectiveness of current practice. It is giving impetus to explore how additional sources of evidence can reduce uncertainty and improve decision-making, including interdisciplinary approaches to conservation planning, incorporating social, economic, political and scientific constraints and evidence (e.g. Parr et al. 2003; Holden et al. 2007; Reed et al. 2009). Facing challenging policy targets and rapid environmental change, conservation practitioners need the most complete evidence base possible to balance the competing demands of protecting and enhancing the natural environment, whilst maintaining the range of ecosystem services that society derives from it.

One area of apparent weakness in much of the evidence base used for conservation is the strong present-future focus. Long-term perspectives are implied in the language used in conservation policy documents, since terms such as 'sustainability', 'heritage' and 'resilience' are concepts with a substantial time dimension. However, this implication is seldom followed up by any substantive use of long-term perspectives in ecological research agendas or conservation practice. Outside the spheres of built heritage and archaeological monuments, past and historical perspectives tend to be marginalized in debate. Although their considerable potential value for helping landscape and ecosystem conservation activities achieve the targets and goals defined in light of neoecological, social and pragmatic constraints is recognized in the academic literature (e.g. Foster et al. 1990, Swetnam et al. 1999, Birks 1996, Willis et al. 2007, Froyd & Willis 2008), they as yet rarely play much part in ecological or management debates. For instance, analysis of the 100 questions of conservation importance identified by Sutherland et al. (2006) found that answering 54 of those

Citation: Davies, A.L. and Bunting, M.J, "Applications of Palaeoecology in Conservation," The Open Ecology Journal, 2010, 3, 54-67. http://benthamopen.com/contents/pdf/TOECOLJ/TOECOLJ-3-2-54.pdf

questions required at least some consideration of processes acting over multiple years or of conditions in the past as well as the present (see Table 1). In practice, and despite the many papers setting a clear rationale for the value of alternative long-term records, datasets used as the evidence base for practice tend to be relatively short-term and rely overwhelmingly on ecological observation and experiment. For example, no dataset used in the Millennium Ecosystem Assessment (2005) is longer than 50 years in duration. This apparent lack of time-depth can generate misleading inferences and potentially lead to inappropriate management activities.

'Long-term' studies, those incorporating data from a time range of fifty years or more, exceed human generations, and are one or more orders of magnitude longer than the typical 3-5 year span of research funding, management planning or political programming. Such datasets matter because many ecological processes 'play out' over decades and centuries (Delcourt & Delcourt 1991, Morecroft et al. 2009). The relatively long life spans of many important taxa (e.g. trees, dwarf shrubs, larger mammals and birds) mean that population consequences of system change may not be fully seen for decades (Willis et al. 2005, Grant & Edwards 2008, Lindbladh et al. 2008). A particularly striking example of this is the concept of 'extinction debt' (Kuussaari et al. 2009).

These issues are recognized in a recent review of the first fifteen years of monitoring at the Environmental Change Network sites, where Morecroft et al. (2009: 2830) write that "trends in physical, chemical and biological parameters are emerging which could not have been detected at an earlier stage. This is still, however, a relatively short time compared to many ecological processes". Longer-term effects were often masked by substantial inter-annual variability during shorter monitoring periods. Results like this suggest that reliance on short timeframe monitoring and experimental data for the evidence base in conservation practice is insufficient. Given the rapid rate of change in landscape impacting human activities and projected climate changes, extending conventional methods to longer time frames using a 'wait and see' approach is not possible either. Conservation practice needs to explore other routes to the necessary evidence base; Pullin & Knight (2003) identify shortcomings in current research practice networks, and others call for increased horizon scanning to improve anticipation of change (Sutherland et al. 2008, Sutherland & Woodroof 2009). While palaeoecology remains an underused resource in conservation, it is clear long-term ecological datasets are an essential resource for predicting the consequences of management interventions and for establishing appropriate strategies in research, policy and practice. They can assist in identifying alternate steady states, quantifying natural variability and identifying thresholds of risk for state change events, and remind us of the probability of future 'ecological surprises'.

Table 1. Summary of Potential for Palaeoecological Data to Contribute to Discussion of the 100 Key Questions for UK Conservation Policy Identified by Sutherland et al. (2006)

	Potential for palaeoecological data to contribute to the discussion	major	minor	minimal
	ECOSYSTEM SERVICES			
1	What are the benefits of protected habitats... relative to non-protected land?	X		
2	What is the role of biodiversity in maintaining specific ecosystem functions?	X		
3	What are the roles of soil biodiversity (...) in ecosystem function, resilience and recovery?			X
4	How does soil biodiversity both influence and respond to above ground biodiversity?			X
5	What is the role of marine biota and benthopelagic coupling in ocean-atmosphere carbon coupling and primary production?			X
6	How can we measure natural capital (...) and integrate such a measure into GDP?		X	
	FARMING			
7	How will CAP reform affect biodiversity at the landscape scale?	X		
8	What are the environmental consequences of farming patterns...?	X		
9	How do farming systems ... compare in terms of their effects on... environmental impacts?	X		
10	How for current agricultural practices affect the conservation value and extent of non-agricultural habitats... and how can detrimental impacts be mitigated?		X	
11	What are the impacts of agricultural activities... on soil biodiversity and function?		X	
12	What are the ecological consequences of changes in upland grazing regimes for biodiversity and soil ecology?	X		
13	What are the impacts on soil and... invertebrates of poaching... and soil compaction at different stocking levels?			X
14	What are the impacts on biodiversity of prophylactic treatment of farm livestock...?			X
15	What lessons can be learnt from agri-environment schemes to optimise their... benefit?			X
16	How does the ecological impact of UK farming compare internationally?		X	

	FORESTRY			
17	What are the environmental benefits of large-scale woodland planting schemes…?			X
18	Where should new woodlands be located?		X	
19	What overall number, age structure and spatial distribution of trees are necessary for the long-term survival of species dependent on ancient/veteran trees?		X	
20	What are the relative benefits for biodiversity of the re-introduction of management to… semi-natural woodlands *vs*.… absence of active management?	X		
21	Why have many woodland birds declined?		X	
22	Which approach to the removal of plantations on ancient woodland sites… yields the greatest biodiversity benefit?			X
	FISHERIES, AQUACULTURE AND MARINE CONSERVATION			
23	What is the biodiversity impact of the harvest of forage fish…?			X
24	What are the ecological impacts of… [inputs] from aquaculture?		X	
25	How important are caged fishes as reservoirs of parasites and pathogens…?			X
26	What are the direct… and indirect… impacts of commercial fishing on cetaceans and seabirds?			X
27	How large should marine protected areas be, and where should they be located…?			X
28	What will be the impact of marine protected areas on wide-ranging species…?			X
29	How important are coastal, estuarine, and fluvial habitats for endangered migratory fish populations…?		X	
30	What is the range of minimal viable population sizes for broadcast spawning marine species?			X
31	How long does it take the seabed to recover from disturbance…?	X		

	Potential for palaeoecological data to contribute to the discussion	major	minor	minimal
	RECREATION AND FIELD SPORTS			
32	What are the impacts of recreational activities on biodiversity?		X	
33	Which ecological principles should guide the choice of… species appropriate for game exploitation?			X
34	What overall impacts do introductions of game species… (including recreational fishing) have on biodiversity?		X	
35	What are the ecological impacts… of a ban on hunting with dogs?			X
	URBAN DEVELOPMENT			
36	How can provision for wildlife be maximised in… urban [sites]?			X
37	What are the consequences for biodiversity of fragmentation by development and infrastructure?		X	
38	What are the ecological impacts on semi-natural habitats and ecosystems of adjacent large developments…?		X	
39	How can sustainable urban drainage systems be optimally designed…?			X
	ALIENS AND INVASIVE SPECIES			
40	What criteria should be used to determine when to intervene to deal with invasive species?		X	
41	How can we manage microbial ecology to control invasive plant pathogens?			X
42	How can we understand better the epidemiology… of diseases within wildlife reservoirs…?			X
43	What are the genetic threats to UK biodiversity posed by introgression from genetically modified organisms…?			X
44	What is the optimal method of managing bracken-dominated habitats…?		X	
45	What are the effects of domestic cats on vertebrate populations…?			X

	POLLUTION			
46	What impact does plastic-derived litter have on the marine environment?			X
47	How can one ameliorate the effects of aerially deposited nitrogen…?		X	
48	What are the critical thresholds for nitrogen and phosphorus inputs into water…?	X		
49	…which [chemicals] are or are likely to become significant environmental problems…?			X
50	What are the long-term impacts of depositing… organic wastes on to agroecosystems?			X
51	How can catchment management be used to reduce diffuse pollution?		X	
52	How will acidification of surface water from rising CO_2 concentrations affect… marine organisms?	X		
53	What are the effects of light pollution… on wildlife…?			X
	CLIMATE CHANGE			
54	Which species are likely to be the best indicators of the effects of climate change on natural communities?	X		
55	Which habitats and species might we lose completely in the UK because of climate change?		X	
56	What will be the ecological impacts of changing agricultural patterns in response to climate change?			X
57	What time lags can be expected between climate change and ecological change?	X		
58	What is the likely relationship between the extent of climate change and the pattern of species extinction?	X		
59	How does climate change interact with other ecological pressures…?	X		
60	How can we increase the resilience of habitats and species to cope with climate change?	X		
61	How well suited is the current UK protected area system for conserving biodiversity… and how can it be enhanced…?		X	
62	How will changes in oceanographic conditions… affect marine ecosystems?	X		
63	What actions are required to recreate the full range of coastal landscapes, habitats and species distributions to compensate for their loss, for example as a result of sea level rise?			X

	Potential for palaeoecological data to contribute to the discussion	**major**	**minor**	**minimal**
	ENERGY GENERATION AND CARBON MANAGEMENT			
64	What are the consequences of biofuel production for biodiversity at field, landscape and regional levels?			X
65	What are the potential impacts of a) terrestrial and b) marine windfarms on biodiversity?			X
66	What are the comparative biodiversity impacts of newly emerging types of renewable energy such as tidal energy?			X
67	How can soil carbon be retained and further carbon sequestered in the soil?		X	
	CONSERVATION STRATEGIES			
68	How can biodiversity action plans be designed to take account of larger scale population processes?		X	
69	How can we best measure favourable conservation status for each of the species and habitats listed within the EU's habitats directive?			X
70	How effective is the current UK protected area network…under current conditions?			X
71	With what precision can we predict the ecological impact of different policy options and the ecological effects of management action?	X		
72	At an international scale, what are the ecological implications of conservation actions and policies adopted within the UK?			X
73	How effective as indicators of overall biodiversity are current indicators…?			X
74	Why are common moths declining and are their declines driving declines in other taxa…?			X
75	What scale and type of land-use change is required to halt the decline in biodiversity…?		X	
76	Are there reliable ways to predict the long-term sustainability of populations of poorly known species (*e.g.* most invertebrates) using … ecological characteristics?		X	

	HABITAT MANAGEMENT AND RESTORATION			
77	What are the costs and benefits of concentrating conservation work on designated sites...?		X	
78	What are the ecological consequences of 'wilding'...as a long-term conservation strategy?	X		
79	What are the consequences of different moorland management techniques... for the uplands...?	X		
80	What measures of habitat condition should we use to measure habitat change in protected areas?		X	
81	How should ditches... be managed...?			X
82	What hedgerow structure and...management produce the greatest wildlife benefits?			X
83	How do recreated habitats differ from their semi-natural analogues?	X		
84	How can we effectively prioritise ...large-scale ecological restoration projects...?			X
85	What is the most appropriate and ecologically sustainable way of dealing with excess nutrients during ...habitat restoration?		X	
86	What are the implications of changing deer densities for agriculture, forestry and biodiversity in different landscape types?			X
87	In reintroductions, does local provenance matter?...			X
	CONNECTIVITY AND LANDSCAPE STRUCTURE			
88	What are the lag times between habitat fragmentation and the loss of species...?	X		
89	Is it better to extend existing habitat patches or create further patches within a landscape?			X
90	How should we manage landscape mosaics for the conservation of diverse taxa that operate on different spatial scales?			X
91	What are the relative merits of different indices of habitat connectivity?...			X
92	What is the value of linear habitats...?			X
93	For [relevant] species... how can 'source' and 'sink' populations.. be identified...?			X
94	How important are core versus peripheral areas in the conservation strategy of a species?		X	
95	How reliant are animal and plant populations in small nature reserves on the maintenance of habitat in surrounding non-protected areas?		X	

	Potential for palaeoecological data to contribute to the discussion	major	minor	minimal
	MAKING SPACE FOR WATER			
96	What have been the consequences of past and present riparian engineering works...?	X		
97	What would be the ecological implications of large-scale rive and floodplain restoration schemes... and would they be... cost-effective...?	X		
98	What are the likely consequences for biodiversity of changes in water quality and sedimentation in rivers?		X	
99	What methods most accurately measure ecological status in the EU water Framework Directive?	X		
100	How can flood control be assisted by... habitat management and restoration...?		X	

In this paper, we draw together previously published palaeoecological case studies to give concrete examples of how a long-term perspective can contribute to ecological debate and conservation management. Recent literature shows increasing awareness of the potential use of palaeoecological data, and by giving specific illustrations of practical application this paper aims to contribute to the transition from awareness of potential towards regular use of long-term data in conservation ecology, and improved communication with conservation ecologists in the design and dissemination of palaeoecological research. We then consider briefly some of the barriers to achieving this transition and means of lowering them.

PALAEOECOLOGY IN THE CONSERVATION TOOL KIT

Palaeoecology is one of a suite of approaches available for acquiring long-term datasets. It forms a time-efficient alternative to long-term monitoring, since data covering several hundred (or thousand) years of change can be acquired from months of research effort. Palaeoecology can be understood as reading the results of natural or historical experiments that were begun a long time ago, and whose evolution is recorded in the sedimentary archives of lakes and mires. Much published palaeoecological research begins by or focuses on reconstructing past ecosystems and system dynamics over time. These reconstructions offer an evidence-based challenge to short-term understandings of 'typical' ecosystem function, and can be particularly useful in disentangling the relative importance of different drivers and disturbance regimes, such as management activities and climate change, in ecosystem development or population dynamics. Site histories contextualize and explain the present status of the site, offer a means of presenting landscape as cultural heritage, and can be of considerable interest to stakeholder groups and an effective tool in engaging local community interest. They can also provide clear grounds to support management decisions; for example, demonstrating that a site has been a raised mire for 3000 years and was only colonized by trees in the last 20 years can be a compelling argument for woodland clearance.

A common misconception is that these techniques only provide generalized background to a site, disconnected from current management concerns, or are solely useful for definition of a restoration target in terms of known past conditions. Whilst palaeoecological data can effectively deliver these outcomes, they also offer insight into processes and system function. Demonstrating that sites formerly had a higher ecological value or which sites are closest to 'pristine' or 'pre-disturbance' status can be used directly to prioritize sites and set targets for highly interventionist conservation activities such as restoration or re-creation of habitats in a minority of situations (Cumming et al. 1994). However, in most cases it is questionable whether selecting fixed points from the past as future goals will result in sustainable ecosystems, particularly in the face of unprecedented and uncertain anticipated environmental changes. Palaeoecology demonstrates the ephemeral nature of community composition over the longer term, and since present environmental conditions and the wider landscape setting are different to those prevailing in the past, faithful re-creation of the past is unrealistic (William & Jackson 2007). Target-setting can however draw on palaeoecological data to define more general measures of ecosystem health relative to present ecosystem status (e.g. estimates of water chemistry, dominant species or relative biodiversity).

Britain has a long tradition of palaeoecological research, partly owing to the combination of climate and geomorphology which means that mires and lakes - and therefore waterlogged sediments containing palaeoecological records - are fairly common landscape elements. Since relatively undisturbed sites often occur in areas of conservation interest, records tend to come disproportionately from such locations. This means that there is already a rich resource of data in the academic literature, generated over the last 40 or so years (e.g. Oldfield 1970, Huntley 1991, Birks 1993, 1996, Brown 2002), which can be mined to explore a variety of issues relevant to the current state and future trajectories of the physical environment. To illustrate these principles and applications of long-term evidence to current questions of conservation importance, and consider the extent to which they influence conservation practice, five UK case studies are presented in response to questions raised by Sutherland et al. (2006).

CASE STUDY 1: THE WATER FRAMEWORK DIRECTIVE (QUESTION 99: SUTHERLAND ET AL. 2006; SEE TABLE 1)

The European Union Water Framework Directive (WFD; European Union 2000) is designed to drive integrated management of surface waters at the catchment scale throughout Europe over several decades, and its main objectives are to prevent further deterioration, provide long-term protection and enhance the status of water resources by achieving 'good ecological quality' in all relevant surface water bodies by 2015. Since its goal is integrated management of fresh water bodies, the WFD emphasizes ecological structure and function, with physical and chemical parameters recorded as supporting elements. The definition of ecological quality is intrinsically historical in character, since quality is "judged by the degree to which present-day conditions deviate from those in the absence of anthropogenic influence, termed reference conditions. Sites in which the various elements correspond totally or almost totally to undisturbed (reference) conditions will be classed as High status" (Bennion & Batterbee 2007: 285).

Identifying reference conditions is a vital first step to compliance with the WFD, and the Directive itself includes hindcasting methods such as palaeolimnology (the study of long-term ecosystem dynamics within lakes and ponds) in the list of possible methods to be used. Member States define river basin districts, identify and characterize the water bodies into ecotypes, and then describe type-specific reference conditions. Since almost all European waters are affected by human activity to some extent, identifying a suite of undisturbed comparison sites ranges from difficult to impossible (e.g. for highly cultural landscapes such as lowland England), depending on the river basin district and ecotype. Long-term monitoring series rarely extend to before major periods of disturbance (e.g. the Industrial Revolution) and even when data series do exist, problems such as a lack of methodological continuity frequently restrict their usefulness. Derivation of site-specific reference conditions from the sediment record therefore offers a valuable addition to the tool kit and, for some sites such as shallow lowland lakes, may be the only way to determine pre-impact conditions (Bennion et al. 2003).

Since the WFD focuses on ecological quality, a multiproxy approach to reconstruction is needed. Bennion and Battarbee (2007) review various ways in which palaeolimnological methods are used to determine physicochemical reference conditions, such as via diatom-inferred Total Phosphorous and pH (e.g. in Scotland: Battarbee et al. 2005, Finland: Kauppila 2006 and Ireland: Leira et al. 2006). These reconstructions are supplemented by a range of indicators representing different parts of the trophic web, including both flora (e.g. diatoms, aquatic plant macrofossils and pigments) and fauna (e.g. chironomids, cladocerans, fish scales), which potentially allow reconstruction of structural and functional characteristics of the ecosystem at points in the past, including the 'pre-human-impact' point required to determine good ecological status under the WFD. Such direct ecological information provides a sound basis for management decisions, and numerous studies illustrate the effectiveness of the approach for assessing ecological change and defining reference conditions (e.g. Meriläinen et al. 2003, Hynynen et al. 2004, Guilizzoni et al. 2006, Taylor et al. 2006).

However, reconstructing the history of the ecosystem is not sufficient to define the baseline reference state. A single point in time has to be selected from the multi-millennial record that is usually available, and it seems sensible that baseline 'pre-impact' dates will vary between ecotypes, reflecting historical and contextual differences. Bennion et al. (2004) chose A.D. 1850 as the reference date for Scottish lochs, since this pre-dates the major changes made by largescale industrialisation and agricultural intensification in the region. Palaeoecological studies from Denmark (e.g. Bradshaw and Rasmussen 2004; Bradshaw et al. 2006), where a long history of agricultural exploitation of most catchments is clear in the lake records, suggest that true preimpact baselines should be sought several millennia ago. Pragmatic and political considerations will also play a part in setting an age for reference conditions; in the densely populated, long settled agricultural lowlands of Europe, setting a target of the state of the ecotype in the preagricultural forests of the early Holocene would be unrealistic and unachievable.

Once a reference date for an ecotype is chosen, it is possible to reduce the costs and increase the speed of the palaeolimnological analyses by using dating methods to identify the location of the reference date within the core, and analyze just two samples, one from this date and one from modern surface sediments (to assess the degree of deviation from the reference state). Whilst we argue above that single reference states are not generally identifiable, WFD target-setting in the UK context requires that a single, measureable reference state be identified. Defining a baseline date allows efficient quantification of this reference state. This 'before and after' sampling can be rapidly applied to a large number of lakes, and the data used to classify sites in terms of degree of deviation from the reference state, which gives a clear basis for planning and costing management action. More detailed analysis of the record from representative lakes in each ecotype can provide supplementary information such as how recently good status was lost and an indication of the primary cause of loss of good status, which can then inform the detailed planning of intervention strategies.

Using the sedimentary record to assess present status offers an advantage over conventional monitoring, since sampling an appropriate depth of sediment integrates the record from several years, offering a good picture of overall conditions. In order to measure the ecosystem health of a lake effectively, conventional monitoring needs to be carried out at several dates during at least one yearly cycle, and therefore the palaeoecological approach reduces the number of visits needed. Assessment and active management of sites classed as having poor ecological status is not just an issue for fresh water systems, although there is no single terrestrial equivalent of the WFD. Management plans and strategies at multiple levels, from the Millennium Ecosystem Assessment to national networks (e.g. Countryside Survey and Environmental Change Network), national and regional Biodiversity Action Plans and assessments of individual designated sites, all give repeated examples of threatened habitats in 'unfavorable' state and in need of restoration (e.g. Britton et al. 2005, Williams 2006). Just as in the freshwater situation, effective planning and restoration action depends on sound understanding of ecosystem processes, particularly system responses to environmental change, and definition of appropriate 'baselines' and targets. Two case studies are used to illustrate how long-term datasets can inform and support terrestrial ecosystem restoration initiatives.

CASE STUDY 2: MOORLAND MANAGEMENT TECHNIQUES (QUESTION 79; SUTHERLAND ET AL. 2006; SEE TABLE 1)

UK moorland ecosystems are of national and international importance for their biodiversity, landscape, cultural heritage and ecosystem services contributions to human livelihoods (Millennium Ecosystem Assessment 2005, Reed et al. 2009). However, they are also threatened by the loss of conservation values and by widespread erosion (e.g. Grieve et al. 1995, Thompson et al. 1995, Higgitt et al. 2001, Holden et al. 2007). It is predicted that peatlands will show heightened sensitivity to disturbance as a result of climatic change over coming decades (Bragg & Tallis 2001, Moore 2002); these threats and the subsequent risk of non-reversible damage are therefore expected to increase without active intervention. Understanding local drivers of change is essential for establishing appropriate responses on a site level. Frank Chambers and colleagues have investigated the causes and process of degradation at several sites on Exmoor (England) and in Wales using pollen, plant macrofossil, peat stratigraphy, radiocarbon-dating and the occurrence of soot particles (fossil fuel-derived pollutants used to date peat accumulated since c.1850). Their work (Chambers et al. 1999, 2007a, 2007b) establishes that degradation is associated with strong directional shifts in the competition balance between dominant taxa and species impoverishment. This involved a switch from mixed, dynamic heather-grass moors to unpalatable and homogeneous grass moors, dominated by Molinia caerulea (purple moor grass), and occurred in the late 19th and early 20th centuries. Causes of this switch include changes in grazing regime, in burning patterns and in atmospheric nitrogen deposition, which may be continuing to favor grass dominance (Chambers et al. 1979, 1999, Holden et al. 2007, Hughes et al. 2007). These data suggest strongly that efforts to 'restore' these degraded blanket bogs by reducing Molinia growth are justified, since the shift in dominance is recent and unprecedented. However, the conservation preference often given to Callunetum needs to be mitigated with the knowledge that its presence may not always be long-standing (Chambers et al. 2007b). In contrast with the loss of heather cover from blanket peat catchments in some parts of the UK over recent centuries (Stevenson & Thompson 1993), in northern England and South Wales Calluna dominated communities are part of the relatively recent shifts towards mono-specific dominance on blanket mires (Atherden 2004, Chambers et al. 2007a).

In contrast with the uniformity of some present-day grass moors and Callunetum, palaeoecological data show that upland moorlands were historically vegetated with a dynamic mosaic of communities, reflecting both local-scale variations in topography and hydrology and spatial heterogeneity in landscape use. Although many moors were, as now, characterized by a small number of dominant taxa, this mosaic offered a wider variety of habitat structure and microclimate within relatively small areas, which probably sustained local biodiversity in many taxonomic groups (e.g. Littlewood et al. 2006). Rather than prioritizing a single dominant or community type, management strategies which aim to ensure structural, if not phyto-taxonomic, diversity within Calluna- or grass-dominated areas contribute to improvements in condition in terms of biodiversity and system function.

CASE STUDY 3: CURRENT SEMI-NATURAL ANALOGUES IN HABITAT RESTORATION (QUESTION 83: SUTHERLAND ET AL. 2006; SEE TABLE 1)

In upland moorland, the replacement of mixed communities by either Callunetum or grass moor is relatively recent. This is not the case for some woodland restoration target areas in northern Scotland, where trees have not been present for centuries to millennia (Tipping 1994). The UK is one of the least densely forested countries in Europe and in Scotland the majority of existing tree cover consists of conifer plantations, interspersed with extensive tracts of moor and peatland. This has given rise to numerous woodland restoration initiatives (Caledonian Partnership 2003, Hobbs 2009). Despite frequent emphasis on human 'degradation' of native habitats (e.g. Darling 1955, McIntosh 2006), existing woods or current soil and climatic criteria are often assumed to provide suitable templates for establishing appropriate locations and composition in woodland restoration (e.g. Towers et al. 2004). Managers in West Glen Affric (NW Scottish Highlands), however, chose to fund research into past vegetation and climate dynamics to assess the limitations of a present-analogue approach and inform future management strategies on tree planting and woodland restoration.

In contrast with the current open, peat-dominated landscape, diverse woodland communities once existed in West Affric, including a mosaic of mixed deciduous woods on alluvial soils and open pine-birch-heaths on peaty ground, with the abundance of pine decreasing westwards (Davies 2003, 2010, Tipping et al. 2006). Current pine-dominated woodland in East Affric thus does not provide a good analogue for past woodland communities in West Affric; woods throughout the valley were more diverse than the present remnants in East Affric (Tipping et al. 1999, Shaw & Tipping 2006). The abundant pine stumps eroding out of peat in West Affric and across NW Scotland are often interpreted as evidence that these areas once supported pinedominated woodland. However, these are an artefact of preservation bias rather than a true reflection of past woodland composition or stability: drier climatic phases allowed episodic, short-lived pine colonization events onto bogs, after which a return to wetter conditions led to the death of the invading trees, renewed growth of peat and preservation of pine stumps (Bridge et al. 1990). Major woodland contraction occurred during a period of climate change between 4400-3800 years ago, which first altered competitive balances between pine, birch, oak and alder, and then led to the widespread failure of tree regeneration throughout West Affric and across NW Scotland (Davies et al. 2004, Tipping et al. 2006; Fig. 1). Given the sensitivity of these woods to past climatic change and predictions of future climatic variability, restoration approaches which rely on recent baselines and current conditions may not generate self-perpetuating or sustainable woods. In response, managers of West Affric continue to work with woodland restoration organizations, but have revised the extent of their plans for woodland growth in recognition of the palaeoenvironmental findings on past species composition and the drivers of woodland decline.

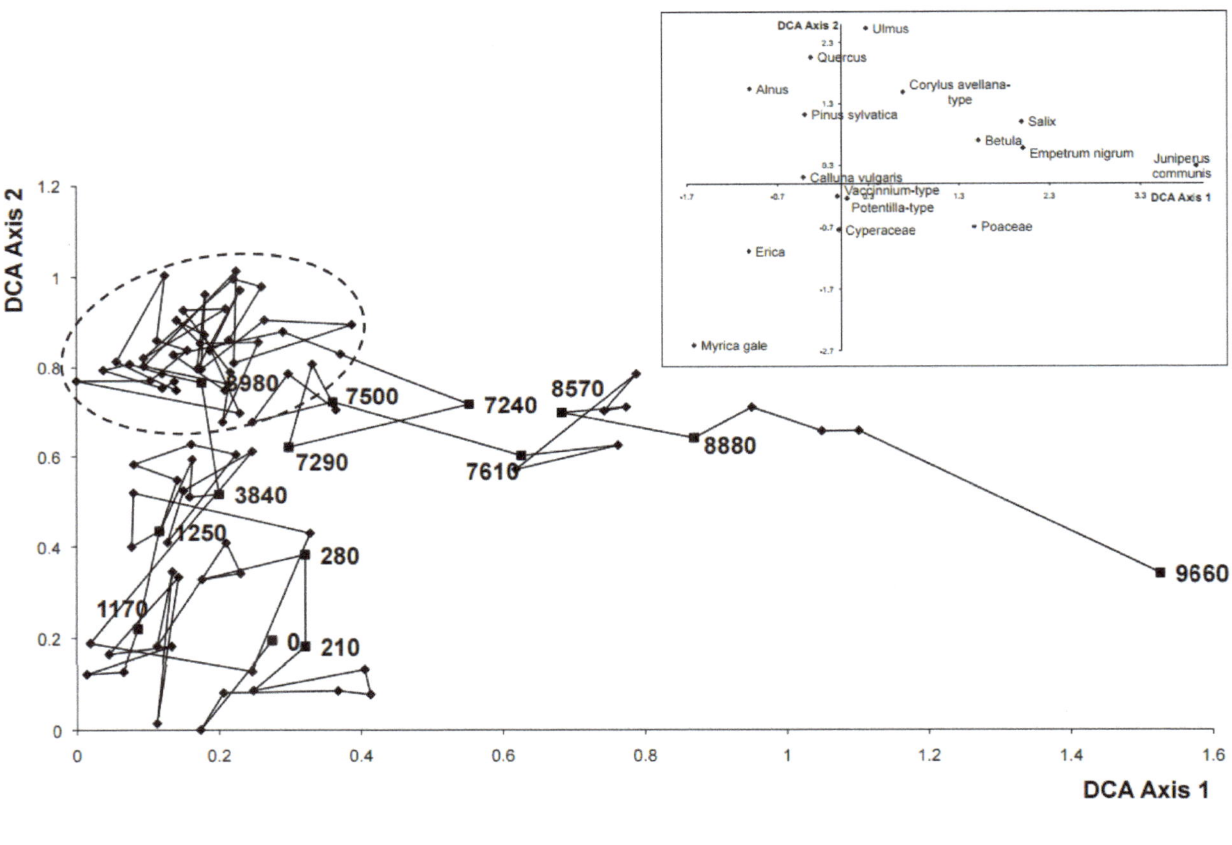

Figure 1. Ordination (detrended correspondence analysis) plot of pollen data from Torran Beithe, an area of blanket peat cover in West Affric (NW Scotland), showing changeable post-glacial migration and succession phase (c.9700-7000 years ago), followed by a period of relative stability during the woodland/heathland mosaic phase (circled), followed by a shift to the present peat-dominated vegetation around 4000 years ago.

CASE STUDY 4: GRAZING IMPACTS & HISTORICAL CONTEXT OF BIODIVERSITY LOSS (QUESTIONS 8 & 12; SUTHERLAND ET AL. 2007; SEE TABLE 1)

Habitat loss and changes in agricultural practice pose major global threats to biodiversity. In particular, agricultural intensification and abandonment threaten many biodiversity, landscape and cultural heritage values that have formed through centuries of interaction with agricultural communities (Henle et al. 2008). This is especially the case in upland areas, where significant reductions in grazing pressure are forecast (Reed et al. 2009). Understanding how species respond is a priority for meeting the challenge of conserving biodiversity in the context of changes in agriculture and climate. A long-term perspective can increase the evidence-base by providing information about the effects of many different forms of agriculture on upland biodiversity, not just modern mechanized, low labor-input farming.

Hanley et al. (2008, 2009) analyzed the effects of changes in resource management and associated socio-economic factors on plant diversity in upland Scotland over the last 400 years using an approach that combined palynological and historical datasets via panel data analysis to pool spatiotemporal information. Rarefaction analysis or palynological richness was used as a proxy for plant diversity, reflecting a combination of diversity and structural (evenness) parameters (Birks & Line 1992, Odgaard 1999). Diversity levels have varied considerably over the last 400 years, partly as a function of changes in land management, indicating that static baselines and 'naturalness' are problematic concepts in conservation and ecology more widely. Changes in livestock grazing pressures brought about by changes in market prices had significant effects on diversity: both higher grazing pressures and abandonment are associated with reductions in plant diversity. Grazing therefore needs to remain a key concern for maintaining diversity.

Numerous sites in this study (Fig. 2) and across a range of European habitats show significant reductions in plant diversity over the last 200 years (e.g. Davies & Dixon 2007, Berglund et al. 2008, Feurdean & Willis 2008). Homogenization is thus not simply the late 20th century trend seen in many repeated ecological surveys (e.g. Britton et al. 2009, Keith et al. 2009); this knowledge can help establish more realistic reference frameworks for conservation. For example, meadow and species-rich grassland diversity has suffered as a result of agricultural intensification and abandonment (Davies & Dixon 2007, Gustavsson et al. 2007, Hanley et al. 2008, cf. Lindbladh 1999). This has implications for the faunal communities that depend on the diminishing range of less intensively managed grassland habitats and on landscape heterogeneity (e.g. Benton et al. 2003, Redpath et al. 2010). Palaeoentomological data further support this threat faced by faunal communities due to habitat fragmentation, as many beetle species have become locally and nationally extinct owing to the loss of their preferred habitat (Whitehouse 2006). Ecosystem rearrangement and habitat fragmentation arising from climate change is likely to exacerbate the extinction risk for many species (e.g. Huntley et al. 2008). This heightens the need to use multiple datasets and 'horizon-scan' (Sutherland et al. 2008) beyond familiar ecological datasets to establish appropriate 'baseline' measures for diversity.

CASE STUDY 5: SEMI-NATURAL WOODLAND MANAGEMENT (QUESTION 20; SUTHERLAND ET AL. 2007; SEE TABLE 1)

Despite recognizing that no habitats in Europe have escaped human influence of some kind, (relative) naturalness continues to influence management decisions (e.g. Angermeier 2000). In reality, however, the current state of natural regeneration is relatively poor in many native woods, with evidence for long-term homogenization, the continued loss of old growth woods and the spread of shade-tolerant and generalist species, partly as a consequence of limited management (Petit et al. 2004, Kirby et al. 2005, Hopkins & Kirby 2007, Keith et al. 2009). This has led some ecologists to question what woodland models are appropriate for conservation and to assess the value of managed disturbance (Amphlett 2003, Quelch 2004, Hall & Stone 2004, Hancock et al. 2005). Applied palaeoecology does not propose going back to past management strategies, but can help assess what disturbance regimes favored regeneration in particular habitats to establish models and management plans that will maintain continuity and conservation values into the future.

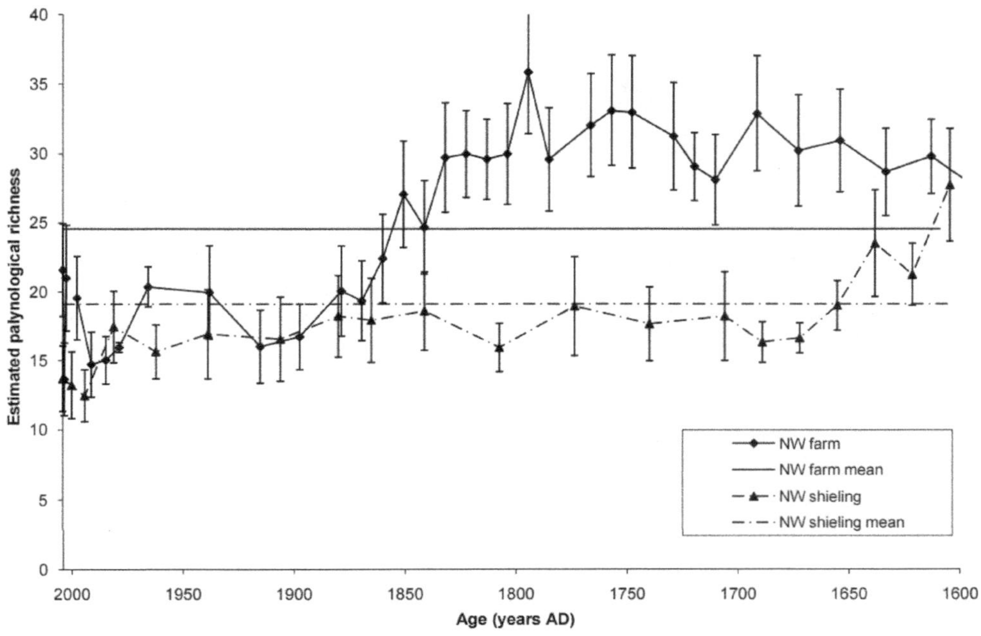

Figure 2. Rarefaction (palynological richness) trends at two sites in NW Scotland over the last 400 years, showing 17th century loss of diversity at shieling (summer farm) site due to woodland loss with the establishment of grazing, and 19th century diversity decline at the farm owing to the eviction of tenant farmers in 1812 during the creation of an extensive sheep farm which became unviable and was then abandoned in the late 19th century.

Pollen studies in high conservation value Atlantic oak woods in N England, W Scotland, N Wales and SW Ireland show that these valued communities have no long history: they have developed since the cessation of timber and grazing management around 100-200 years ago (Edwards 1986, Mitchell 1988, 1990, Birks 1993, Sansum 2004). The current composition and age structure of the canopy of these woods, especially the dominance of oak, reflect past management effects rather than present conditions owing to the slow rate of maturation in populations with a long generation time. Past management limited shrub growth and maintained a more open, disturbance-adapted ground flora, while species selection and repeated disturbance reduced tree-species diversity. Although bryophytes are important for current woodland conservation value, they are poor indicators of the antiquity of the observed woodland communities and structure. The persistence of the valued, rich bryophyte floras stems from the presence of 'semiwoodland' habitats and microrefugia (e.g. streamsides, crags), and the fact that some form of tree cover was always present. 'Natural' features, such as deadwood accumulation and limited numbers of shade-tolerant understorey species, have been scarce or absent for centuries owing to this management history. Consequently, as indicated by ecological experiments and monitoring, disturbance is required to maintain current values and herbivore exclusion alone is unlikely to result in widespread tree regeneration (Palmer et al. 2004). Equally, the results indicate that human intervention need not be damaging or run contrary to conservation objectives which prioritise regeneration and structural heterogeneity (Sansum 2004).

DISCUSSION

These case studies show that palaeoecological data can make contributions to a range of conservation activities, from defining targets to supporting the development of effective management plans for achieving those targets. In combination with shorter-term datasets, they allow scientists to investigate how ecological processes develop over decades, centuries and millennia, and understand the past conditions whose lagged consequences are still manifest in the modern landscape. Why, therefore, does palaeoecology continue to fall outside the 'horizons' being scanned to improve anticipation of needs and threats that we face on a global scale?

Although many papers by palaeoecologists have been published in ecological journals, giving both UK and international examples of how palaeoecological data might contribute to ecology and conservation (e.g. Birks 1996, Brown 2002, Willis et al. 2005, 2007, Willis & Birks 2006), these are mostly cited by palaeoecologists and it may be that there is still a simple lack of awareness among the ecological community of the potential of palaeoecology to address ecologically interesting questions with sufficient precision to warrant attention (Froyd & Willis 2008). Several factors could contribute to this. Over the last few decades in Britain academic palaeoecologists have generally found employment in Geography or Archaeology departments, rather than Biology or Ecology, meaning that many graduates in disciplines leading to conservation-related jobs have barely encountered palaeoecological ideas during their studies.

Time to become familiar with the field and its techniques is a luxury that few researchers, policy-makers or practitioners would lay claim to, but, in general, limited attention is given to communication with non-specialists. Many palaeoecological studies, especially older ones or book-form syntheses, focus on large-scale and long-term patterns which are less directly applicable to conservation questions, and tend to be cited only as introductory information, rather than seen as relevant to ecological results or practices. In addition, palaeoecological data are conventionally presented in the form of multivariate stratigraphic diagrams, which are information-dense, require skilled understanding to correctly interpret, and are unfamiliar to many ecologists. Ordination diagrams are just as complex and information-dense as stratigraphic diagrams, but are a more familiar method of displaying multivariate data for an ecological audience (Fig. (1), e.g. Bradshaw et al. 2005). The increasing number of papers that utilise alternative presentation methods (e.g. Lindbladh, et al. 2007, Hanley et al. 2008, van Leeuven et al. 2008) and incorporate ecological baselines (e.g. Gillson & Duffin 2007) to illustrate the conservation implications of long-term datasets may improve communication by making the information more accessible.

In addition to differences in the methods used to present data, developing ways of working around differences in the ways that ecology and palaeoecology represent ecosystems is a significant challenge, which can only be addressed through interdisciplinary cooperation. In palaeoecology a strong current area of activity is quantitative reconstruction of past vegetation from pollen records. Presenting reconstructions as vegetation cover estimates (Hellman et al. 2008) or even as mapped distributions (Bunting & Middleton 2009) improves legibility for the non-specialist, although they can hide uncertainties.

When long-term data are used in conservation research, old maps, documents and historical narratives remain more popular owing to their relative accessibility to non-historians (e.g. Holl & Smith 2007), although this may lead to uncritical and inappropriate use of datasets that are inherently biased towards changeable sets of human values (e.g. Bowman et al. 2010, Szabó 2009). Even within the discipline of ecology, changing methodologies, the challenges of relocating sites for repeat monitoring to build on historical data and changing taxonomies all make historical documents such as old plant lists or surveys difficult to integrate with current data and concerns (Ross et al. 2010).

This paper presents clear examples indicating how palaeoecology can contribute to conservation practice, but argues that this potential is only rarely realized. It is notable that only one of the case studies presented above involves publications where practitioners are co-authors (Stevenson & Thompson 1993, Chambers et al. 1999, 2007a, 2007b). Furthermore, there is only one example where long-term data have an established role in policy and practice: the Water Framework Directive. This example probably has the least 'information transfer' problem, since a single pre-industrial baseline has been defined as representative of 'favorable' condition, and results can be presented in terms of difference of present from baseline. As indicated above, this simple, easily measured approach may not be universally appropriate, but it is feasible, affordable and straight-forward to implement, which is highly valued in the practical context of the WFD. This single baseline approach contrasts markedly with the terrestrial examples considered, where palaeoecological data emphasise the undesirability of defining an inflexible single 'baseline' for dynamic ecosystems. Recent debates about vegetation structure and regeneration mechanisms in naturalistic woodlands has elicited both single discipline and joint responses (e.g. Bradshaw et al. 2003, Hodder et al. 2005, Whitehouse & Smith 2010). This illustrates how the different approaches to ecology can be used together to address common questions, when there is sufficient incentive and common interest.

There is a growing risk that, by relying on the relatively small numbers of current communities considered to be 'natural' or in 'favorable condition', or defined within commonly used classification systems (e.g. National Vegetation Classification communities), conservation protection, management targets and objectives may become too narrow. This could lead to fixed ideas of what a site 'should' look like, excluding natural extremes of variation, and is likely to exacerbate homogenization and reduce habitat resilience to change (e.g. Wilkinson 2001). Trends towards homogenization (e.g. Britton et al. 2009, Keith et al. 2009) are of particular concern when environmental change is predicted to be rapid and pressures on the landscape are high. Diversity seems to be an important factor in enabling resilience, and trends towards a smaller number of communities suggest increasing vulnerability to future losses of biodiversity and of habitats. By revealing unexpected past ecosystem states, such as the diversity of woodland in West Affric and the absence of pure Callunetum in many parts of the uplands, palaeoecological data can support acceptance of a wider range of possible future ecosystem states. Overcoming the temptation to set targets tightly in terms of specific community compositions that are easily measured and focusing on fostering a diversity of sustainable, resilient habitats and ecosystem processes, rather than preferred compositions, across the landscape will be challenging. The longer term palaeoecological record shows clearly that communities are transient; whatever humans do or however climate changes, new communities will form as species respond individualistically to a wide suite of dynamic environmental factors.

Learning from the ecological surprises of the past has the potential to improve anticipation of future shifts and increase the effectiveness of conservation practice. However, just as conservation practice would benefit from drawing on the broadest possible evidence base, there is also an onus on palaeoecologists to understand the conservation concerns, such as homogenization, to identify the ways in which their data can inform this and other current debates, and to find appropriate data formats and outlets to contribute to the debate. Palaeoecologists would thus benefit from increasing their awareness of ecological issues in order to refocus their efforts and develop research and publications that address current needs in ecological research, practice and policy.

There are structural barriers to developing the interdisciplinary, intersectoral networks needed to integrate effectively palaeoecological approaches into conservation thinking, which are exacerbated by time and resource pressures. In addition to requiring opportunities for palaeo- and neo-ecologists to exchange views, both in person and through the literature, overcoming resistance to rethinking established perceptions in ecology and conservation may remain an ongoing challenge (Carrion & Fernandez 2009). It remains to be seen whether the interdisciplinary and knowledge exchange impetus brought about by environmental uncertainty will speed changes in thought and action on both sides of the broad ecological community.

REFERENCES

1. Amphlett, A (2003) Contexts, developing ideas and emerging issues in the conservation management of the RSPB Abernethy Forest Reserve. Botanical Journal of Scotland, 55, 135-148.
2. Angermeier, PL (2000) The natural imperative for biological conservation. Conservation Biology, 14, 373-381.
3. Atherden, M (2004) Ecological change on the North York Moors during the last millennium and their impact on the landscape. In Society, Landscape and Environmental in Upland Britain, ID Whyte & AJL Winchester (Eds), Society for Landscape Studies supplementary series 2, 1-9.
4. Battarbee, RW, Monteith, DT, Juggins, S, Evans, CD, Jenkins, A, Simpson, GL (2005) Reconstructing pre-acidification pH for an acidified Scottish loch: a comparison of palaeolimnological and modelling approaches. Environmental Pollution, 137, 135-149.
5. Bennion, H & Battarbee, RW (2007) The European Union Water Framework Directive: opportunities for palaeolimnology. Journal of Palaeolimnology, 38, 285-295.
6. Bennion, H, Fluin, J, Simpson, GL (2004) Assessing eutrophication and reference conditions for Scottish freshwater lochs using subfossil diatoms. Journal of Applied Ecology, 4, 124-138
7. Bennion, H, Simpson, G, Hughes, M, Phillips, G & Fozzard, I (2003) The role of palaeolimnology in identifying reference conditions and assessing ecological status of lakes. In: Ruoppa, M, Heinonen, P, Pilke, A, Rekolainen, S, Toivo, H, Vuoristo, H (Eds) How to assess and monitor ecological quality in freshwaters. Tema Nord, 547. Nordic Council of Ministers, Copenhagen, 57-63.
8. Benton, TG, Vickery, JA & Wilson, JD (2003) Farmland biodiversity: is habitat heterogeneity the key? Trends in Ecology & Evolution, 18, 182-188.
9. Berglund, BE, Gaillard, M-J, Bjorkman, L & Persson, T (2008) Long-term changes in floristic diversity in southern Sweden: palynological richness, vegetation dynamics and land-use, Vegetation History and Archaeobotany, 17, 573-583.
10. Birks, HJB (1993) Quaternary palaeoecology and vegetation science - current contributions and possible future developments. Review of Palaeobotany and Palynology, 79, 153-177.
11. Birks, HJB (1996) Contributions of Quaternary palaeoecology to nature conservation. Journal of Vegetation Science, 7, 89-98.
12. Birks, HJB & Line, JM (1992) The use of rarefaction analysis for estimating palynological richness from Quaternary pollen-analytical data. The Holocene, 2, 1-10.
13. Bowman, DMJS, Prior, LD & Williamson, G (2010)

The roles of statistical inference and historical sources in understanding landscape change: the case of feral buffalo in the freshwater floodplains of Kakadu National Park. Journal of Biogeography, 37, 195-199.

14. Bradshaw EG, Nielsen AB & Anderson NJ (2006) Using diatoms to assess the impacts of prehistoric, pre-industrial and modern land-use on Danish lakes. Regional Environmental Change, 6, 17-24.

15. Bradshaw EG & Rasmussen P (2004) Using the geological record to assess the changing status of Danish lakes. Geological Survey of Denmark Greenland Bulletin, 4, 37-40.

16. Bradshaw, RHW, Hannon, GE & Lister, AM (2003) A long-term perspective on ungulate-vegetation interactions. Forest Ecology & Management, 181, 267-280.

17. Bradshaw, RHW, Wolf, A & Moller, PF (2005) Long-term succession in a Danish temperate deciduous forest. Journal of Biogeography, 28, 157-164.

18. Bragg, OM & Tallis, JH (2001) The sensitivity of peat-covered upland landscapes. Catena, 42, 345-360.

19. Bridge, MC, Haggart, BA & Lowe, JJ (1990) The history and palaeoclimatic significance of subfossil remains of Pinus sylvestris in blanket peats from Scotland. Journal of Ecology, 78, 77-99.

20. Britton, AJ, Beale, CM, Towers, W & Hewison, RL (2009) Biodiversity gains and losses: Evidence for homogenisation of Scottish alpine vegetation. Biological Conservation, 142, 1728-1739.

21. Britton, AJ, Pearce, ISK & Jones, B (2005) Impacts of grazing on montane heath vegetation in Wales and implications for the restoration of montane areas. Biological Conservation 125, 515-524.

22. Brown, AG (2002) Learning from the past: palaeohydrology and palaeoecology. Freshwater Biology, 47, 817-829.

23. Bunting, MJ & Middleton, R (2009) The Multiple Scenario Approach - a pragmatic method for past vegetation mosaic reconstruction. The Holocene, 19, 799-803.

24. Caledonian Partnership (2003) Restoring Natura Forest Habitats. Highland Birchwoods, Munlochy. Carrión, JS & Fernández, S (2009) The survival of the 'natural potential vegetation' concept (or the power of tradition). Journal of Biogeography, 36, 2202-2203.

25. Chambers, FM, Dresser, PQ & Smith, AG (1979) Radiocarbon dating evidence on the impact of atmospheric pollution on upland peats. Nature, 282, 829-831.

26. Chambers, FM, Mauquoy, D & Todd, PA (1999) Recent rise to dominance of Molinia caerulea in environmentally sensitive areas: new perspectives from palaeoecological data. Journal of Applied Ecology, 36, 719-733.

27. Chambers, FM, Mauquoy, D, Cloutman, EW, Daniell, JRG & Jones, PS (2007a) Recent vegetation history of Drygarn Fawr (Elenydd SSSI), Cambrian Mountains, Wales: implications for conservation management of degraded blanket mires. Biodiversity and Conservation, 16, 2821-2846.

28. Chambers, FM, Mauquoy, D, Gent, A, Pearson, F, Daniell, JRG & Jones, PS (2007b) Palaeoecology of degraded blanket mire in South Wales: data to inform conservation management. Biological Conservation, 137, 197-209.

29. Cumming, BF, Davey, KA, Smol, JP & Birks, HJB (1994) When did acidsensitive Adirondack Lakes (New York, USA) begin to acidify and are they still acidifying? Canadian Journal of Fisheries and Aquatic Sciences, 51, 1550-1568.

30. Darling, FF (1955) West Highland Survey: An Essay in Human Ecology. Oxford University Press, Oxford. Davies, A (2003) Carnach Mor and Camban: woodland history and land-use in alluvial settings. In: Tipping, RM (Ed) The Quaternary of Glen Affric and Kintail. Quaternary Research Association, London, 75- 83.

31. Davies, AL & Dixon, P (2007) Reading the pastoral landscape: palynological and historical evidence for the impacts of long-term grazing on Wether Hill, Cheviot foothills, Northumberland. Landscape History, 29, 35-45.

32. Davies, AL (2010) Palaeoecology, management, and restoration in the Scottish Highlands. In: Hall, M. (Ed) Restoration and history: the search for a useable environmental past, Routledge, Abingdon, 74- 86.

33. Davies, AL, Tisdall, E & Tipping, R (2004) Holocene climatic variability and human settlement in the Scottish Highlands: fragility and robustness. In: Housley, RA & Coles, G (Eds) Atlantic Connections and Adaptations: economies, environments and subsistence in lands bordering the North Atlantic. Oxbow Books, Oxford, 2-11.

34. Delcourt, HR & Delcourt, PA (1991) Quaternary Ecology: a Palaeoecological Perspective. Chapman & Hall, London. Edwards, ME (1986) Disturbance histories of four Snowdonian woodlands and their relation to Atlantic bryophyte distributions. Biological Conservation, 37, 301-320.

35. European Union (2000) Directive 2000/60/EC of the European Parliament and of the Council of 23 October 2000 on establishing a framework for community action in the field of water policy. Journal of the European Community, L327, 1-72.

36. Feurdean, A & Willis, KJ (2008) Long-term variability of Abies alba in NW Romania: implications for its conservation management. Diversity and Distributions, 14, 1004-1017.

37. Foster, DR, Schoonmaker, PK & Pickett, STA (1990) Insights from paleoecology to community ecology. Trends in Ecology and Evolution, 5, 119-122.

38. Froyd, CA & Willis, KJ (2008) Emerging issues in biodiversity & conservation management: The need for a palaeoecological perspective. Quaternary Science Reviews, 27, 1723-1732.

39. Gillson, L & Duffin, KI (2007) Thresholds of potential concern as benchmarks in the management of African savannahs. Philosophical Transactions of the Royal Society of London Series B-Biological Sciences, 362,

309-319.

40. Grant, M & Edwards, M (2008) Conserving idealized landscapes: past history, public perception and future management in the New Forest (UK). Vegetation History & Archaeobotany, 17, 551-562.

41. Grieve, IC, Davidson, DA & Gordon, JE (1995) Nature, extent and severity of soil-erosion in upland Scotland. Land Degradation and Rehabilitation, 6, 41-55.

42. Guilizzoni P, Lami A, Manca M, Musazzi S, Marchetto A (2006) Palaeoenvironmental changes inferred from biological remains in short lake sediment cores from the Central Alps and Dolomites. Hydrobiologia, 562, 167-191.

43. Gustavsson, E, Lennartsson, T & Emanuelsson, M (2007) Land use more than 200 years ago explains current grassland plant diversity in a Swedish agricultural landscape. Biological Conservation, 138, 47- 59.

44. Hall J & Stone D (2005) Overall biodiversity and the spatial patterns of Atlantic oakwoods. Botanical Journal of Scotland, 57, 115-118.

45. Hancock, M, Egan, S, Summers, R, Cowie, N, Amphlett, A, Rao, S, & Hamilton, A (2005) The effect of experimental prescribed fire on the establishment of Scots pine Pinus sylvestris seedlings on heather Calluna vulgaris moorland. Forest Ecology and Management, 212, 199-213.

46. Hanley, N, Davies, A, Angelopoulos, K, Hamilton, A, Ross, A, Tinch, D & Watson, F (2008) Economic determinants of biodiversity change over a 400 year period in the Scottish uplands. Journal of Applied Ecology, 45, 1557-1565.

47. Hanley, N, Tinch, D, Angelopoulos, K, Davies, A, Barbier, EB & Watson, F (2009) What drives long-run biodiversity change? New insights from combining economics, palaeoecology and environmental history. Journal of Environmental Economics and Management, 57, 5-20.

48. Hellman, S, Gaillard, MJ, Broström, A & Sugita, S (2008) The REVEALS model, a new tool to estimate past regional plant abundance from pollen data in large lakes: validation in southern Sweden. Journal of Quaternary Science, 23, 21-42.

49. Henle, K, Alard, D, Clitherow, J, Cobb, P, Firbank, L, Kull, T, McCracken, D, Moritz, R FA, Niemela, J, Rebane, M, Wascher, D, Watt, A & Young, J (2008) Identifying and managing the conflicts between agriculture and biodiversity conservation in Europe - A review. Agriculture Ecosystems & Environment, 124, 60-71.

50. Higgitt, DL, Warburton, J & Evans, ML (2001) Sediment transfer in upland environments. In: Higgitt, DL & Lee, EM (Eds) Geomorphological Processes and Landscape Change: Britain in the last 1000 years, Blackwell, Oxford, 190-214.

51. Hobbs, R (2009) Woodland restoration in Scotland: ecology, history, culture, economics, politics and change. Journal of Environmental Management, 90, 2857-2865.

52. Hodder, KH, Bullock, JM, Buckland, PC & Kirby, KJ (2005) Large herbivores in the wildwood and modern naturalistic grazing systems. English Nature Research Report Number 648, Peterborough.

53. Holden, J, Shotbolt, L, Bonn, A, Burt, TP, Chapman, PJ, Dougill, AJ, Fraser, EDG, Hubacek, K, Irvine, B, Kirkby, MJ, Reed, MS, Prell, C, Stagl, S, Stringer, LC, Turner, A & Worrall, F (2007) Environmental change in moorland landscapes. Earth-Science Reviews, 82, 75-100.

54. Holl, K & Smith, M (2007) Scottish upland forests: History lessons for the future. Forest Ecology and Management, 249, 45-53. Hopkins, JJ & Kirby, KJ (2007) Ecological change in British broadleaved woodland since 1947. Ibis, 149, 29-40.

55. Hughes, PDM, Lomas-Clarke, SH, Schulz, J & Jones, P (2007) The declining quality of late-Holocene ombrotrophic communities and the loss of Sphagnum austinii (Sull. ex Aust.) on raised bogs in Wales. The Holocene, 17, 613-625.

56. Huntley, B (1991) Historical lessons for the future. In The Scientific Management of Temperate Communities for Conservation, Spellerberg, IF, Goldsmith, FB & Morris, MG (Eds), Blackwell Scientific Publications, Oxford, 473-503.

57. Huntley, B, Collingham, YC, Willis, SG & Green, RE (2008) Potential Impacts of Climatic Change on European Breeding Birds. PLoS ONE, 3, e1439.

58. Hynynen, J, Palomäki, A, Meriläinen, JJ, Witick, A, Mäntykoski, K (2004) Pollution history and recovery of a boreal lake exposed to a heavy bleached pulping effluent load. Journal of Paleolimnology, 32, 351-374.

59. Kauppila, T (2006) Sediment-based study of the effects of decreasing mine water pollution on a heavily modified, nutrient enriched lake. Journal of Paleolimnology, 35, 25-37.

60. Keith, SA, Newton, AC, Morecroft, MD, Bealey, CE & Bullock, JM (2009) Taxonomic homogenization of woodland plant communities over 70 years. Proceedings of the Royal Society B-Biological Sciences 276, 3539-3544.

61. Kirby, KJ, Smart, SM, Black, HIJ, Bunce, RGH, Comey, PM & Smithers, RJ (2005) Long term ecological change in British woodland (1971- 2001). English Nature Research Report 653, Peterborough.

62. Kuussaari, M, Bommarco, R, Heikkinen, RK, Helm, A, Krauss, J, Liundborg, R, Ockinger, R, Partel, M, Pino, J, Roda, F, Stefanescu, C, Teder, T, Zobel, M & Steffan-Dewenter, I (2009) Extinction debt: a challenge for biodiversity conservation. Trends in Ecology and Evolution, 24, 564-571.

63. Leira, M, Jordan, P, Taylor, D, Dalton, C, Bennion, H, Rose, N, Irvine, K (2006) Assessing the ecological status of candidate reference lakes in Ireland using palaeolimnology. Journal of Applied Ecology, 43, 816-827.

64. Lindbladh, M (1999) The influence of former land-use on vegetation and biodiversity in the boreo-nemoral zone of Sweden. Ecography, 22, 485-498.

65. Lindbladh, M, Brunet, J, Hannon, G, Niklasson, M,

Eliasson, Eriksson, G & Ekstrand, A (2007) Forest history as a basis for ecosystem restoration - a multidisciplinary case study in a south Swedidh temperate landscape. Restoration Ecology, 15, 284-295.

66. Lindbladh, M, Niklasson, M, Karlsson, M, Bjorkman, L & Churski, M (2008) Close anthropogenic control of Fagus sylvatica establishment and expansion in a Swedish protected landscape: implications for forest history and conservation. Journal of Biogeography, 35, 682-697.

67. Littlewood, NA, Pakeman, RJ & Woodin, SJ (2006) The response of plant and insect assemblages to the loss of Calluna vulgaris from upland vegetation. Biological Conservation, 128, 335-345.

68. McIntosh, B (2006) Native pinewoods in Scotland: perspectives on policy and management. Forestry, 79, 303-307.

69. Meriläinen, JJ, Hynynen, J, Palomäki, A, Mäntykoski, K, Witick, A (2003) Environmental history of an urban lake: a palaeolimnological study of Lake Jyväsjärvi, Finland. Journal of Paleolimnology, 30, 387- 406.

70. Millennium Ecosystem Assessment (2005) Ecosystems and Human WellBeing: Current State and Trends. Findings of the Condition and Trends Working Group. Island Press; Vol. 1.

71. Mitchell, FJG (1988) The vegetational history of the Killarney oakwoods, SW Ireland: evidence from fine spatial resolution pollen analysis. Journal of Ecology, 76, 415-436.

72. Mitchell, FJG (1990) The impact of grazing and human disturbance on the dynamics of woodland in S.W. Ireland. Journal of Vegetation Science, 1, 245-254.

73. Moore, PD (2002) The future of cool temperate bogs, Environmental Conservation, 29, 3-20.

74. Morecroft, MD, Bealey, CE, Beaumont, DA, Benham, S, Brooks, DR, Burt, TP, Critchley, CNR, Dick, J, Littlewood, NA, Monteith, DT, Scott, WA, Smith, RI, Wlamsley, C & Watson, H (2009) The UK Environmental Change Network: emerging trends in the composition of plant and animal communities and the physical environment. Biological Conservation, 142, 2814-2832.

75. Odgaard, BV (1999) Fossil pollen as a record of past biodiversity. Journal of Biogeography, 26, 7-17.

76. Oldfield, F (1970) The ecological history of Blelham Bog, National Nature Reserve. In: Walker, R & West, RG (Eds) Studies of the vegetational history of the British Isles, Cambridge University Press, Cambridge, 141-157.

77. Palmer, SCF, Mitchell, RJ, Truscott, AM & Welch, D (2004) Regeneration failure in Atlantic oakwoods: the roles of ungulate grazing and invertebrates. Forest Ecology and Management, 192, 251-265.

78. Parr, TW, Sier, ARJ, Battarbee, RW, Mackay, A & Burgess,J (2003) Detecting environmental change: science and society - perspectives on long-term research and monitoring in the 21st century. Science of the Total Environment, 310, 1-8.

79. Petit, S, Howard, DC & Stuart, RC (2004) A national perspective on recent changes in the spatial characteristics of woodland in the British landscape. Landscape and Urban Planning, 69, 127-135.

80. Pullin, AS & Knight, TM (2003) Support for decision making in conservation practice: An evidence-based approach. Journal for Nature Conservation, 11, 83-90.

81. Quelch, PR (2005) Structure and utilisation of the early oakwoods. Botanical Journal of Scotland, 57, 99-105.

82. Redpath, N, Osgathorpe, LM, Park, K & Goulson, D (2010) Crofting and bumblebee conservation: The impact of land management practices on bumblebee populations in northwest Scotland. Biological Conservation, 143, 492-500.

83. Reed, MS, Bonn, A, Slee, W, Beharry-Borg, N, Birch, J, Brown, I, Burt, TP, Chapman, D, Chapman, PJ, Clay, GD, Cornell, SJ, Fraser, EDG, Glass, JH, Holden, J, Hodgson, JA, Hubacek, K, Irvine, B, Jin, N, Kirkby, MJ, Kunin, WE, Moore, O, Moseley, D, Prell, C, Price, MF, Quinn, CH, Redpath, S, Reid, C, Stagl, S, Stringer, LC, Termansen, M, Thorp, S, Towers, W & Worrall, F (2009) The future of the uplands. Land Use Policy, 26 (Suppl 1), S204-S216.

84. Ross, LC, Woodin, SJ, Hester, AJ, Thompson, DBA & Birks, HJB (2010) A test of plot relocation methods in a re-visitation study: how important is spatial heterogeneity? Plant Ecology and Diversity, in press. Sansum, PA (2004) Argyll oakwoods: use and ecological change, 1000 to 2000 AD - a palynological-historical investigation. Botanical Journal of Scotland, 57, 83-97.

85. Shaw, H & Tipping, R (2006) Recent pine woodland dynamics in east Glen Affric, northern Scotland, from highly resolved palaeoecological analysis. Forestry, 79, 331-310.

86. Stevenson, AC & Thompson, DBA (1993) Long-term changes in the extent of heather moorland in upland Britain and Ireland: palaeoecological evidence for the importance of grazing. The Holocene, 3, 70-76.

87. Sutherland, WJ, Armstrong-Brown, S, Armsworth, PR, Brereton, T, Brickland, J, Campbell, CD, Chamberlain, DE, Cooke, AI, Dulvy, NK, Dusic, NR, Fitton, M, Freckleton, RP, Godfray, HCJ, Grout, N, Harvey, HJ, Hedley, C, Hopkins, JJ, Kift, NB, Kirby, J, Kunin, WE, Macdonald, DW, Marker, B, Naura, M, Neale, AR, Oliver, T, Osborn, D, Pullin, AS, Shardlow, MEA, Showler, DA, Smith, PL, Smithers, RJ, Solandt, JL, Spencer, J, Spray, CJ, Thomas, CD, Thompson, J, Webb, SE, Yalden, DW & Watkinson, AR (2006) The identification of 100 ecological questions of high policy relevance in the UK. Journal of Applied Ecology, 43, 617-627.

88. Sutherland, WJ, Bailey, MJ, Bainbridge, IP, Brereton, T, Dick, JTA, Drewitt, J, Dulvy, NK, Dusic, NR, Freckleton, RP, Gaston, KJ, Gilder, PM, Green, RE, Heathwaite, AL, Johnson, SM, Macdonald, DW, Mitchell, R, Osborn, D, Owen, RP, Pretty, J, Prior, SV, Prosser, H, Pullin, AS, Rose, P, Stott, A, Tew, T, Thomas, CD, Thompson, DBA, Vickery, JA, Walker, M, Walmsley, C, Warrington, S, Watkinson, AR, Williams, RJ, Woodroffe, R & Woodroof, HJ (2008) Future novel threats and opportunities facing UK biodiversity identified by horizon scanning. Journal of Applied Ecology, 45, 821-833.

89. Sutherland, WJ & Woodroof, HJ (2009) The need for environmental horizon scanning. Trends in Ecology & Evolution, 24, 523-527.

90. Swetnam, TW, Allen, CD & Betancourt, JL (1999) Applied historical ecology: Using the past to manage for the future. Ecological Applications, 9, 1189-1206.

91. Szabó, P (2009) Open woodland in Europe in the Mesolithic and in the Middle Ages: can there be a connection? Forest Ecology and Management, 257, 2327-2330.

92. Taylor, D, Dalton, C, Leira, M, Jordan, P, Chen, G, León-Vintró, L, Irvine, K, Bennion, H & Nolan, T (2006) Recent histories of six productive lakes in the Irish Ecoregion based on multiproxy palaeolimno-logical evidence. Hydrobiologia, 571, 237-259.

93. Thompson, DBA, MacDonald, AJ, Marsden, JH & Galbraith, CA (1995) Upland heather moorland in Great Britain: a review of international importance, vegetation change and some objectives for nature conservation. Biological Conservation, 71, 163-178.

94. Tipping, R (1994) The form and fate of Scotland's woodlands. Proceedings of the Society of Antiquaries of Scotland, 124, 1-54.

95. Tipping, R, Buchanan, J, Davies, A & Tisdall, E (1999) Woodland biodiversity, palaeo-human ecology and some implications for conservation management. Journal of Biogeography, 26, 33-43.

96. Tipping, R, Davies, A & Tisdall, E (2006) Long-term woodland dynamics in West Glen Affric, northern Scotland. Forestry, 79, 351-359.

97. Towers, W, Hall, J, Hester, AJ, Malcolm, A & Stone, D (2004) The potential for native woodland in Scotland: the native woodland model. Natural Heritage Management Series. Scottish Natural Heritage, Battleby.

98. van Leeuwen, JFN, Froyd, CA, van der Knaap,WO, Coffey, EE, Tye, A & Willis, KJ (2008) Fossil pollen as a guide to conservation in the Galápagos. Science, 322, 1206.

99. Whitehouse, NJ (2006) The Holocene British and Irish ancient forest fossil beetle fauna: implications for forest history, biodiversity and faunal colonisation. Quaternary Science Reviews, 25, 1755-1789.

100. Whitehouse, NJ & Smith, D (2010) How fragmented was the British Holocene wildwood? Perspectives on the "Vera" grazing debate from the fossil beetle record. Quaternary Science Reviews, 29, 539- 553.

101. Wilkinson, DM (2001) An open place? A contribution to the discussion on trees on British mires. In: Whild, S, Meade, R & Daniels, J (Eds) Management of water and trees on raised bogs. English Nature Research Reports 407, English Nature, Peterborough, 19-20.

102. Williams, JM (Ed) (2006) Common Standards Monitoring for Designated Sites: First Six Year Report. Peterborough, JNCC. Williams, JW & Jackson, ST (2007) Novel climates, no-analog communities, and ecological surprises. Frontiers in Ecology and the Environment, 5, 475-482.

103. Willis, KJ & Birks, HJB (2006) What is natural? The need for a long-term perspective in biodiversity conservation. Science, 314, 1261-1265.

104. Willis, KJ, Gillson, L, Brncic, TM & Figueroa-Rangel, BL (2005) Providing baselines for biodiversity measurement. Trends in Ecology & Evolution, 20, 107-108.

105. Willis, KJ, Gillson, L & Knapp, S (2007) Biodiversity hotspots through time: an introduction. Philosophical Transactions of the Royal Society B, 362, 169-174.

SECTION II

CHAPTER 7 COMBINING STATE-AND-TRANSITION SIMULATIONS AND SPECIES DISTRIBUTION MODELS TO ANTICIPATE THE EFFECTS OF CLIMATE CHANGE

Brian W. Miller[1,2], Leonardo Frid[3], Tony Chang[4], Nathan Piekielek[5], Andrew J. Hansen[4], Jeffrey T. Morisette[6]

[1] Natural Resource Ecology Laboratory, Colorado State University, Fort Collins, CO, 80523-1499, USA

[2] Department of the Interior North Central Climate Science Center, Fort Collins, CO, 80523-1499, USA

[3] Apex Resource Management Solutions Ltd., Bowen Island, BC V0N 1G1, Canada

[4] Montana State University, Bozeman, MT, 59717, USA

[5] The Pennsylvania State University, University Park, PA, 16802, USA

[6] U.S. Geological Survey, Department of the Interior North Central Climate Science Center, Fort Collins, CO, 80523-1499, USA

ABSTRACT

State-and-transition simulation models (STSMs) are known for their ability to explore the combined effects of multiple disturbances, ecological dynamics, and management actions on vegetation. However, integrating the additional impacts of climate change into STSMs remains a challenge. We address this challenge by combining an STSM with species distribution modeling (SDM). SDMs estimate the probability of occurrence of a given species based on observed presence and absence locations as well as environmental and climatic covariates. Thus, in order to account for changes in habitat suitability due to climate change, we used SDM to generate continuous surfaces of species occurrence probabilities. These data were imported into ST-Sim, an STSM platform, where they dictated the probability of each cell transitioning between alternate potential vegetation types at each time step. The STSM was parameterized to capture additional processes of vegetation growth and disturbance that are relevant to a keystone species in the Greater Yellowstone Ecosystem—whitebark pine (*Pinus albicaulis*). We compared historical model runs against historical observations of whitebark pine and a key disturbance agent (mountain pine beetle, *Dendroctonus ponderosae*), and then projected the simulation into the future. Using this combination of correlative and stochastic simulation models, we were able to reproduce historical observations and identify key data gaps. Results indicated that SDMs and STSMs are complementary tools, and combining them is an effective way to account for the anticipated impacts of climate change, biotic interactions, and disturbances, while also allowing for the exploration of management options.

Citation: Brian W. Miller, Leonardo Frid, Tony Chang, Nathan Piekielek, Andrew J. Hansen, Jeffrey T. Morisette. Combining state-and-transition simulations and species distribution models to anticipate the effects of climate change. AIMS Environmental Science, 2015, 2(2): 400-426. doi: 10.3934/environsci.2015.2.400

INTRODUCTION

There is a challenging dichotomy in the resource management community. On the one hand, there is a need to know and understand fundamental physical processes at a very specific and local level. On the other, there is also a need to know the larger context within which the system works. This is particularly true for integrating climate change into natural resource management practices. Ecological restoration and management actions require an understanding of local conditions (such as species competition, disturbance conditions, micro topography, and a given jurisdiction's management options), but climate impacts are often measured and modeled at a coarser scale and imply the need to consider the wider landscape. Recent work has called for integrating various methods in order to overcome limitations of individual tools and the challenges associated with climate change adaptation [1].

Species distribution modeling (SDM) has proven useful for estimating the potential impacts of climate change and other abiotic factors on species distributions. Many SDMs are popular due to their ability to generate complex predictions without high computational requirements; however, there has been increasing recognition of the limitations of SDMs related to interspecific interactions, dispersal, equilibria, stochastic events, and fundamental niche space [2,3]. In addition to their inherent limitations, there are some specific questions that are not appropriate or possible within the framework of correlative modeling but may be of keen interest to natural resource managers. For example, what would happen to a species if the budget for a given restoration project was doubled or cut in half?

State-and-transition simulation models (STSMs) provide a spatial and quantitative framework to explore "what if" scenarios, which can be used to explore both management options and evaluate the sensitivity of the system to specific parameterizations or assumptions [4]. STSMs have been recognized as useful tools for incorporating the effects of multiple disturbances, biotic interactions, and management scenarios, but lack statistically robust techniques for relating climate data to species distributions. Together, STSMs and SDMs are well-equipped to account for the various facets of natural resource management challenges, at both regional and local scales, and combine them in a spatial simulation framework.

Our case study of whitebark pine in the Greater Yellowstone Ecosystem serves as a proof-of-concept for combining stochastic simulation models (STSMs) and correlative models (SDMs), and sets the stage for future exploration of management options. This research offers two novel contributions:

- accounting for climate change and other dynamics through a combination of SDMs and STSMs; and
- outlining an approach for validating STSMs based on recent developments in agent-based modeling, which remains an active area of exploration.

BACKGROUND

Species Distribution Models

Species distribution models (SDMs) refer to a variety of empirical methods that associate species with the attributes of their preferred bioclimatic environment, also sometimes referred to as "habitat" [5,6]. SDMs evolved over the last several decades from our conceptual understanding of niche theory combined with the increased availability of geospatial datasets and geographic information systems with which to quantify bioclimatic attributes and their variation across space. SDMs typically represent individual species presence (and/or absence) at the spatial resolution of commonly available climate data (usually ~ 1 km or coarser) at temporal resolutions of decades or longer (often 30 year averages). More recently, interest in the ecological impacts of climate change has spurred rapid development and growth in application of SDMs because they are relatively easy to train on contemporary conditions and apply to projected future conditions. Future SDM projections often support commonly held assumptions about climate change and species extinctions and range shifts, although their temporal transferability (i.e., ability to be calibrated for one time period and make predictions for another) is rarely considered [7].

There are numerous challenges to producing actionable science using SDMs including that: they often represent unknown mechanistic relationships between species and their environments [8]; empirical methods do not always capture the direct effects of climate on species in ecologically meaningful ways [9]; there is often a mismatch of spatial and temporal resolution between SDMs and management action [10]; and SDMs rarely consider climate change impacts to communities of species and functional types whereas management is often charged with preserving assemblages of species and ecosystem function [11]. Numerous refinements of SDMs have been proposed [12]; however, the above challenges suggest that there remains a need to develop new methods that better produce scientific information that is useful to natural resource managers who are engaged in climate adaptation planning.

State-and-Transition Models

State-and-transition models originated as conceptual models that represented groups of vegetation communities and the shifts between them [13]. The definition of states often depends on the modeling objectives and data availability, but can generally be thought of as suites of vegetation communities that have distinct functional groups, ecosystem processes, and structure [14]. Transitions include natural events, management interventions, or a combination of both [15]. State-and-transition models are typically represented using box and arrow diagrams, in which boxes or nested boxes represent vegetation phases and states, and arrows represent the transitions between them.

These conceptual models remain at the heart of more recent quantitative computer-based state-and-transition models, or state-and-transition simulation models (STSMs, reviewed by Daniel and Frid [4]). In STSMs, transitions can be deterministic (e.g., growth, aging) or probabilistic (e.g., fire, invasion), and can be aspatial or spatially explicit. Spatially explicit STSMs are analogous to joint cellular automata-Markov models[16], hybrid Markov-cellular automaton models [17], and spatio-temporal Markov chains [18]. Bestelmeyer et al. [19] note that spatial context is important for conceptual state-and-transition models because spatial dynamics such as contagion, feedbacks between patches, spatial patterns in historical legacies, and variation in soils, topography, and climate can all affect the likelihood and location of transitions. These spatial dynamics can similarly be incorporated into spatially explicit STSMs.

STSMs can be used to compare and evaluate different resource management scenarios [20,21,22,23], and can incorporate climate effects[24,25,26,27,28]. Here we present a novel approach to incorporating the effects of climate on vegetation into STSMs that is spatially explicit and based on correlative models of habitat suitability.

Study Area

Whitebark pine (*Pinus albicaulis*) is an iconic member of the subalpine forest community in the western North American mountains and is highly valued for the ecosystem services it provides. It is considered a keystone species because it has strong influence on ecosystem properties. Whitebark pine (WBP) is able to tolerate the harsh conditions typical of mountainous terrain and creates forest in high-elevation locations that would otherwise be shrubs and herb. By colonizing open subalpine patches, it provides safe sites for subalpine fir (*Abies lasiocarpa*) and other species resulting in greater forest cover. In addition to providing habitat for many wildlife species the unusual reproductive strategy of producing large crops of cones with large nutritious seeds every few years provides a rich food source to the threatened Grizzly bear and other wildlife species. In addition to these positive effects on high-elevation ecosystems, the species has strong social values. The unique umbrella shape, large size, and grizzled appearance developed over its centuries-long lifespan evokes strong emotional appeal to people visiting these mountain haunts.

The Greater Yellowstone Ecosystem (GYE), which includes Yellowstone National Park, Grand Teton National Park, and a number of state and federally managed forests, is a mid- to high-latitude region in the Northern Rocky Mountains of western North America. Conifers are dominant in the range, with forest types composed of *Pinus contorta*, *Abies lasiocarpa*, *Pseudotsuga menziesii*, *Pinus albicaulis*, *Juniperus scopulorum*,*Pinus flexis* and *Picea engelmannii*, although the deciduous hardwood *Populus tremuloides*, is also wide spread. Plateaus and lowlands are dominated by species of *Artemisia tridentata* and open grasslands of mixed composition. The GYE study area encompasses 150,700 km^2 with an elevational gradient from 522-4,206 m that represents 14 surrounding mountain ranges [29].

WBP is present in the GYE from below 2,100 m to nearly 3,300 m, an elevation range that spans the montane and subalpine zones [30]. WBP is subdominant to other conifer species in the lower portion of its distribution and is dominant in many locations at upper treeline [31]. The WBP population in the GYE has been particularly hard hit in recent years: in some areas whitebark pine mortality has exceeded 95% of cone bearing trees (DBH > 15 cm) [32] due to factors related to warming climate, mountain pine beetle (*Dendroctonus ponderosae*), and an exotic fungal pathogen (*Cronartium ribicola*) associated with white pine blister rust [33]. Beyond the current forest die-off, resource managers are concerned because the area of suitable habitat for WBP in the GYE is projected to decline dramatically in the coming century due to projected climate change [29,34,35,36]. Consequently, the US Fish and Wildlife Service listed the WBP on the U.S. candidate species list [37]. Yet questions remain as to how best to manage WBP under these various threats.

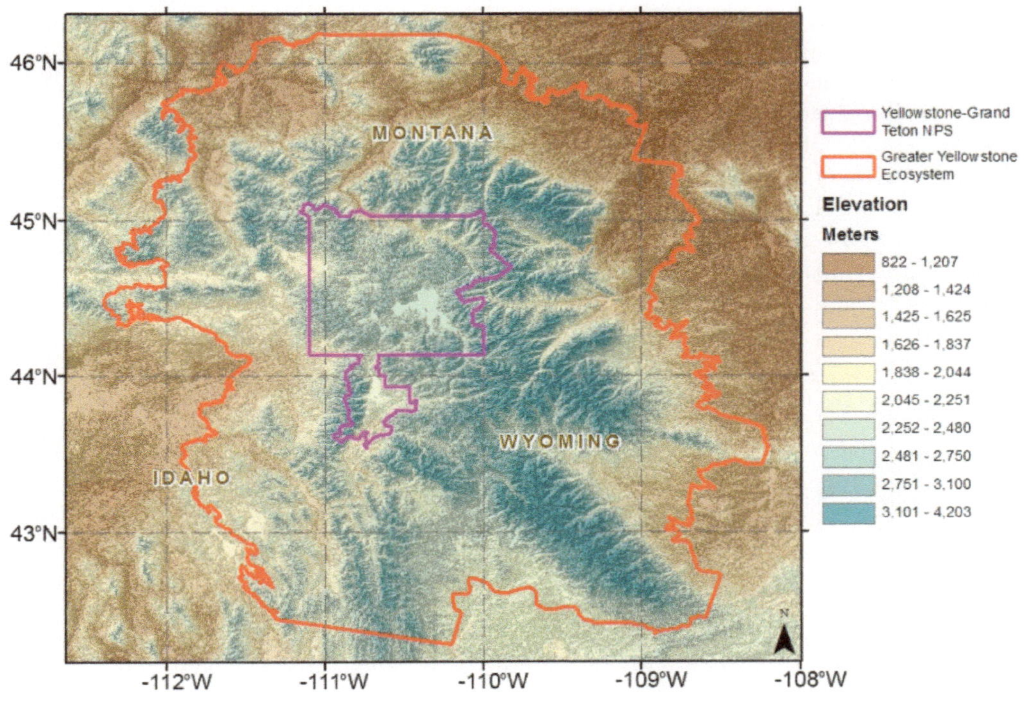

Figure 1. Study area, highlighting the approximate extent of the Greater Yellowstone Ecosystem.

METHODS

Species Distribution Modeling

For the SDM portion of the research, we used the *VisTrails: SAHM* software package to preprocess data, select predictor variables, compare model algorithms, produce model diagnostics, and capture data and workflow provenance. *VisTrails* is a free open source scientific workflow software [38] that has been customized for SDM using a set of add-on tools that comprise the Software for Assisted Habitat Modeling (*SAHM*)[39].

Field observations of adult tree presences and absences were compiled from the Forest Inventory and Analysis (FIA) program, Whitebark/Limber Pine Information System [40], and long term monitoring plots established by the National Park Service Greater Yellowstone Inventory and Monitoring Network [41]. "Adult" class WBP were selected for modeling based on a recorded diameter at breast height (DBH > 20 cm). WBP within central Montana are reported to reach 100 years of age at approximately 8-12 m in height with DBH 15-20 cm. Given previous silvicultural studies, it was assumed that 20 cm DBH for WBP represent adult class individuals for the GYE, with potential to reproduce. First, 2,545 WBP observations from the Forest Inventory and Analysis (FIA) [42] program were assembled. FIA plots are located on a regular gridded sampling design with one plot at approximately every 2, 500 forested hectares, with swapped and fuzzed exact plot locations within 1.6 km to protect privacy. Gibson et al. [43] found that model accuracy was not dramatically affected by data fuzzing, but to provide the most spatial accuracy, this study culled FIA field points where measured elevation were > 300 m different than a 30 m USGS digital elevation model. In order to generate probability of occurrence surfaces for competitor species, separate species distribution models for spruce-fir [1] and lodgepole pine were constructed using 2,489 FIA plots.

Probability surfaces were generated using a random forest algorithm. Random forest is an ensemble learning technique that generates independent random classification trees using a subset of the total predictor variables and classifies a bootstrap random subsample of the data. The predictor variables were 30-year means (1950□1980) of Parameter-elevation Regressions on Independent Slopes Model (PRISM) variables and Thornthwaite-based dynamic water balance model variables [29]. We used a maximum correlation filter of 0.75 (all pairs of variables) to avoid collinearity issues, and physiologically relevant variables were determined by expert analysis and literature review [29]. A total of eight predictor variables were used to determine the probability of WBP occurrence: minimum temperature January, vapor pressure deficit March, precipitation April, snow water equivalent April, maximum temperature July, actual evapotranspiration July, potential evapotranspiration August, precipitation September. Atmospheric CO2 concentrations were not in the list of available predictors, but are noteworthy because they could affect tree physiology and response to warming [44].

Accuracy for the model was evaluated by calculating the receiver operator characteristic curve (AUC). Although AUC alone does not provide an explicit description of commission and omission error rates within a model, it does serve as an index of how likely a model can discriminate a presence versus an absence [45]. As a general rule of thumb, an AUC of 0.5 indicates performance no better than random and 1.0 indicates perfect model prediction. AUC measures above approximately 0.7 are generally considered to be good and above 0.9 excellent [46]. After fitting, our WBP model reported an AUC value of 0.94, displaying high specificity and sensitivity. In addition to examining the AUC for WBP, the out-of-bag (OOB) error estimation was also examined and found a rate of commission (13.1%) and omission (10.9%) for ensemble bagged trees, demonstrating low test error. Evaluation of AUC for the other species displayed good/excellent skill for present day climate conditions.

A post-hoc comparison of habitat niche fit with seedling class WBP (< 2.54 cm DBH, n = 497) was also constructed, displaying a spatial distribution for WBP analogous to adults (Figure 2). Binary classification of presence and absence based on a probability threshold, where sensitivity and specificity were equal, generated near equivalent distribution maps for adults and seedlings despite seedlings presenting lower occurrence probabilities as a result of lower empirical sample prevalence. Comparison of predicted presence distributions across predictor variables again verified similar environmental gradients for both adults and seedlings (Figure 3). This post-hoc analysis rationalizes use of the "adult" fit distribution probabilities under future climate for state and transition simulations as potential suitable habitats for dispersed recruit colonization (details below) due to its more complete empirical representation of the population on the landscape.

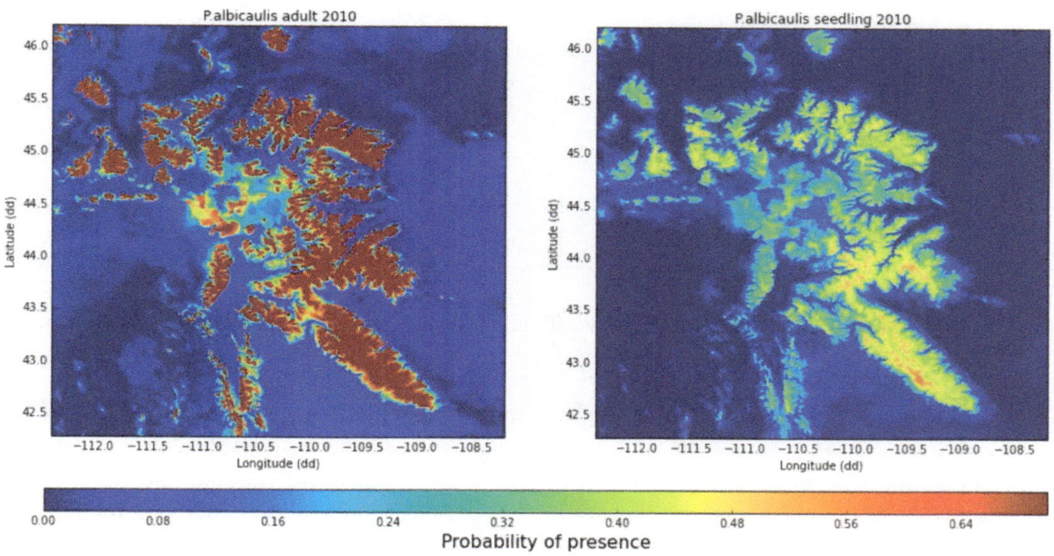

Figure 2. Modeled suitable bioclimatic envelope for adult WBP (938 presence, 1631 absence; accuracy rate of about 93%) (left panel), and for seedlings (497 presence, 2072 absence; accuracy rate of about 84%) (right panel).

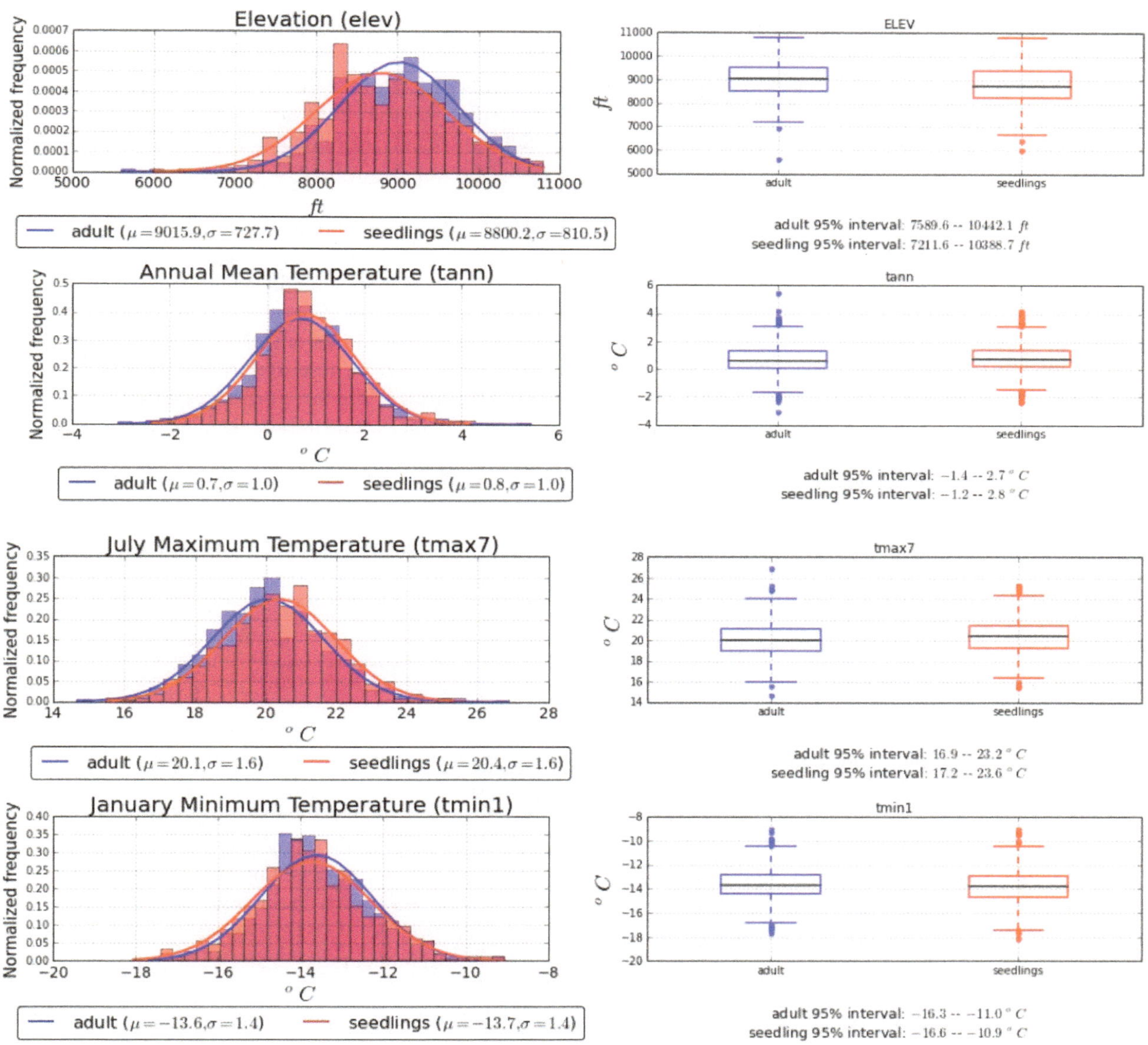

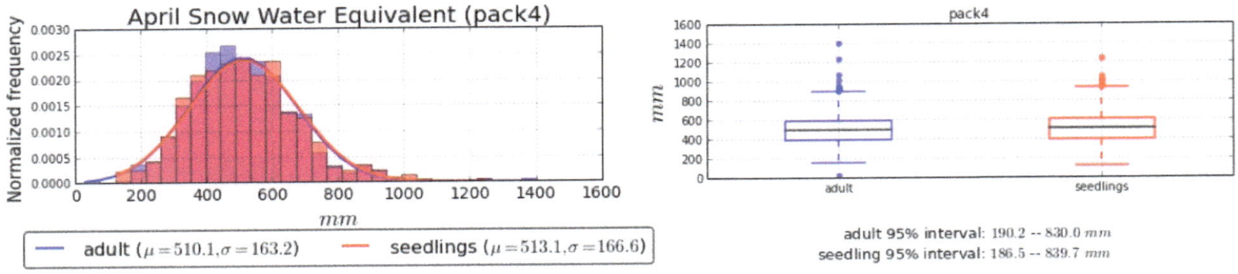

Figure 3. Binary classification of modeled presence for WBP adults (blue) and seedlings (red) display similar environmental gradient distributions.

Following model fitting and applying the model to a historic climate period (1950□1980 and 1980-2010, respectively), we projected our SDM under two global climate models (GCMs) and two carbon concentration scenarios. Using a Bias-Correction Spatial Disaggregation (BCSD) approach, an archive of 9 statistically downscaled CMIP5 climate projections for the conterminous United States at 30-arc-second spatial resolution was assembled by the NASA Center for Climate Simulation NEX-DCP30. For this analysis, two GCMs that represented the greatest and least change in total WBP habitat in the GYE by the year 2099 (CNRM-CM5 [46] and HadGEM2-AO [47], respectively), and two representative concentration pathway (RCP) scenarios were used to project WBP occurrence probabilities. RCP 4.5 was the first, representing increased radiative forcing until stabilization of greenhouse emissions between 2040 and 2050. RCP 8.5 was the second, representing the "business as usual" scenario, with uncontrolled radiative forcing with stabilization by 2099. Using this approach, we sought to demonstrate the "bookends" of range for projected WBP probable habitat under the high uncertainty and variability of global climate models [48]. Occurrence probability surfaces were constructed for the 2040, 2070, and 2099 climate projections.

State-and-Transition Modeling

We implemented the STSM in the *ST-Sim* modeling platform [49]. *ST-Sim*'s graphical user interface streamlines the process of defining distributions and probabilities, managing model inputs and outputs, viewing results, and creating and comparing alternative simulation scenarios.

Within *ST-Sim*, we defined state-and-transition pathways for each potential vegetation type (or stratum) in the model. Strata and state classes were defined based on vegetation dynamics models developed for the area as part of the Landscape Fire and Resource Management Planning Tools Project (LANDFIRE) [50]; specifically, we created three forest types (WBP, lodgepole, and spruce-fir) based on the dominant species in three LANDFIRE biophysical settings (BPS) (northern rocky mountain subalpine woodland and parkland, rocky mountain lodgepole pine forest, and rocky mountain subalpine mesic-wet spruce-fir forest and woodland). We also included an alpine stratum as a potential location for WBP colonization, and a shrub-herb stratum for forest that underwent a mortality event. The pathway diagrams for each forest stratum were then updated based on available information in the literature on species life histories. We initialized the model with multiple sources of spatial data. The extent of the spruce-fir and lodgepole forest types were based on spatial data from LANDFIRE, and WBP was based on a previously published dataset that was generated with Landsat ETM+ imagery [51]. Cells that were classified as WBP by the LANDFIRE biophysical settings data, but not by the Landenburger et al. dataset were reclassified as spruce-fir because there was greater overlap in its BPS description than lodgepole. The location of state classes within each biophysical setting were spatially randomized in proportion to the LANDFIRE mapped state class distributions.

The entire simulated landscape covers just over 15 million hectares centered on the GYE, at a spatial resolution approximately 64 hectares per cell. The simulation was initiated in the year 1920 and run for 30 years to allow the model (specifically, tree populations) to stabilize. The historical period covers 1950-2010, and the model was then projected to 2100. The model included transitions for reproduction, aging, mortality, and competition. Modeled disturbances included replacement fire, mixed fire, disease, insect pathogens, and wind/weather. Parameters for these and other probabilistic transitions were derived from a variety of published sources (see Supplementary Material). Deterministic parameters (age-based transitions) were based on LANDFIRE pathway diagrams and life histories of the three main tree species. Fire size, seed dispersal distances, and mountain pine beetle outbreak size distributions were estimated based on the Monitoring Trends in Burn Severity (MTBS) dataset, U.S. Forest Service species information, and published literature [52].

The relationship between probabilities of mountain pine beetle infection and temperature was estimated using a modified logistic function:

$$p = \frac{L}{(1 + e^{((-\mu + M)/a)})} \qquad (1)$$

where L is the maximum probability, μ is the 2-year moving average annual minimum temperature anomaly, M is the mean annual minimum temperature anomaly for 1999-2009, and a is adjusted so the resulting probability matches observed beetle mortality for the same period. In order to calculate annual temporal multipliers, this probability was then divided by the historical rates of insect and disease infection from the LANDFIRE model (0.003). This relationship was used for both historical model runs (1950-2010) and model projections (2010-2100) in order to vary the probability of mountain pine beetle through time. Other probabilistic transitions were held constant through time.

Unfortunately, data was insufficient to parameterize blister rust in detail. There is a wide range of estimates of blister rust infection rates across locations within the GYE [31], and little information on the size of blister rust outbreaks. Moreover, the mortality of WBP with blister rust is often associated with other disturbances (mountain pine beetle, fire). Modeling blister rust is further complicated by the complex relationship between blister rust and climate (temperature and humidity both influence the spread of blister rust), and the fungus' dependence on multiple hosts. As a result, we did not explore the influence of blister rust on WBP beyond including it in the model and ensuring that the modeled prevalence of infection and mortality matched observations from the GYE.

The establishment of new forest stands after disturbance was dependent on dispersal distances (from the literature), and on habitat suitability as defined by SDM output data. SDM output for each period (1950-1980, 1980-2010, 2010-2040, 2040-2070, 2070-2100) and each forest type (WBP, spruce-fir, lodgepole pine) were used as spatial multipliers to dictate probabilities of forest establishment (Figure 4). If establishment failed, the cell returned to the open state in the shrub-herb stratum.

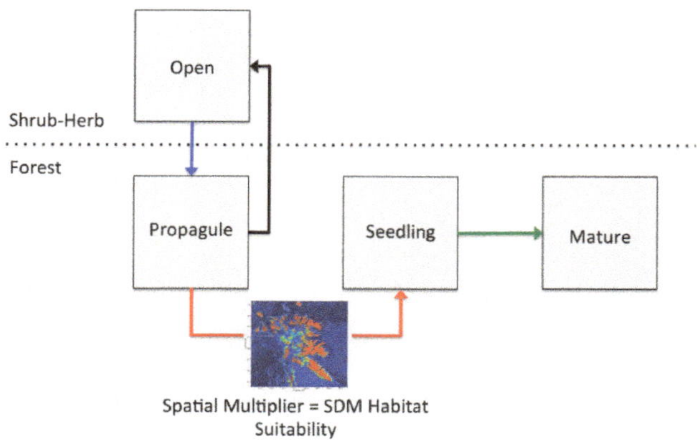

Figure 4. Simplified pathway diagram of the integration of species distribution model output into the state-and-transition simulation model, including dispersal (blue arrow), depletion of the seedbank or mortality of juvenile trees (black arrow), successful establishment (whose probability is characterized by the SDM output, red arrow), and aging (green arrow). Other transitions (e.g., MPB infection, blister rust infection, fire) were omitted from the figure due to space considerations and because each forest stratum had a unique pathway diagram.

We verified realistic model function and then evaluated model outputs to ensure they corresponded to real-world patterns (similar to techniques used in agent-based modeling [53,54]). In particular, verification involved model debugging and parameter calibration, and validation included sensitivity analysis and pattern evaluation. We calibrated the equation that dictated the relationship between mountain pine beetle and minimum annual temperature anomaly (Equation 1) to match observations (a similar approach to Frid et al. [55,56]). In particular, we compared observed levels of beetle and blister rust infection in GYE to results from simulations using four logistic curves with different maxima for the probability of beetle infection.

Evaluation of simulation models remains a significant challenge [57,58,59]. Our evaluation effort was limited by the availability of independent datasets, but we implemented sensitivity analysis and pattern evaluation [60] to check the validity of our model. We sought to reproduce the recent large-scale dynamics of WBP population and disturbance. We compared mountain pine beetle induced WBP mortality using different stratum shift (dead forest stands to open shrub-herb) probabilities to observations from the U.S. Forest Service aerial detection survey. Modeled results were subset by the area flown in each year of the aerial survey.

To test the robustness of our model to parameter uncertainty, we performed sensitivity analysis by running the model for the historical period with different values for parameters that might change substantially with climate or had lower certainty (were unknown or estimated, see Supplementary Material). In particular, we varied the probabilities of alpine colonization by WBP, replacement fire, and spruce-fir replacement of WBP stands by +/−50% [54,61]. We ran 40 Monte Carlo simulations (iterations or model runs) for each perturbation, and compared the means of our primary outcome variables (WBP population, and mountain pine beetle mortality and infection at year 2010) to those from the original model specification. The sensitivity index [61] was calculated as:

$$S_x = \frac{dx/x}{dp/p} \tag{2}$$

where p is the value of the independent variable, dp is the value for a change of p, x is the value of the dependent variable, and dx is the corresponding change in x in response to the change in p.

After model verification and validation we ran the STSM using a no climate change scenario and multiple scenarios that included combinations of two GCMs (CNRM-CM5 and HadGEM2-AO) and two RCPs (4.5 and 8.5) to capture the range of change in projected WBP habitat suitability. We compared temporal patterns and values of key outcome variables at 2100 for these four climate scenarios and a no climate change scenario. The no climate change scenario used SDM output from 1950-1980 and the mean probability of mountain pine beetle infection for the same period.

RESULTS

Verification and Validation

Calibration of the relationship between annual minimum temperature anomaly and mountain pine beetle infection indicated that 0.15 was a reasonable estimate of the maximum probability of beetle infection. This value produced estimates of beetle infection and mortality, and blister rust infection that were comparable to observations from the literature (Figure 5); about half of sub-watersheds in the GYE had high to complete mortality from mountain pine beetle [62], and the proportion of infected WBP trees in 2010 was around 20% [63].

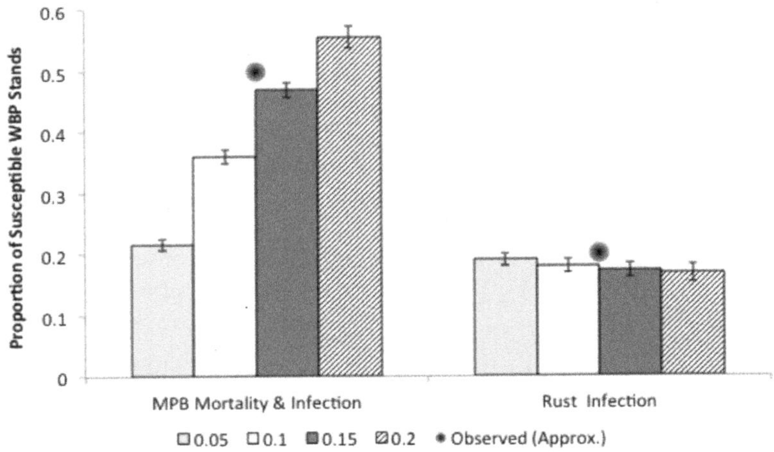

Figure 5. Results of model calibration using four different maxima for the probability of mountain pine beetle (MPB) infection (bars) as compared to observations (points). Columns represent the means of 40 Monte Carlo simulations and error bars the standard deviations.

We also tested the sensitivity of key model outputs (total WBP, live mature WBP, beetle kill WBP, and beetle infected WBP) to changes in parameters with less certainty (alpine colonization, replacement fire, and spruce-fir competition) (Table 1). Perturbations of all of the three parameters produced significant differences in one of the three main model outputs. Reducing (by 50%) the time required for spruce-fir forest to replace WBP produced the largest sensitivity index values.

Table 1. Sensitivity analysis results

Parameter	Default Value	Perturbation Value	Total WBP		Live Mature WBP		Beetle Kill WBP		Beetle Infected WBP	
			Hectares (% Change)	Sensitivity	Hectares (% Change)	Sensitivity	Hectares (% Change)	Sensitivity	Hectares (% Change)	Sensitivity
Alpine colonization (probability)[(a)]	0.0001	0.0002	**561,060** (0.30)	0.30%	**163,524** (1.14)	1.14%	146,815 (−0.05)	N/A	6,838 (−0.68)	N/A
Replacement fire (probability)[(b)]	0.001–0.003	0.0015–0.0045	**586,759** (4.90)	9.79%	161,757 (0.05)	N/A	**142,137** (−3.24)	−6.48%	6,674 (−3.06)	N/A
		0.0005–0.0015	**531,778** (−4.93)	9.86%	**162,647** (0.60)	−1.19%	**152,048** (3.51)	−7.02%	6,923 (0.56)	N/A
Spruce-Fir Competition (years)	350	175	**459,347** (−17.88)	35.76%	**123,146** (−23.83)	47.67%	**103,335** (−29.65)	59.31%	**4,818** (−30.01)	60.03%
		525	**567,269** (1.41)	2.83%	**164,816** (1.94)	3.88%	**150,577** (2.51)	5.01%	**6,537** (−5.05)	−10.10%

Outcome values are means of 40 model runs measured at year 2010. These values are compared to baseline model output using t tests; significant differences ($p < 0.05$ for the two-tailed distribution) are shown in bold, and sensitivity values are reported for statistically significant differences. Percent changes from the baseline model are in parentheses.
[(a)]Alpine colonization could not be decreased further or increased by a smaller increment due to the limitations of the software (i.e., precision was limited to four decimal places)
[(b)]The probability of replacement fire varied by state class; default and perturbation values for replacement fire are listed as ranges due to space considerations

In terms of pattern evaluation, the relationship between the probability of beetle infection and annual minimum temperature anomaly for the historical period were similar to simulated beetle survival published elsewhere [64, Figure 2A in 65]. Additionally, the final model specification (value of 0.1 for the probability of a stratum shift from dead forest to open shrub-herb) closely matched temporal patterns of mountain pine beetle induced WBP mortality from the aerial detections survey data, especially when compared to model specifications using other stratum shift values (0.05 and 0.0333) and a no climate change scenario (Figure 6). Model output for the total area of WBP in National Forest and National Park Service lands in 2010 (536,664 hectares) was also comparable to the observed area of WBP dominant stands (531,999 hectares) [31].

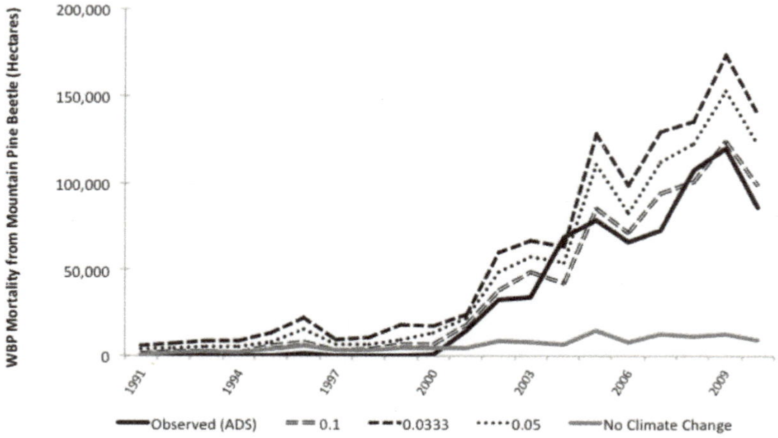

Figure 6. Comparison of mountain pine beetle mortality for four different model specifications (three different probabilities of the shift from dead forest to open shrub-herb, and a no climate change scenario).

Projections

The area of WBP in 2100 was lower for all climate change projections as compared to the no climate change scenario (Figure 7), although this difference does not become pronounced until mid- to late-century (Figure 8). The no climate change scenario resulted in less spruce-fir and more lodgepole as compared to the GCM/RCP scenarios.

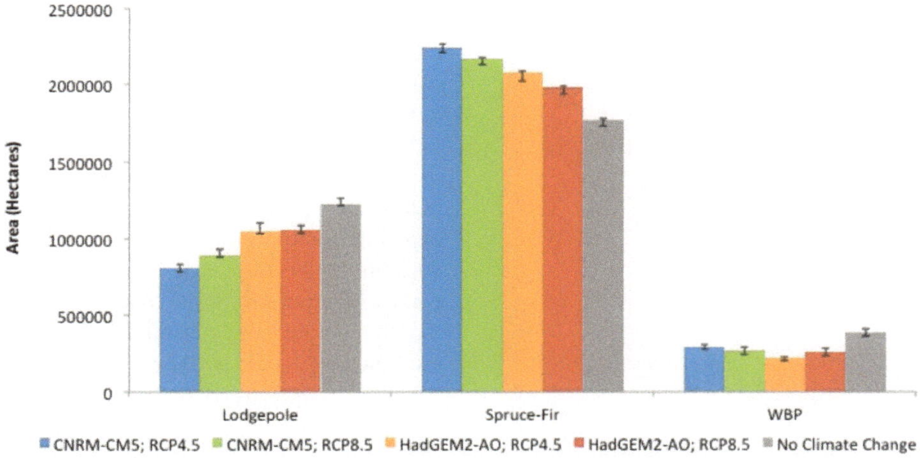

Figure 7. Mean areas of three forest types in 2100 under five different climate scenarios. Error bars represent the 95% percentile range for 40 Monte Carlo simulations.

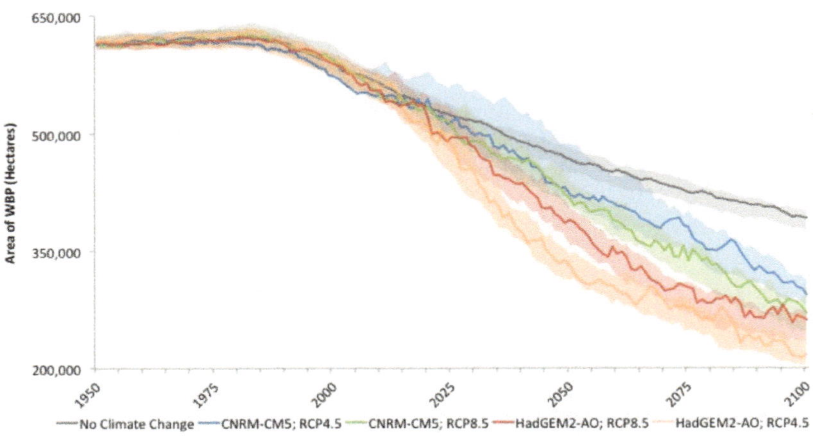

Figure 8. Projected area of WBP forest under five different climate scenarios. Lines represent the means of 40 Monte Carlo simulations and shaded regions represent the 95% percentile range.

Despite the appearance of diminishing mountain pine beetle mortality over time and projections of low mountain pine beetle mortality in 2100 (Figure 9), beetle infected and killed WBP stands accounted for about 40% of susceptible WBP in all climate scenarios at the end of the century. Moreover, by 2100 late seral stage WBP was nearly absent from the landscape in all climate projections except for the no climate change scenario.

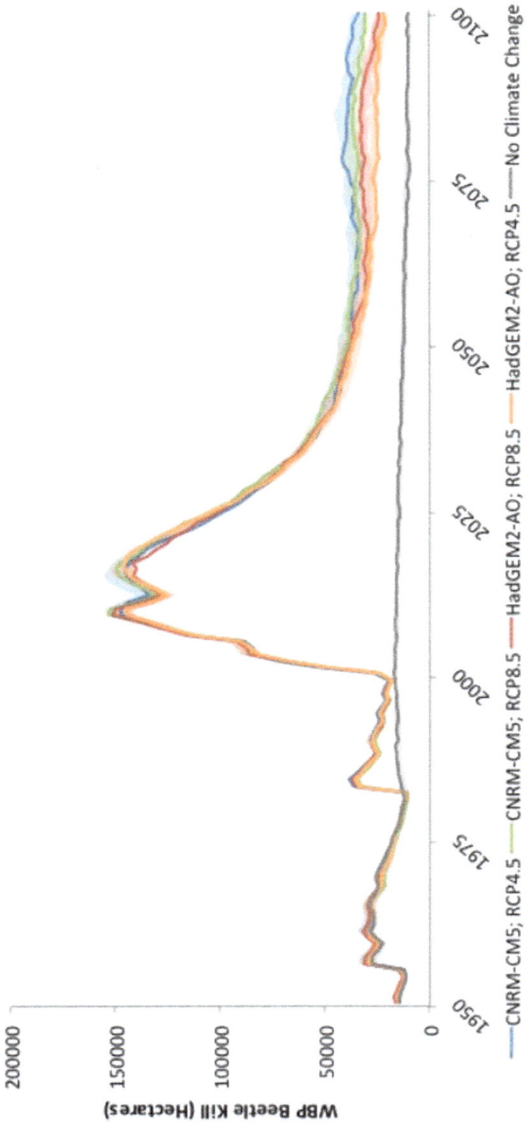

Figure 9. Projected area of WBP mortality from mountain pine beetle under five different climate scenarios. Lines represent the means of 40 Monte Carlo simulations and shaded regions represent the 95% percentile range.

DISCUSSION

Results of the sensitivity analysis suggest that some of the less certain model parameters had a substantial effect on model output. The time for WBP stands to transition to spruce-fir forest had the largest effect. More detailed information on this dynamic would likely improve model reliability. Estimates for seral (as opposed to climax) WBP communities for this parameter range from 100-200 years [66]. Although the 350-year replacement time used in the model produced relatively conservative estimates of WBP persistence, a more nuanced approach could link spruce-fir competition to other factors (e.g., elevation). It is possible that this and other influential parameters, such as the rate of WBP colonization of the alpine zone and variation in the probability of stand replacing fires with climate, could be estimated based on fire, reforestation, and climatic patterns that have been reconstructed from fossil records [30,67].

Another limitation of this modeling effort was the use of output from SDMs of adult trees as input for seedling establishment. In other words, the spatial multiplier files used in our model represent the probability of the presence of adult trees, but dictated the probability of seedling establishment. However, these probability surfaces were based on 30-year moving averages of climate variables, so, depending on the age of the adult trees, these data may indeed represent the climate conditions during establishment. Moreover, our comparison of the presence and absence locations of mature and seedling WBP suggests that the distribution of seedlings is very similar to that of adults. In sum, without additional data points for the presence and absence of seedlings to parameterize more robust seedling SDMs, and given the overlap in WBP adult and seedling distributions it seems reasonable to have used the available SDM output for seedling establishment within the STSM.

Despite these limitations, the model produced results that closely matched historical point estimates and temporal patterns of WBP and mountain pine beetle activity in the GYE, indicating realistic model function. The counterintuitive finding of less spruce-fir under no climate change is likely the result of mountain pine beetle activity; the no climate change scenario had lower beetle infection probabilities, less beetle mortality, and thus fewer opportunities for spruce-fir to colonize cells vacated by beetle-killed lodgepole and WBP. Future work could incorporate SDM output of historic and projected mountain pine beetle habitat suitability in order to provide more nuanced and spatially explicit infection probabilities.

Output for key WBP variables were comparable across GCMs and RCPs. All scenarios showed a substantial decline in WBP forest, and especially late seral stage forest. The differential effects of climate projections may have been attenuated by similar levels of mountain pine beetle activity across climate scenarios. Across all climate projections, the probability of mountain pine beetle infection reaches a maximum by early- to mid-century.

Overall, this modeling effort integrated a wide range of available information on WBP ecology, reproduced observed patterns of WBP mortality in the GYE, and set the stage for future exploration of WBP management scenarios. The sensitivity analyses and findings also point to the importance of accounting for climate, disturbances, biotic interactions, and habitat suitability together. The approach outlined here for combining the SDMs and STSMs provides a strong foundation for exploring alternative approaches to managing resources under the combined pressures of climate change, insects, disease, and interspecific competition.

CONCLUSION

This research demonstrated the benefits of integrating correlative and stochastic simulation models. In particular, we combined SDM's ability to produce statistically robust estimates of the relationship between climate and habitat with STSM's ability to account for disturbances and biotic interactions. The resulting model reproduced observed patterns of WBP in the GYE.

Another important outcome from this research is the identification of data gaps and opportunities for the integration of additional datasets and modeling approaches. Our framework for model validation not only served to corroborate model function, but also identified important research needs, including the ability of WBP to colonize alpine zones, the rate of spruce-fir replacement of WBP across different parts of the landscape, and changes in fire regime. Looking to the historical and paleontological record could help address these shortcomings.

This model and the software platform offer a basis for exploring resource management scenarios such as thinning to protect high value trees from fire, prescribed fires, and the application of pesticides and pheromones to deter mountain pine beetles [31]. Ideally, the specific management scenarios would be carefully developed in conjunction with resource managers. Scenarios could then be implemented in *ST-Sim* by altering transition probabilities to further explore parameter space and model sensitivity, setting management targets for a given outcome (e.g., area of beetle mortality), or setting limits on expenditures for a given action (e.g., pesticide application). Moreover, these scenarios could be applied to specific management units (e.g., National Forests, National Parks), locations (e.g., < 10 km from roadways), and time periods. The costs of implementing management actions or achieving particular targets could also be tracked through time.

STSMs can integrate a variety of data types and sources to create spatially explicit, flexible, and verifiable representations of ecological dynamics. When paired with SDMs, they offer an especially powerful approach for anticipating the effects of climate change, and ultimately, exploring options for managing species in the face of an uncertain future.

ACKNOWLEDGMENTS

We are grateful to Rick Lawrence for sharing data that were used to help initialize the STSM. We also thank Catherine Jarnevich and an anonymous reviewer for their helpful comments on previous drafts of this paper. This research was funded by the U.S. Geological Survey's North Central Climate Science Center. Any use of trade, product, or firm names is for descriptive purposes only and does not imply endorsement by the U.S. Government.

REFERENCES

1. Miller BW, Morisette JT (2014) Integrating research tools to support resource management under climate change. *Ecol Soc* 19: 41.
2. Guisan A, Thuiller W (2005) Predicting species distribution: Offering more than simple habitat models. *Ecol Lett* 8: 993-1009.
3. Sinclair SJ, White MD, Newell GR (2010) How useful are species distribution models for managing biodiversity under future climates. *Ecol Soc* 15: 8.
4. Daniel CJ, Frid L (2012) Predicting landscape vegetation dynamics using state-and-transition simulation models, In: Kerns BK, Shlisky AJ, Daniel CJ, technical editors, Proc First Landscape State-and-Transition Sim Model Conf, June 14–16, 2011, Portland, OR, Gen Tech Rep PNW-GTR-869, Portland, OR: U.S. Department of Agriculture, Forest Service, Pacific Northwest Research Station, 5-22.
5. Franklin J (2009) Mapping species distributions: Spatial inference and prediction. Cambridge, UK: Cambridge University Press.
6. Franklin J (2013) Species distribution models in conservation biogeography: Developments and challenges. *Div Distr* 19: 1217-1223.
7. Dobrowski SZ, Thorne JH, Greenberg JA, et al. (2011) Modeling plant ranges over 75 years of climate change in California, USA: Temporal transferability and species traits. *Ecol Monogr* 81: 241-257.
8. Bell DM, Bradford JB, Lauenroth WK (2014) Early indicators of change: Divergent climate envelopes between tree life stages imply range shifts in the western United States. *Global Ecol Biogeog* 23: 168-180.
9. Stephenson NL (1998) Actual evapotranspiration and deficit: Biologically meaningful correlates of vegetation distribution across spatial scales. *J Biogeogr* 25: 855-870.
10. Benito-Garzón M, Ha-Duong M, Frascaria-Lacoste N, et al. (2013) Habitat restoration and climate change: Dealing with climate variability, incomplete data, and management decisions with tree translocations. *Restor Ecol* 21: 530-536.
11. Schwartz MW (2012) Using niche models with climate projections to inform conservation management decisions. *Biol Cons* 155: 149-156.
12. Iverson LR, Prasad AM, Matthews SN, et al. (2011) Lessons learned while integrating habitat, dispersal, disturbance, and life-history traits into species habitat models under climate change. *Ecosystems* 14: 1005-1020.
13. Westoby M, Walker B, Noy-Meir I (1989) Opportunistic management for rangelands not at equilibrium. *J Range Manage* 42: 266-274.
14. Bestelmeyer BT, Brown JR, Havstad KM, et al. (2003) Development and use of state-and-transition models for rangelands. *J Range Manage* 56: 114-126.
15. Stringham TK, Krueger WC, Shaver PL (2001) States, transitions, and thresholds: Further refinement for rangeland applications. Special Report Number 1024, Oregon State University Agricultural Experiment Station: 1-15.
16. Parker DC, Manson SM, Janssen MA, et al. (2003) Multi-agent systems for the simulation of land-use and land-cover change: A review. *Ann Assoc American Geogr* 93: 314-337.
17. Li H, Reynolds JF (1997) Modeling effects of spatial pattern, drought, and grazing on rates of rangeland degradation: A combined Markov and cellular automaton approach, In: Quattrochi DA, Goodchild MF, editors, Scale in Remote Sensing and GIS, Boca Raton, FL: Lewis Publishers,211-230.

18. Balzter H, Braun PW, Köhler W (1998) Cellular automata models for vegetation dynamics. *Ecol Model* 107: 113-125.

19. Bestelmeyer BT, Goolsby DP, Archer SR (2011) Spatial perspectives in state-and-transition models: A missing link to land management. *J Appl Ecol* 48: 746-757.

20. Forbis TA, Provencher L, Frid L, et al. (2006) Great Basin land management planning using ecological modeling. *Environ Manage* 38: 62-83.

21. Provencher L, Forbis TA, Frid L, et al. (2007) Comparing alternative management strategies of fire, grazing, and weed control using spatial modeling. *Ecol Model* 209: 249-263.

22. Frid L, Wilmshurst JF (2009) Decision analysis to evaluate control strategies for Crested Wheatgrass (Agropyron cristatum) in Grasslands National Park of Canada. *Invasive Plant Sci Manage* 2: 324-336.

23. Costanza JK, Hulcr J, Koch FH, et al. (2012) Simulating the effects of the southern pine beetle on regional dynamics 60 years into the future. *Ecol Model* 244: 93-103.

24. Keane RE, Holsinger LM, Parsons RA, et al. (2008) Climate change effects on historical range and variability of two large landscapes in western Montana, USA. *Forest Ecol Manage* 254: 375-389.

25. Strand EK, Vierling LA, Bunting SC, et al. (2009) Quantifying successional rates in western aspen woodlands: Current conditions, future predictions. *Forest Ecol Manage* 257: 1705-1715.

26. Halofsky JE, Hemstrom MA, Conklin DR, et al. (2013) Assessing potential climate change effects on vegetation using a linked model approach. *Ecol Model* 266: 131-143.

27. Hemstrom MA, Halofsky JE, Conklin DR, et al. (2014) Developing climate-informed state-and-transition models, In: Halofsky JE, Creutzburg MK, Hemstrom MA, editors, Integrating Social, Economic, and Ecological Values Across Large Landscapes, Gen Tech Rep PNW-GTR-896, Portland, OR: U.S. Department of Agriculture, Forest Service, Pacific Northwest Research Station, 175-202.

28. Yospin GI, Bridgham SD, Neilson RP, et al. (2014) A new model to simulate climate change impacts on forest succession for local land management. *Ecol Appl* 25: 226-242.

29. Chang T, Hansen AJ, Piekielek N (2014) Patterns and variability of projected bioclimatic habitat for Pinus albicaulis in the Greater Yellowstone Area. *PLoS ONE* 9: e111669.

30. Whitlock C (1993) Postglacial vegetation and climate of Grand Teton and southern Yellowstone National Parks. *Ecol Monogr* 63: 173-198.

31. Greater Yellowstone Coordinating Committee Whitebark Pine Subcommittee (2011) Whitebark Pine Strategy for the Greater Yellowstone Area. US Dept of Agriculture, Forest Service, Forest Health Protection, and Grand Teton National Park.

32. Gibson K, Skov K, Kegley S, et al. (2008) Mountain pine beetle impacts in high-elevation five-needle pines: current trends and challenges. US Department of Agriculture, Forest Service, Forest Health Protection.

33. Larson ER, Kipfmueller KF (2012) Ecological disaster or the limits of observation. *Reconciling modern declines with the long-term dynamics of whitebark pine communities. Geogr Compass* 6: 189-214.

34. Romme WH, Turner MG (1991) Implications of global climate change for biogeographic patterns in the Greater Yellowstone Ecosystem. *Conserv Biol* 5: 373-386.

35. Bartlein PJ, Whitlock C, Shafer SL (1997) Future climate in the Yellowstone national park region and its potential impact on vegetation. *Conserv Biol* 11: 782-792.

36. Schrag AM, Bunn AG, Graumlich LJ (2008) Influence of bioclimatic variables on tree-line conifer distribution in the Greater Yellowstone Ecosystem: Implications for species of conservation concern. J Biogeogr 35(4): 698-710.

37. U.S. Fish and Wildlife Service (2011) Whitebark pine to be designated a candidate for endangered species protection. Available from: http://www.fws.gov/Wyominges/Pages/Species/Findings/2011_WBP.html

38. Freire J (2012) Making computations and publications reproducible with VisTrails. *Comput Sci Eng* 14: 18-25.

39. Morisette JT, Jarnevich CS, Holcombe TR, et al. (2013) VisTrails SAHM: Visualization and workflow management for species habitat modeling. *Ecography* 36: 129-135.

40. Lockman IB, DeNitto GA, Courter A, et al. (2007) WLIS: The whitebarklimber pine information system and what it can do for you. In: Proceedings of the Conference Whitebark Pine: A Pacific Coast Perspective. Ashland, OR: US Department of Agriculture, Forest Service, Pacific Northwest Region, 146-147.

41. Greater Yellowstone Whitebark Pine Monitoring Working Group (2011) Interagency Whitebark Pine Monitoring Protocol for the Greater Yellowstone Ecosystem, Version 1.1. Greater Yellowstone Coordinating Committee, Bozeman, MT. Available from: http://fedgycc.org/documents/GYE.WBP.MonitoringProtocol.V1.1June2011.pdf

42. Smith WB (2002) FIA: Forest inventory and analysis: a national inventory and monitoring program. *Environ Pollut* 116: S233-S242.

43. Gibson J, Moisen G, Frescino T, et al. (2014) Using publicly available forest inventory data in climate-based models of tree species distribution: Examining effects of true versus altered location coordinates. *Ecosystems* 17: 43-53.

44. Drake BG, Gonzàlez-Meler MA, Long SP (1997) More efficient plants: A consequence of rising atmospheric CO_2. *Annu Rev Plant Biol* 48: 609-639.

45. Lobo JM, Jiménez-Valverde A, Real R (2008) AUC: A misleading measure of the performance of predictive distribution models. *Global Ecol Biogeogr* 17: 145-151.

46. Voldoire A, Sanchez-Gomez E, Salas y Mélia, et al. (2013) The CNRM-CM5. *1 global climate model: Description and basic evaluation. Clim Dynam* 40:

2091-2121.

47. Martin GM, Bellouin N, Collins WJ, et al. (2011) The HadGEM2 family of met office unified model climate configurations. *Geosci Model Dev* 4: 723-757.

48. Deser C, Phillips AS, Alexander MA, et al. (2014) Projecting North American climate over the next 50 years: Uncertainty due to internal variability. *J Climate* 27: 2271-2296.

49. Apex Resource Management Solutions (2014) ST-Sim state-and-transition simulation model software. Available from: www.apexrms.com/stsm

50. Rollins MG (2009) LANDFIRE: A nationally consistent vegetation, wildland fire, and fuel assessment. *Int J Wildland Fire* 18: 235-249.

51. Landenburger L, Lawrence RL, Podruzny S, et al. (2008) Mapping regional distribution of a single tree species: Whitebark Pine in the Greater Yellowstone Ecosystem. *Sensors* 8: 4983-4994.

52. Aukema BH, Carroll AL, Zhu J, et al. (2006) Landscape level analysis of mountain pine beetle in British Columbia, Canada: Spatiotemporal development and spatial synchrony within the present outbreak. *Ecography* 29: 427-441.

53. An L, Linderman M, Qi J, et al. (2005) Exploring complexity in a human-environment system: An agent-based spatial model for multidisciplinary and multiscale integration. Ann Assoc American Geogr 95, 54-79.

54. Miller BW, Breckheimer I, McCleary AL, et al. (2010) Using stylized agent-based models for population-environment research: a case study from the Galápagos Islands. *Popul Environ* 31: 401-426.

55. Frid L, Hanna D, Korb N, et al. (2013) Evaluating alternative weed management strategies for three Montana landscapes. *Invasive Plant Sci Manage* 6: 48-59.

56. Frid L, Holcombe T, Morisette JT, et al. (2013) Using state-and-transition modeling to account for imperfect detection in invasive species management. *Invasive Plant Sci Manage* 6: 36-47.

57. Windrum P, Fagiolo G, Moneta A (2007) Empirical validation of agent-based models: Alternatives and prospects. *J Artif Soc Soc Simulat* 10: 8.

58. Oreskes N, Shrader-Frechette K, Belitz K (1994) Verification, validation, and confirmation of numerical models in the earth sciences. *Science* 263: 641-646.

59. Oreskes N (1998) Evaluation (not validation) of quantitative models. *Environmental Health Perspectives* 106: 1453-1460.

60. Grimm V, Revilla E, Berger U, et al. (2005) Pattern-oriented modeling of agent-based complex systems: Lessons from ecology. *Science* 310: 987-991.

61. Jørgensen SE (1986) Fundamentals of Ecological Modelling. Amsterdam: Elsevier.

62. Macfarlane WW, Logan JA, Kern WR (2009) Using the landscape assessment system (LAS) to assess mountain pine beetle-caused mortality of whitebark pine, Greater Yellowstone Ecosystem, 2009. Project report prepared for the Greater Yellowstone Coordinating Committee, Whitebark Pine Subcommittee, Jackson, WY.

63. Shanahan E, Irvine KM, Roberts D, et al. (2014) Status of whitebark pine in the Greater Yellowstone Ecosystem: A step-trend analysis comparing 2004-2007 to 2008-2011. Natural Resource Technical Report NPS/GRYN/NRTR—2014/917. Fort Collins, CO: National Park Service.

64. Régnière J, Bentz BJ (2007) Modeling cold tolerance in the mountain pine beetle, Dendroctonus pondersoae. *J Insect Physiol* 53: 559-572.

65. Logan JA, Macfarlane WW, Wilcox L (2010) Whitebark pine vulnerability to climate-driven mountain pine beetle disturbance in the Greater Yellowstone Ecosystem. *Ecol Appl* 20: 895-902.

66. Keane RE (2001) Successional dynamics: Modeling an anthropogenic threat, In: Tomback DF, Arno SF, Keane RE, editors, Whitebark Pine Communities: Ecology and Restoration. Washington, DC: Island Press, 159-192.

67. Whitlock C, Millspaugh, SH (2001) A paleoecologic perspective on past plant invasions in Yellowstone. *West N Am Naturalist* 61: 316-327.

68. LANDFIRE (Landscape Fire and Resource Management Planning Tools Project) dataset. Available online at www.landfire.gov.

69. Teste FP, Lieffers VJ, Landhausser SM (2011) Viability of forest floor and canopy seed banks in Pinus contorta var. *latifoia (Pinaceas) forests after a mountain pine beetle outbreak. American Journal of Botany* 98: 630-637.

70. MTBS (Monitoring Trends in Burn Severity) dataset. Available from: at www.mtbs.gov.

71. USFS (United States Forest Service), Species life history information. Available from: www.fs.fed.us.

72. Johnson EA, Fryer GI (1996) Why Engelmann spruce does not have a persistent seed bank. *Canadian Journal of Forest Research,* 26: 872-878.

73. Hatala JA, Crabtree, RL (2009) Modelling the spread of blister rust in the Greater Yellowstone Ecosystem. *Nutcracker Notes, Whitebark Pine Ecosystem Foundation* 16: 17-18.

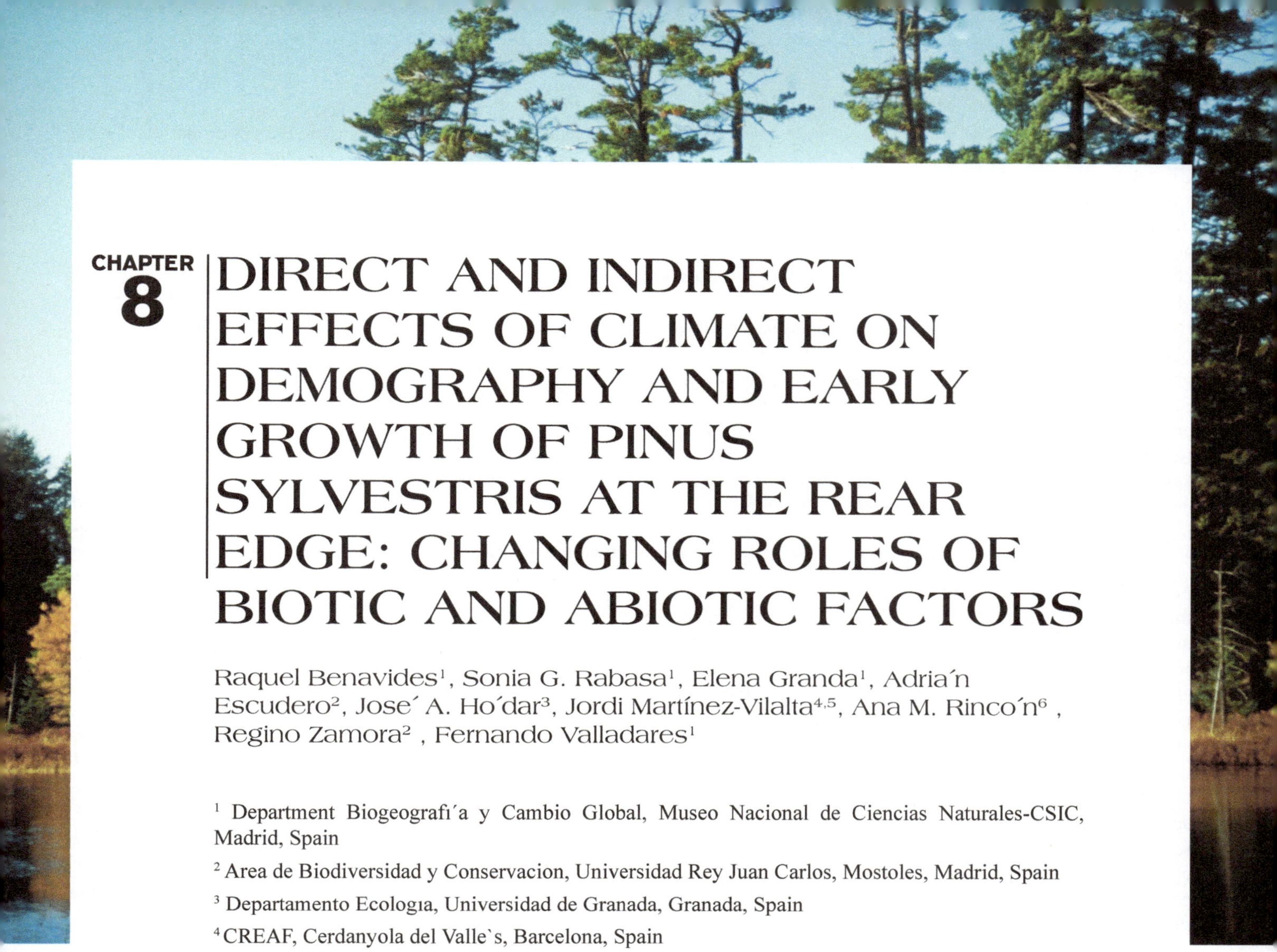

CHAPTER 8

DIRECT AND INDIRECT EFFECTS OF CLIMATE ON DEMOGRAPHY AND EARLY GROWTH OF PINUS SYLVESTRIS AT THE REAR EDGE: CHANGING ROLES OF BIOTIC AND ABIOTIC FACTORS

Raquel Benavides[1], Sonia G. Rabasa[1], Elena Granda[1], Adria´n Escudero[2], Jose´ A. Ho´dar[3], Jordi Martínez-Vilalta[4,5], Ana M. Rinco´n[6], Regino Zamora[2], Fernando Valladares[1]

[1] Department Biogeografı´a y Cambio Global, Museo Nacional de Ciencias Naturales-CSIC, Madrid, Spain

[2] Area de Biodiversidad y Conservacion, Universidad Rey Juan Carlos, Mostoles, Madrid, Spain

[3] Departamento Ecologıa, Universidad de Granada, Granada, Spain

[4] CREAF, Cerdanyola del Valle`s, Barcelona, Spain

[5] Universitat Autonoma Barcelona, Cerdanyola del Valle`s, Barcelona, Spain

[6] Departamento Proteccio´n Vegetal, Instituto de Ciencias Agrarias-CSIC, Madrid, Spain

ABSTRACT

Global change triggers shifts in forest composition, with warming and aridification being particularly threatening for the populations located at the rear edge of the species distributions. This is the case of Scots pine (*Pinus sylvestris*) in the Mediterranean Basin where uncertainties in relation to its dynamics under these changing scenarios are still high. We analyzed the relative effect of climate on the recruitment patterns of Scots pine and its interactions with local biotic and abiotic variables at different spatial scales. Number of seedlings and saplings was surveyed, and their annual shoot growth measured in 96 plots located across altitudinal gradients in three different regions in the Iberian Peninsula. We found a significant influence of climate on demography and performance of recruits, with a non-linear effect of temperature on the presence of juveniles, and a positive effect of precipitation on their survival. Abundance of juveniles of *P. sylvestris* that underwent their first summer drought was skewed towards higher altitudes than the altitudinal mean range of the conspecific adults and the optimum elevation for seedlings' emergence. At local level, light availability did not influence juveniles' density, but it enhanced their growth. Biotic interactions were found between juveniles and the herb cover (competition) and between the number of newly emerged seedlings and shrubs (facilitation). Results also highlighted the indirect effect that climate exerts over the local factors, modulating the interactions with the pre-existing vegetation that were more evident at more stressful sites. This multiscale approach improves our understanding of the dynamics of these marginal populations and some management criteria can be inferred to boost their conservation under the current global warming.

Citation: Benavides R, Rabasa SG, Granda E, Escudero A, Hódar JA, Martínez-Vilalta J, et al. (2013) Direct and Indirect Effects of Climate on Demography and Early Growth of Pinus sylvestris at the Rear Edge: Changing Roles of Biotic and Abiotic Factors. PLoS ONE 8(3): e59824. doi:10.1371/journal.pone.0059824

INTRODUCTION

Climate is the primary filter that determines where a specific species can thrive [1], [2]. Unlike other factors involved in plant species distributions, it is global, permanently active and spatially continuous. Hence, the geographical variation of climate on predictable gradients such as altitude and latitude allows the analysis of the responses of species along ample ranges of environmental variation. Specifically, the study of plant populations' dynamics along climatic gradients has a great interest to predict likely responses to climate warming. Some studies have already raised the alarm about the upwards and polewards shifts of plant species ranges during the twentieth century as a consequence of the global warming [3]–[7]. By contrast, other authors have shown demographic compensations or species benefiting from increases in temperature or aridity [8]–[11].

The research of causal links between climate and plant population responses is particularly interesting for analyzing the dynamics of marginal populations [12]. Species populations at the rear edge of their distribution area are considered an important source of natural history information under past environments [13], essential to estimate their evolutionary potential [14] and to establish proper management guidelines under current uncertainties. Preservation of these populations necessarily depends on the success of the current recruitment under the ongoing changing conditions. Juveniles of woody species are more susceptible to climate, and more specifically to extreme events [15], and respond quicker than adults to environmental changes [16]. Therefore, shifts in the recruitment pattern may reflect subtle differences in climate and their potential future responses. For instance, several authors have already reported increments in elevation for recruitment rates of some tree and shrub species compared to the mean range of the conspecific adult individuals [9], [17]–[19], combined with a replacement of species at lower areas [16], [18], [20]. This seems a response to a direct effect of the changing climate on recruitment over seedling and sapling performance (with higher stress caused by heat and/or drought), and indirect over the performance and mortality of the adult trees, the potential parents [17], [20]–[22]. Some of these effects have been evaluated on rear edge populations, but unfortunately on isolated populations. Furthermore, most of these works did not take into consideration the extreme complexity and heterogeneity of what are called marginal populations [18], [20]–[21].

In the Iberian Peninsula, Scots pine (*Pinus sylvestris* L.) is a clear example of a boreal species at the rear edge of a large distribution area, and, obviously, this pine has to face ecological conditions different from those at the centre of its distribution range. Studies from central and northern Europe pointed out low temperature as the main abiotic limiting factor for the recruitment [23]. Meanwhile, in Mediterranean areas the summer drought is the factor that mainly limits seedling survival and growth [24]. However, Iberian Scots pine populations face very different climatic conditions within this region. In addition to the expected latitudinal trend across the different mountain ranges where the species occurs in this rear edge (i.e. from Pyrenees to Sierra Nevada at the southernmost end of its distribution), there are finer scale gradients nested within each mountain range, which reproduce rear, optimum and leading edge conditions making the interpretation of what is currently occurring in the so-called rear edge extremely complex. At these finer scales, non-climate factors (both biotic and abiotic) can be at least as much relevant for species distribution as climate [25] by modulating the direct effect of regional climate on individuals [2].

Taking into account the predictions in Mediterranean areas in terms of increasing aridity and temperature [26], the population dynamics and mainly the recruitment of this pine may be even more strongly conditioned by the spatial heterogeneity at the local scale [12]. Several works have focused on the study of the spatial heterogeneity at microhabitat scale in order to disentangle the main factors influencing emergence and/or survival in Mediterranean areas for species at their rear edge [20], [27]–[31]. Most of these studies highlighted the decisive role that plant-plant interactions had over the recruitment patterns, specially the nurse role of shrubs ameliorating the harsh climatic conditions imposed by summer drought [24], [32]. Indeed, they all supported the non-random effect that local environmental conditions (biotic and abiotic factors) exert over the juveniles' survival and growth, leading to aggregated patterns in *safe sites* [31]. Particularly, studies addressing the recruitment of *P. sylvestris* in its southernmost distribution area have only considered these local factors neglecting the variability across the landscape [20], [24], [29], [33] (but see [22]). Thus, a critical question remaining is to explore whether juveniles of *P. sylvestris* respond to local scale conditions similarly across climatic and nested altitudinal gradients in different rear edge regions.

In the present study, we aimed to find out whether there are generalizable patterns at different spatial scales in the recruitment of Scots pine populations in its rear edge. Specifically, we explored the relative effect that climate and local factors may have over recruitment across several mountain ranges, surveying the complete altitudinal range of this pine species within them. The local variables comprised both biotic (cover of herbs, shrubs and adult individuals) and abiotic variables (slope, rockiness, light availability), describing not only conditions of the habitat (i.e. at the stand structural level), but also microhabitat conditions (i.e. at microsite scale). Moreover, we tested the effects of climate on local interactions, including plant-plant interactions (among juveniles and the pre-existing vegetation) to figure out whether climate exerts a multi-path impact over recruitment, not only directly over demography, but also indirectly interacting with local factors (at community level).

MATERIALS AND METHODS

Study Sites

The field work was conducted in three forests across a latitudinal gradient in Spain (Fig. 1). The northernmost stand was located in Arcalís, in Central Pyrenees (42° 22′ N, 1° 06′ E), with an altitudinal range from 670 to 1600 m. The central location was placed in Valsaín, on the north-facing slopes of Sierra de Guadarrama (40° 49′ N, 4° 01′ W), between 1130 to 1900 m. Both forests are pure Scots pine stands, with scattered individuals of other species mostly at lower altitudes, mainly *Quercus ilex* and *Q. pyrenaica* at Valsaín, and *Q. ilex*, *Q. humilis* and some planted *Pinus nigra* at Arcalís. These forests have had a long history of harvest and management. In Valsaín, the forest is managed under a group shelterwood system since long ago, always relying on natural regeneration, but no harvest has been accomplished in the sampled areas for the last 15 years. In the forest in Arcalís, logging activities took place in the past but have not been practiced since the last 30 years. In fact, current pine population exhibits natural regeneration and an uneven age and size structure.

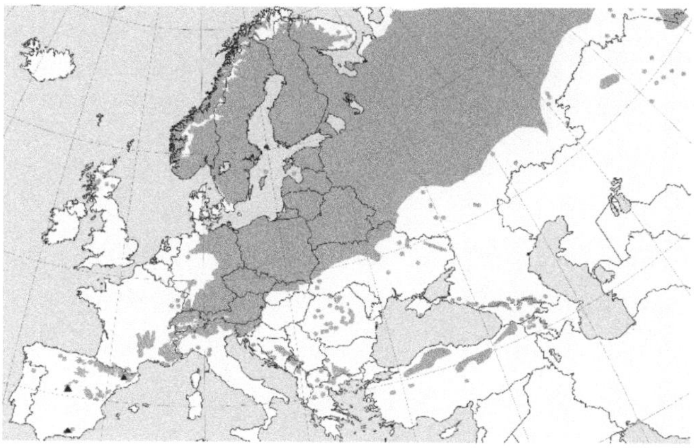

Figure 1. Location of the sampling sites (black triangles). Dark grey denotes distribution range of Scots pine (base map: www.euforgen.org).

The third location is Barranco del Espinar at Sierra Nevada National Park, the southernmost distribution area for Scots pine throughout its overall range (37° 06′ N, 3° 27′ W). The native *P. sylvestris* individuals are present from 1400 to 2100 m and coexist with other tree species, such as *Taxus baccata*, *Acer opalus* ssp. *granatense* or *Quercus ilex* at low tree density (around 114 trees ha^{-1}). The area was traditionally grazed by livestock (goats and cattle), and currently the area is only grazed by increasing populations of wild ungulates, namely Spanish ibex (*Capra pyrenaica*). No specific permits were required for the described field studies, as the locations were not privately owned and there was no involvement of endangered or protected species. The authorities responsible of these areas were informed and they expressed their consent to this sampling.

Sampling and Data

We conducted the field work in late spring-early summer 2010 and 2011. We established six (in Valsaín) and five (in the other two sites) transects at different elevational levels, covering the whole altitudinal range of *P. sylvestris* in each site. The sampling design consisted of six plots at each altitudinal level, separated at least 100 m. Three plots were haphazardly located under fully canopied stands, and the other three in more open areas to include contrasting conditions within each forest and altitude, avoiding ravines, boulders or other geomorphologic elements which could bias our measurements. In the center of every plot we established a 4×4 m frame and counted the number of seedlings recently emerged (from current year) and individuals older than a year and till 1.5 m high (hereafter saplings) in every 1 m^2 subplot (16 within a plot). We estimated their age by counting the terminal bud scars (internodes) along the main stem, included the buried part, and we sorted them into four different age classes: 0) for seedlings; 1) for individuals between 1 and 2 years; 2) for 3 to 5 year-old individuals; and 3) for those over 5 years. The age of the individuals included in the last category was not fully dated due to the difficulty to distinguish scars in the field as individuals grow and become woodier, but it was verified that they exceeded 5 years. We also measured the shoot growth of saplings from previous years (2007, 2008 and 2009). Whenever juvenile density was under five individuals per plot, we enlarged the census area to a final plot of 6×6 m. In summary, we assessed the recruitment analysing three processes: emergence (inferred by seedlings abundance), seedling survival (inferred by the transition from seedlings to saplings) and primary growth of the saplings [34]. In early summer, the abundance of seedlings would mean *sensu stricto* the early survival of seedlings [34], but in the present study it has been considered a proxy of emergence as juveniles had not still coped with their main bottleneck: the summer drought.

The recorded variables represented biotic and abiotic environmental conditions assessed at different scales. At broad scale, we translated the variation in elevation and latitude into climatic variables using the Climatic Digital Atlas of the Iberian Peninsula [35].

At plot level (i.e. at habitat scale), we measured the slope and the light availability (GSF, global site factor), obtained by a hemispherical photograph taken at the center of each plot [36]. In terms of biotic conditions, all adult trees (individuals of any species with diameter at breast height -dbh- over 7.5 cm) within a 10 m radius around the regeneration plot were mapped. Afterwards we calculated the basal area for every plot as an index of competition with adult trees. This radius size was consistent with that used in other tree competition studies [37]–[38].

At subplot level (microhabitat scale), we visually estimated the percentage of rocks, shrubs and herbs in every 1 m^2. Finally, we assessed a potential fecundity index for each subplot considering the conspecific adult trees previously mapped, together with those bigger than 20 cm of dbh located between 10 m and 20 m from the plot centre. This index had two components (see [38], [39]): one related to the number of seeds that an adult tree can produce (seed rain), which is proportional to the tree size (dbh); and other that accounted for the seed dispersal capacity, which was assumed to decline exponentially as the distance to a specific tree increases.

Statistical Analyses and Zero-Inflated Models (ZIP)

To evaluate the effects of climate on demography and performance of juveniles of *P. sylvestris* we carried out mixed models. The density of individuals, both seedlings and pooled saplings (age class was neglected due to the low figures in some categories), and the mean annual shoot growth were our response variables; and variables describing the environment (biotic and abiotic) at different scales were the explanatory ones (see list in Table S1). The interactions between temperature and the biotic factors (cover of co-occurring vegetation) and the quadratic term of temperature was included to consider non-monotonic responses along the gradient.

The density of seedlings and saplings per subplot (per m^2) followed a Zero-Inflated Poisson distribution (ZIP), which accounts for an excess of zeros avoiding the underestimation of the number of zeros and the overestimation of large count occurrence [40], [41]. This ZIP consists of a binomial distribution model representing the occurrence (observing or not recruits), and a Poisson distribution model, conditional on the first, representing the abundance (number of recruits), both processes evaluated at the same time. A complete review of ZIP models for count data can be found in literature (e.g. [42], [43]). To fit models with ZIP distributions we used the SAS procedure NLMIXED (SAS Institute Inc. 2008. SAS/STAT 9.2) that allows the optimization of our customized likelihood function.

We added random components into the intercepts of the linear predictors of the models [41]due to the hierarchical structure of the data, with likely different level of correlation among observations from the same site, transect or plot (nested design). The statistical procedure used in this study recommends the use of just one random effect due to the complexity that the likelihood function reaches and problems detected with the convergence of the model. Thus, we assumed site as a fixed effect, considering that this variable only counted with three levels. Regarding transect and plot, we preliminary fit the model including both effects separately, and we chose the model which harboured more variability (i.e. plot) using the Akaikés Information Criterion (AIC). Concerning fixed effects, we selected the best model following the principle of parsimony to find the simplest model that was not significantly worse than any more complicated one in terms of AIC.

Previously, in order to avoid multi-collinearity problems we checked the correlation among variables (Table S1). This analysis showed high correlation among potential fecundity index, basal area and GSF. Since Gómez-Aparicio [44] showed that in Mediterranean areas recruitment was more correlated to the best patches for emergence and survival than to seed rain -site instead of seed limitation-, and since competition among adults and juveniles is expected to be mainly related to light availability, we kept GSF as a proxy of the overstorey structure and discarded the other two predictors for ZIP models.

For analyzing mean primary growth, we considered the same random and fixed variables (coupled with age class as a fixed factor) than those used for the juveniles' density; and again the selection criteria followed the principle of parsimony using the AIC as a measure of model fit. This analysis was also implemented using the SAS procedure NLMIXED, considering a normal distribution of the data after a log-transformation.

RESULTS

Site Characteristics and Regeneration Distribution

In terms of climate, the three sites were quite representative of the distribution area of Scots pine in Spain (Figure S1), but for precipitation in SN (p<0.001). The altitudinal range of the three sites was quite different, but there was no significant difference in terms of temperature among them.

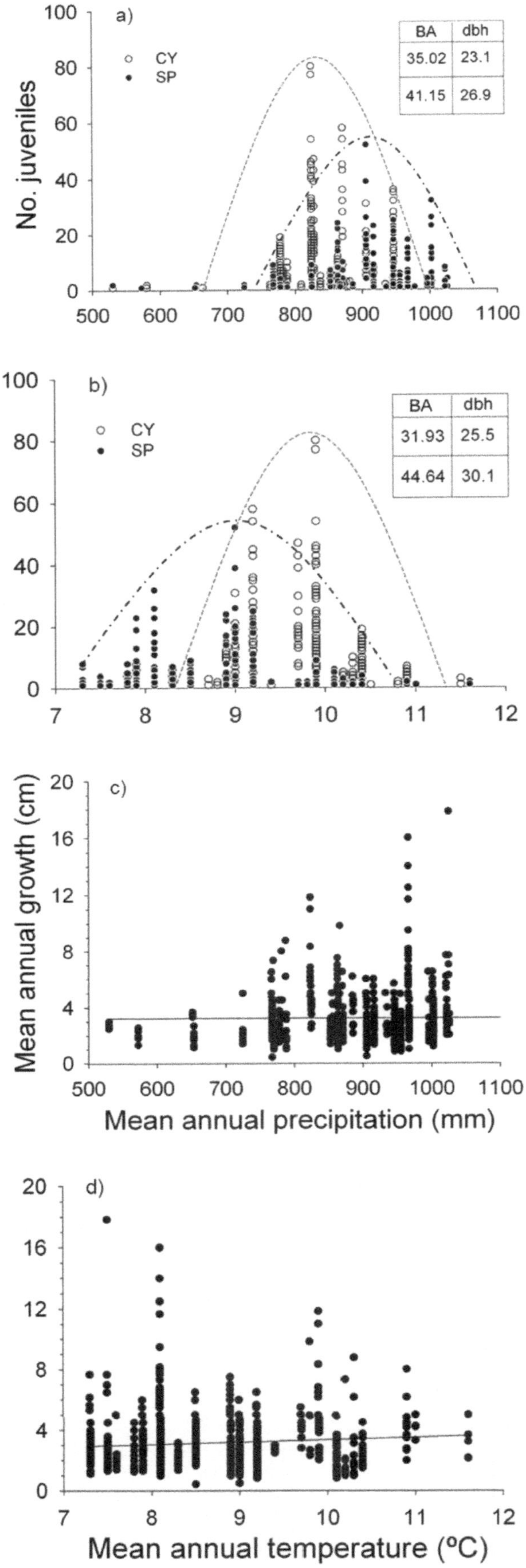

Figure 2. Distribution of current-year seedlings (CY) and juveniles older than a year (SP) according to a) the annual mean precipitation; and b) the mean annual temperature (with their envelope lines for

each cohort). Distribution of mean growth according to c) the mean annual precipitation and d) mean annual temperature. (BA: mean basal area (m^{2}ha^{-1}); dbh: mean diameter at breast height (cm); both calculated within an interval around the optimum annual mean precipitation (±50 mm) and temperature (±0.5 °C), respectively).

a)

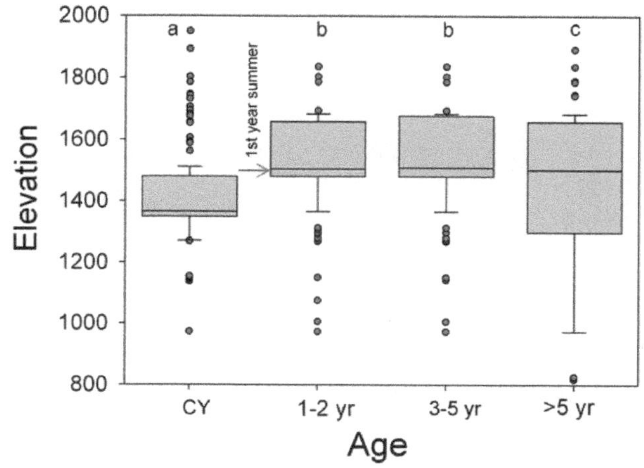

The presence of juveniles was different among sites (p<0.001) (Table 1). Across the altitudinal gradients and sites, the distribution of juveniles varied non-linearly. The optimum climatic conditions for pooled seedlings appeared around 823 mm and 9.9°C; while for pooled saplings the presence was higher at lower temperatures and higher precipitation, i.e. around 905 mm and 9°C (Fig. 2). These climatic divergences were the result of the altitudinal gap between the optima of both cohorts (Fig. 3), showing a higher frequency of seedlings and saplings at 1362 m and 1490 m respectively (optimum altitude weighted by site).

b)

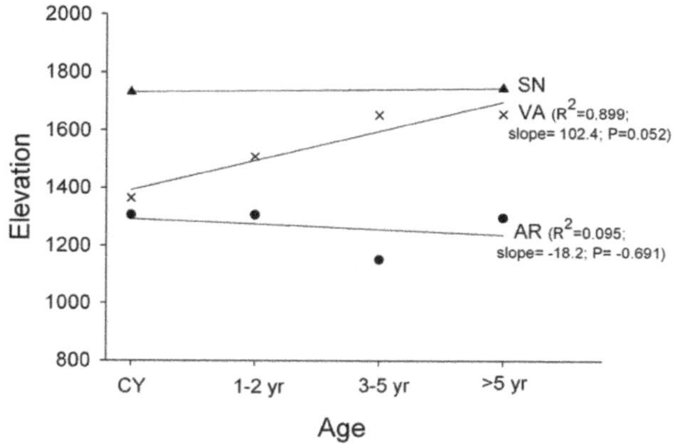

Figure 3. a) Altitudinal distribution of juveniles according to their age class and; b) altitudinal distribution of their medians according to each site. Different letters indicate significant differences (Kruskal-Wallis test)

Table 1. Descriptive variables measured in the sampling plots in the three study sites

	Valsaín	Arcalís	Sierra Nevada
Latitude	40°49′ N	42°22′N	37°06′N
Longitude	4°01′W	1°06′E	3°27′W
Altitudinal range (m)	1138–1837	672–1597	1425–1989
T range (°C)	7.3–10.9	7.3–11.6	8.7–11.5
Mean T (°C)	9.3 (1.3)	9.6 (1.6)	9.9 (1.0)
P range (mm)	762.7–1025.4	572.3–955.1	529.7–867.0
Mean P (mm)	873.6 (91.0)	827.2 (109.7)	691.4 (102.7)
Mean slope (%)	15.7(10.3)	45.1 (19.8)	35.0 (12.9)
Mean cover shrubs (%)	7.5 (15.8)	25.7 (26.2)	26.7 (28.1)
Mean cover herbs (%)	39.0 (32.7)	28.2 (26.5)	20.3 (21.5)
Mean cover rocks (%)	2.5 (7.2)	9.7 (15.3)	14.0 (20.0)
Mean tree density (pines ha^{-1})	298.9 (255.4)	603.7 (473.6)	165.5 (117.1)
Mean basal area (m^2 ha^{-1})	36.8 (22.7)	31.8 (13.8)	19.7 (14.4)
Mean dbh of adult pines (cm)	36.7 (11.8)	22.9(5.7)	37.8 (27.3)
Mean GSF	0.39(0.24)	0.27 (0.12)	0.43 (0.28)
No. observed CY	3794	77	45
No. observed SP	1211	197	7
No. plots with zero recruits	1	11	19

Standard deviations are shown in brackets.
T: mean annual temperature, P: annual mean precipitation; dbh: diameter at breast height; GSF: global site factor, CY: current-year seedlings, SP: juveniles older than a year.
doi:10.1371/journal.pone.0059824.t001

Factors Affecting Recruitment

Table 2 shows the variables included in the final ZIP models. It is remarkable that these models showed the significant variables affecting the occurrence (in particular the "no event", i.e. the absence) and abundance of juveniles independently.

Table 2. Zero-Inflated Poisson Mixed Models for the occurrence (absence)* and abundance of recently emerged seedlings and juveniles older than a year (saplings)

		Occurrence Model*				Abundance Model			
	Variables	Estimate	SE	t value	Pr>|t|	Estimate	SE	t value	Pr>|t|
Seedlings	Constant	78.5909	32.5935	2.41	0.0179	−84.6586	11.1953	−7.56	<.0001
(AIC = −12038)	Site 2	−4.1053	1.4245	−2.88	0.0049	4.5493	0.4974	9.15	<.0001
(EF = 0.742)	Site 3	1.5625	1.41	1.11	0.2707	2.6094	0.5332	4.89	<.0001
(ΔAIC = 125)	T	−15.7198	6.8578	−2.29	0.0242	17.3094	2.3224	7.45	<.0001
	T^2	0.7892	0.3609	2.19	0.0313	−0.9152	0.1217	−7.52	<.0001
	P	-	-	-	-	-	-	-	-
	Slope	-	-	-	-	0.03153	0.01165	2.71	0.0081
	Herbs	-	-	-	-	−0.03711	0.01675	−2.22	0.0292
	Shrubs	-	-	-	-	-	-	-	-
	Rocks	-	-	-	-	-	-	-	-
	GSF	-	-	-	-	-	-	-	-
	Herbs xT	-	-	-	-	0.004191	0.00173	2.42	0.0174
	Shrubs x T	0.002273	0.001005	2.26	0.0261	-	-	-	-
	GSF x T	-	-	-	-	-	-	-	-
Saplings	Constant	5.2785	1.2245	4.31	<.0001	−32.0442	10.2046	−3.14	0.0023
(AIC = −1646)	Site 2	−5.1948	1.1335	−4.58	<.0001	1.2807	0.949	1.35	0.1805
(EF = 0.598)	Site 3	−4.6633	1.1484	−4.06	0.0001	1.229	0.9268	1.33	0.1882
(ΔAIC = 87)	T	-	-	-	-	7.0072	2.1234	3.3	0.0014
	T^2	-	-	-	-	−0.4299	0.1147	−3.75	0.0003
	P	-	-	-	-	0.00431	0.001508	2.86	0.0053
	Slope	−0.03114	0.02128	−1.46	0.1468	−0.01973	0.01113	−1.77	0.0796
	Herbs	0.01801	0.005788	3.11	0.0025	−0.0734	0.0203	−3.62	0.0005
	Shrubs	-	-	-	-	-	-	-	-
	Rocks	-	-	-	-	-	-	-	-
	GSF	-	-	-	-	-	-	-	-
	Herbs xT	-	-	-	-	0.007874	0.002344	3.36	0.0011
	Shrubs x T	-	-	-	-	-	-	-	-
	GSF x T	-	-	-	-	-	-	-	-

Every variable introduced in the models is listed, but only the estimates for the variables included in the final model are shown.
Site 1: Valsaín, Site 2: Arcalís, Site 3: Sierra Nevada. T: mean annual temperature, P: annual mean precipitation, GSF: Global site factor; EF: modelling efficiency; ΔAIC: difference in AIC referred to saturated model.
*The parameters estimated in the binary part are referred to the probability *P* (finding no recruit) assessment, being 1-*P* the probability of presence.
doi:10.1371/journal.pone.0059824.t002

The selected model for seedlings showed a clear quadratic effect of mean annual temperature with significant values for both parts of the model (occurrence and abundance). The difference among sites (site factor) in terms of presence and abundance was also significant. The slope affected positively, and the percentage of herbaceous plants negatively to the density of seedlings (abundance model). We observed that temperature also interacted with other variables. Thus, the interaction between temperature and the cover of shrubs was significant (occurrence model), with an increasing probability of observing seedlings in places with a high cover of shrubs as annual temperature increased (Fig. 4a). Temperature also interacted with the cover of herbs, increasing its negative effect at the extreme low and high temperatures (abundance model) (Fig. 4b).

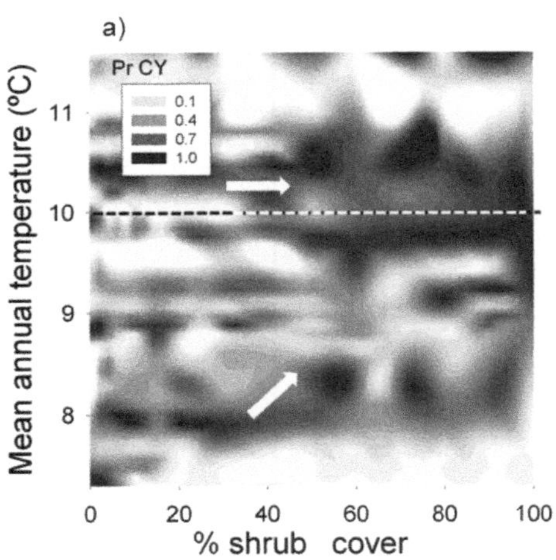

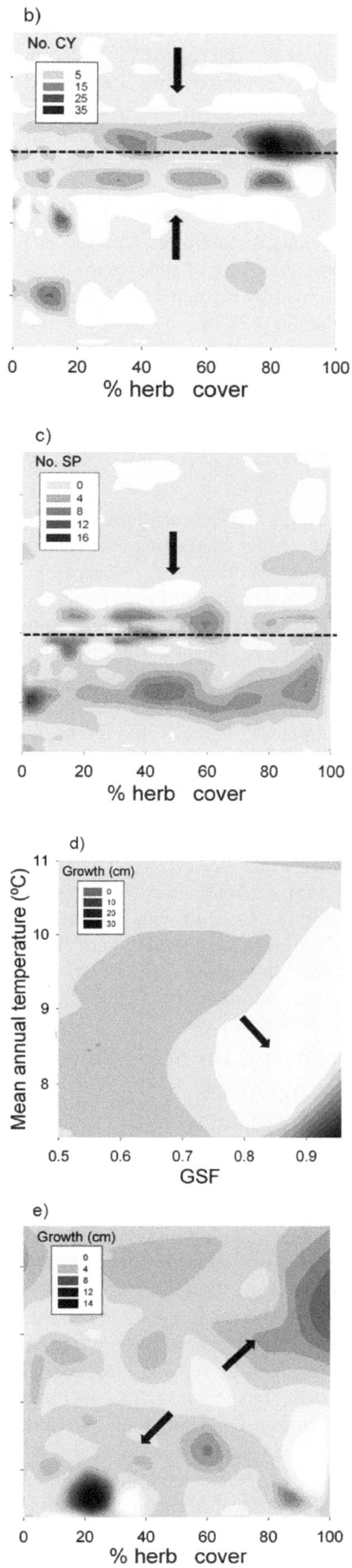

Figure 4. Significant interactions among mean annual temperature and: a) cover of shrubs affecting the probability of current-year seedlings' presence (Pr CY); b) herb cover affecting the abundance of seedlings (No. CY); c) herbs over the abundance of saplings (No. SP); d) GSF affecting mean growth of SP; and e) herbs on mean growth of saplings. Dotted line shows the mean annual temperature at which the number of CY (a, b) and SP (c) were more frequent, respectively.

The final model for saplings also included the site as a significant factor (occurrence model) and the climatic variables (abundance model): annual temperature again showed a quadratic significant effect and annual precipitation had a positive effect. The cover occupied by herbs affected negatively in both parts of the model, and interacted with temperature being less important at lower temperatures (Fig. 4c).

Factors Affecting Sapling Growth

Mean values of growth seemed to stay approximately constant with temperature and precipitation (Table 3; Figs. 2c, 2d), but it can be observed that the number of juveniles and the dispersion of the data was lower under more stressful conditions (higher temperatures and lower precipitation) (Figs. 2c, 2d). High light availability (GSF), rockiness and a low herb cover led to significant greater shoot elongations (Table 3). In addition, GSF and herbs interacted with annual temperature. The positive effect of light on growth was higher at low temperatures (Fig. 4d); whereas growth showed a more complex response when temperature interacted with herbs: at low temperatures juveniles grew more with low percentage of herbaceous, but also at high temperatures with high cover of herbs (Fig. 4e). Finally, the age of the saplings was also significant, showing higher mean growth for juveniles over 5 years.

Table 3. No Linear Mixed Model for the mean shoot growth of juveniles

Variable	Estimate	SE	t value	Pr>|t|	AIC	ΔAIC	EF
Constant	2.1146	0.4303	4391	<.0001	3717.2	209.6	0.407
Age class 1	−0.5550	0.1477	−3.76	0.0005			
Age class 2	−1.4442	0.1556	−9.28	<.0001			
T	-	-	-	-			
T^2	-	-	-	-			
P	-	-	-	-			
Slope	-	-	-	-			
Herbs	−0.09288	0.02941	−3.16	0.0029			
Shrubs	-	-	-	-			
Rocks	0.01521-	0.006872	2.21	0.0321			
GSF	16.0222	4.2561	3.76	0.0005			
Herbs x T	0.01051	3.11	0.0033	0.0033			
Shrubs x T	-	-	-	-			
GSF x T	−1.2034	0.4558	−2.64	<.0001			

Every variable introduced in the models is listed, but only the estimates for the variables included in the final model are shown.
T: mean annual temperature; P: annual mean precipitation; GSF: global site factor; Age class 1: 1–2 years old; Age class 2: 3–5 years old; Age class 3: >5 years old; ΔAIC: AIC increment referred to saturated model; EF: modelling efficiency.
doi:10.1371/journal.pone.0059824.t003

DISCUSSION

Effects of Climate on Demography

In the present study we have accomplished, for the first time, a nested survey of the recruitment pattern of Scots pine along altitudinal gradients at three different mountain ranges throughout the southernmost limit of its distribution area. Despite the different altitudinal range and characteristics of each site, the altitudinal gradient within each mountain range reproduced the rear, optimum and leading edges of the study populations under similar temperature ranges which led to a similar distribution pattern of juveniles in the different rear edge regions. Thus, the results showed a clear non-linear effect of temperature on the distributions of both cohorts, with an optimum temperature range where juveniles were more frequent, above and below which the number dropped gradually similarly to results in previous studies [9], [17], [19]. These higher frequencies of juveniles tallied with altitudinal levels that were above the mean elevation of the overall range of the adult stands (1330 m), though the differences among seedlings and saplings were remarkable. Regarding the former, the optimum of their distribution is close to the mean altitudinal range of adults. Meanwhile, the optimum for saplings' density was over 100 m skewed towards higher altitudes (lower temperatures and higher precipitation) compared to seedlings. In fact, considering the age classes of juveniles, there was an abrupt transition towards higher altitudes after the first year. This transition from seedlings to sapling is acknowledged as the main bottleneck for Scots pine recruitment with reported mortality rates around 80–100% after the first summer drought [15], [27], [31]. Thus, our data suggest that first summer survival may principally occur at locations with higher precipitation and lower temperature than those where more seedlings emerged. Likewise, other studies have already highlighted the positive role of water availability (related with precipitation and temperature) for the survival of Scots pine seedlings in Mediterranean areas [27], [31].

Effect of Climate on Plant-Plant Relationships

The surrounding structure of adult trees have a great relevance on regeneration patterns, in terms of seed source and mitigation of heat stress conditions (positive effects); or in terms of competition for light (negative effect) [27], [33], [39]. Thus, the direction of the net effect of overstorey structure on recruitment will depend on species-specific trade-offs or on the specific limiting factor within a system. Previous studies pointed out the light availability (as a direct consequence of the overstorey structure) as one of the best predictors for recruit distribution[29], [45], [46]. In our study case, we found no significant net effect of light availability on the abundance of seedlings or saplings. This might be due to the negligible relative effect of seed rain distribution in pure stands, or the heat stress amelioration provided by adult trees partially offsetting the lack of light. Nevertheless, our results did reveal the critical positive effect that light availability had over the shoot growth of saplings, as an indicator of their current performance and potential success within the stand [30], [45]. This effect was more marked at sites with lower temperatures, where the stress caused by heat and drought dropped. Many works have investigated plant-plant interactions along gradients of environmental severity[47]. In Mediterranean areas, compelling evidence about the nurse effect exerted by shrubs on recruits has already been reported by different studies [27], [32], [46], [48]. Likewise, our results showed a significant effect of climate (temperature) on the interaction between seedlings and shrubs: the probability of finding them was higher near shrubs as mean temperature increased, supporting the idea of stress amelioration underneath. Surprisingly, we did not detect any effect of the shrub presence in relation to saplings, as previous research has reported in Sierra Nevada [24], [27]. This lack of significance may suggest either that the benefits provided by the nurse plant may not overcome its own resource uptake [49]; or may be due to different lifespan of shrubs and tree juveniles that may hide interactions in the past. Additionally, the specific shrub species in each site and their peculiarities, not considered in the present study (e.g. allelopathic effects, architecture, herbivory appeal) may play a crucial role in fading or neglecting this nursing effect [32], [50].

The herbaceous layer constitutes a physical barrier preventing seeds from contacting the mineral soil [24], and competes with seedlings for resources [51]. This is in accordance with the negative effect that herb cover had on seedling emergence and survival in our study. This effect varied with temperature, shrinking at less stressful sites, i.e. at intermediate temperatures for emergence (seedlings); and at lower temperatures for survival (saplings), precisely where precipitation was higher. This might indicate that the competition is harder under harsher conditions, subscribing evidence that negative relationships do not always predominate under less stressful conditions [52]. Cover of herbs significantly hampered the mean growth of saplings, likely due to competition for water as well [51], [33], being less evident in places with cooler temperatures (less stressful conditions). However, at warmer sites, growth increased with a higher cover of herbs, suggesting a stimulation of shoot elongation in more stressful sites, in order to escape from competition in the same way as shade conditions do [53], [54].

Complex Scenarios for Scots Pine: Considerations for Management

Recruitment of Scots pine in the Iberian Peninsula is episodic, especially dependant on the climatic conditions during critical stages of seedling establishment or the availability of *safe sites*. In Mediterranean areas, rare wet summers entail the opportunity for high sporadic establishment [55]; whereas episodic events of severe drought occasionally cause massive seedling mortality [56]. Since these extreme events are predicted to become more frequent[26], it seems crucial to work on the preservation or promotion of these *safe sites*. The results of the present study showed that seedlings' emergence in southern populations of *P. sylvestris* booms at warmer temperatures than survival, and thus subsequent seedlings' success would rely on higher water availability and less heat-stress. Based on our results and in accordance with other studies, the partial removal of the herbaceous vegetation may ameliorate the competition for the scarce water [33], [57] and consequently boost seedling survival and growth. Moreover, the promotion of shrubs as nurse plants, not only may mitigate the summer stress for seedlings [24], [48], but also protect juveniles against mechanic damages such as grazing [50]. Nevertheless, management in Mediterranean areas should not only be based on silvicultural manuals. Silvicultural strategies must be tailor-made according to the idiosyncrasy of each place. Our results showed differences in the recruitment pattern among the three sites, probably due to multiple and diverse reasons. For instance, Sierra Nevada represents the most stressful site in climatic terms, with low density of adult trees and an important potential damage to juveniles by wild animals' browsing [58]. Therefore, operations aimed at the avoidance of the senescence of the existing adult trees and the promotion of the protective shrub layer may help to preserve stands with similar characteristics. The northernmost location, Arcalís, harbours other peculiarities. Galiano, Martínez-Vilalta & Lloret [20] suggested that management abandonment in the last decades (with consequent high tree density), might have favored a poorer performance of adult trees under extreme events, constraining partially the recruitment in this area. Thus, a sustainable thinning may ameliorate conspecific competition for water and improve the performance of potential parents [59]. Finally, in the third stand, Valsaín, recruitment does not seem to be compromised. This may surely be linked to the higher precipitation records, but also to the regular management that so far has guaranteed the presence of mature and vigorous trees as seed source and provided appropriate size gaps for seedling establishment.

CONCLUSION

The consideration of environmental variables at different scales constitutes a step forward to tackle the heterogeneity and captures the potential variability of recruitment across the landscape. Particularly interesting is to consider different elevation gradients (three sites in this study) within the overall rear edge area of any species, because they in turn reproduced the leading and rear edge conditions. Thus, our results have shown the general incidence of climate directly and indirectly on recruitment from demographic, performance and community standpoints, despite the idiosyncrasy of each site. This pattern may be useful for disentangling the potential effect of climate change on Scots pine recruitment and for establishing generalizable criteria for management in these marginal populations. Moreover, the consideration and assimilation of local biotic and abiotic interactions is essential to reach sustainable management for populations of *P. sylvestris*, but also to improve our understanding of the dynamics of these marginal populations. Therefore, we advocate further studies at individual and at community level, both in the field or in green houses, in order to understand the non-linear responses of communities to combined effects (climatic and non-climatic) and to deepen into their influence over the performance of the juveniles of these marginal populations.

SUPPORTING INFORMATION

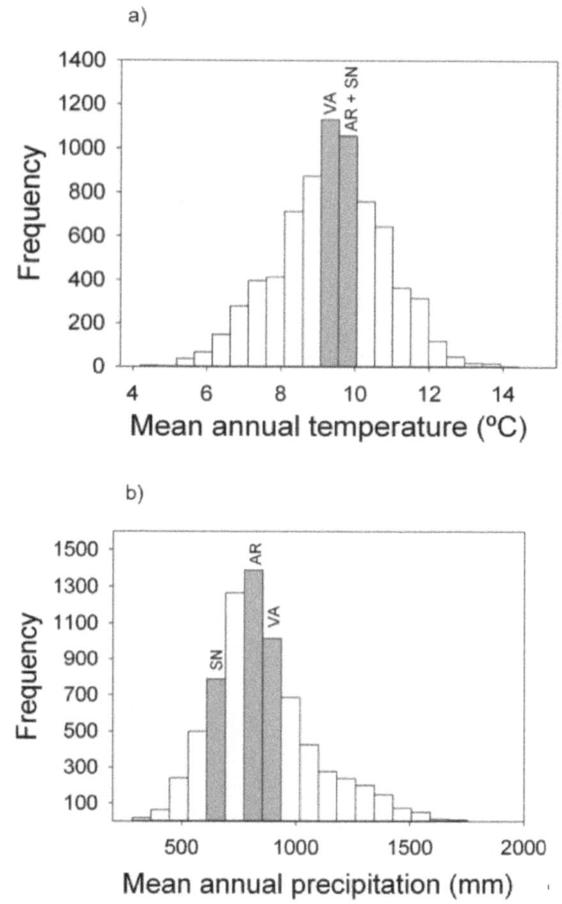

Figure S1. Histograms of frequencies of the plots belonging to the III Spanish National Forest Inventory with *Pinus sylvestris* as a dominant tree, according to their mean annual temperature and precipitation [35]. The grey bars show where our sampling sites belong (AR: Arcalís, VA: Valsaín; SN: Sierra Nevada).

Table S1. Spearman correlation coefficients among the variables measured in the study plots (T: mean annual temperature; P: annual mean precipitation; GSF: global site factor; FI: potential fecundity index; BA: basal area)

	T	P	Slope	Shrubs	Herbs	Rocks	GSF	FI	BA
T	1,000	,040	,085**	,341**	-,322**	,155**	,137**	-,215**	-,405**
P	-	1,000	-,031	-,259**	,014	-,101**	,004	,063**	,203**
Slope	-	-	1,000	,315**	-,046	,255**	-,042	-,020	-,118**
Shrubs	-	-	-	1,000	-,172**	,230**	,147**	-,214**	-,349**
Herbs	-	-	-	-	1,000	-,127**	,057*	-,145**	,055*
Rocks	-	-	-	-	-	1,000	,235**	-,287**	-,420**
GSF	-	-	-	-	-	-	1,000	**-,548****	**-,539****
FI	-	-	-	-	-	-	-	1,000	**,641****
BA	-	-	-	-	-	-	-	-	1,000

ACKNOWLEDGMENTS

We thank all the people that made this intensive fieldwork possible, especially to R. Freire, B. Santamaría, J. Martínez, D. López and R. Ruíz-Puche. We are also very grateful to R. Manson for the statistical support.

AUTHOR CONTRIBUTIONS

Conceived and designed the experiments: RB SGR EG AE JMV JAH AMR RZ FV. Performed the experiments: RB SGR EG JAH AMR RZ. Analyzed the data: RB. Contributed reagents/materials/analysis tools: RB SGR EG AE JMV JAH AMR RZ FV. Wrote the paper: RB AE FV.

REFERENCES

1. Keddy P (1992) Assembly and response rules: two goals for predictive community ecology. Journal of Vegetation Science 3: 157–164. doi: 10.2307/3235676
2. Pearson RG, Dawson TP (2003) Predicting the impacts of climate change on the distribution of species: are bioclimate envelope models useful? Global Ecology & Biogeography 12: 361–371. doi: 10.1046/j.1466-822x.2003.00042.x
3. Wardle P, Coleman MC (1992) Evidence for rising upper limits of four native New Zealand forest trees. New Zealand Journal of Botany 30: 303–314. doi: 10.1080/0028825x.1992.10412909
4. Kullman L (2002) Rapid recent range-margin rise of tree and shrub species in the Swedish Scandes. Journal of Ecology 90 (1): 68–77. doi: 10.1046/j.0022-0477.2001.00630.x
5. Lloyd AH, Fastie CL (2003) Recent changes in tree line forest distribution and structure in interior Alaska. Ecoscience 10 (2): 176–185.
6. Peñuelas J, Boada M (2003) A global change-induced biome shift in the Montseny mountains (NE Spain). Global Change Biology 9 (2): 131–140. doi: 10.1046/j.1365-2486.2003.00566.x
7. Beckage B, Osborne B, Gavin DG, Pucko C, Siccama T, et al. (2008) A rapid upward shift of a forest ecotone during 40 years of warming in the Green Mountains of Vermont. Proceedings of the National Academy of Sciences of the United States of America 105: 4197–4202. doi: 10.1073/pnas.0708921105
8. Wang T, Zhang Q-B, Ma K (2006) Treeline dynamics in relation to climatic variability in the central Tianshan Mountains, northwestern China. Global Ecology and Biogeography 15(4): 406–415. doi: 10.1111/j.1466-822x.2006.00233.x
9. Lenoir J, Gegout JC, Pierrat JC, Bontemps JD, Dhote JF (2009) Differences between tree species seedling and adult altitudinal distribution in mountain forests during the recent warm period (1986-2006). Ecography 32(5): 765–777. doi: 10.1111/j.1600-0587.2009.05791.x
10. Doak DF, Morris WF (2010) Demographic compensation and tipping points in climate-induced range shifts. Nature 467(7318): 959–962. doi: 10.1038/nature09439
11. Gimeno TE, Camarero JJ, Granda E, Pías B, Valladares F (2012) Enhanced growth of*Juniperus thurifera* under a warmer climate is explained by a positive carbon gain under cold and drought. Tree Physiology 32: 326–336. doi: 10.1093/treephys/tps011
12. Hampe A, Jump AS (2011) Climate Relicts: Past, Present, Future. Annual Review of Ecology, Evolution, and Systematics 42: 313–333. doi: 10.1146/annurev-ecolsys-102710-145015
13. Horsák M, Chytrý M, Pokryszko BM, Danihelka J, Ermakov N, et al. (2010) Habitats of relict terrestrial snails in southern Siberia: lessons for reconstructions of paleoenvironments of full-glacial Europe. Journal of Biogeography 37: 1450–1462. doi: 10.1111/j.1365-2699.2010.02280.x
14. Hampe A, Petit RJ (2005) Conserving biodiversity under climate change: the rear edge matters. Ecology Letters 8(5): 461–467. doi: 10.1111/j.1461-0248.2005.00739.x
15. Castro J, Zamora R, Hódar JA, Gómez JM (2005) Alleviation of summer drought boosts establishment success of *Pinus sylvestris* in a Mediterranean mountain: an experimental approach. Plant Ecology 181: 191–202. doi: 10.1007/s11258-005-6626-5
16. Lloret F, Peñuelas J, Prieto P, Llorens L, Estiarte M (2009) Plant community changes induced by experimental climate change: seedling and adult species composition. Perspectives in Plant Ecology Evolution and Systematics 11: 53–63. doi: 10.1016/j.ppees.2008.09.001
17. Gworek JR, Wall SBV, Brussard PF (2007) Changes in biotic interactions and climate determine recruitment of Jeffrey pine along an elevation gradient. Forest Ecology and Management 239(1-3): 57–68. doi: 10.1016/j.foreco.2006.11.010
18. Peñuelas J, Ogaya R, Boada M, Jump AS (2007) Migration, invasion and decline: changes in recruitment and forest structure in a warming-linked shift of European beech forest in Catalonia (NE Spain). Ecography 30(6): 829–837. doi: 10.1111/j.2007.0906-7590.05247.x
19. Jurado E, García JF, Flores J, Estrada E, Gónzalez H (2011) Abundance of seedlings in response to elevation and nurse species in Northeastern Mexico. Southwestern Naturalist 56(2): 154–161. doi: 10.1894/f03-mlk-18.1
20. Galiano L, Martínez-Vilalta J, Lloret F (2010) Drought-induced multifactor decline of Scots pine in the Pyrenees and potential vegetation change by the expansion of co-occurring oak species. Ecosystems 13(7): 978–991. doi: 10.1007/s10021-010-9368-8
21. Jump AS, Hunt JM, Peñuelas J (2007) Climate relationships of growth and establishment across the altitudinal range of *Fagus sylvatica* in the Montseny Mountains, northeast Spain. Ecoscience 14(4): 507–518. doi: 10.2980/1195-6860(2007)14[507:crogae]2.0.co;2
22. Vilá-Cabrera A, Martínez-Vilalta J, Vayreda J, Retana J (2011) Structural and climatic determinants of demographic rates of Scots pine forests across the Iberian Peninsula. Ecological Applications 21: 1162–1172. doi: 10.1890/10-0647.1
23. Ryyppö A, Iivonen S, Rikala R, Sutinen ML, Vapaavuori E (1998) Responses of Scots pine seedlings to low root zone temperature in spring. Physiologia

Plantarum 102(4): 503–512. doi: 10.1034/j.1399-3054.1998.1020404.x

24. Castro J, Zamora R, Hódar JA, Gómez JM (2002) Use of shrubs as nurse plants: A new technique for reforestation in Mediterranean mountains. Restoration Ecology 10: 297–305. doi: 10.1046/j.1526-100x.2002.01022.x
25. Lo YH, Blanco JA, Kimmins JP (2010) A word of caution when planning forest management using projections of tree species range shifts. The Forestry Chronicle 86: 312–316. doi: 10.5558/tfc86312-3
26. Christensen JH, Hewitson B, Busuioc A, Chen A, Gao X et al.. (2007) Regional climate projections. In: Solomon S, Qin D, Manning M, Chen Z, Marquis M et al., editors; Cambridge and New York. Cambridge University Press. pp. 847–943.
27. Castro J, Zamora R, Hódar JA, Gómez JM (2004) Seedling establishment of a boreal tree species (*Pinus sylvestris*) at its southernmost distribution limit: consequences of being in a marginal Mediterranean habitat. Journal of Ecology 92: 266–277. doi: 10.1111/j.0022-0477.2004.00870.x
28. Pardos M, Montes F, Cañellas I (2008) Spatial dynamics of natural regeneration in two differently managed *Pinus sylvestris* stands before and after silvicultural intervention using replicated spatial point patterns. Forest Science 54(3): 260–272.
29. Barbeito I, Fortin M-J, Montes F, Canellas I (2009) Response of pine natural regeneration to small-scale spatial variation in a managed Mediterranean mountain forest. Applied Vegetation Science 12: 488–503. doi: 10.1111/j.1654-109x.2009.01043.x
30. Matías L, Gómez-Aparicio L, Zamora R, Castro J (2011) Effects of resource availability on plant recruitment at the community level in a Mediterranean mountain ecosystem. Perspectives in Plant Ecology Evolution and Systematics 13(4): 277–285. doi: 10.1016/j.ppees.2011.04.005
31. Quero JL, Herrero A, Zamora R (2011) Linking stochasticity to determinism of woody plant recruitment in a mosaic landscape: A spatially explicit approach. Basic and Applied Ecology 12(2): 161–171. doi: 10.1016/j.baae.2010.11.007
32. Gómez-Aparicio L, Zamora R, Gómez JM, Hódar JA, Castro J, et al. (2004) Applying plant facilitation to forest restoration: A meta-analysis of the use of shrubs as nurse plants. Ecological Applications 14(4): 1128–1138. doi: 10.1890/03-5084
33. Barbeito I, LeMay V, Calama R, Canellas I (2011) Regeneration of Mediterranean*Pinus sylvestris* under two alternative shelterwood systems within a multiscale framework. Canadian Journal of Forest Research 41: 341–351. doi: 10.1139/x10-214
34. Jordano P, Herrera CM (1995) Shuffling the offspring-uncoupling and spatial discordance of multiple stages in vertebrate seed dispersal. Ecoscience 2: 230–237.
35. Ninyerola M, Pons X, Roure JM (2005) Atlas Climático Digital de la Península Ibérica. Metodología y aplicaciones en bioclimatología y geobotánica. Bellaterra: Universidad Autónoma de Barcelona.
36. Valladares F, Guzmán B (2006) Canopy structure and spatial heterogeneity of understory light in an abandoned Holm oak woodland. Annals of Forest Science 63(7): 749–761. doi: 10.1051/forest:2006056
37. He FL, Duncan RP (2000) Density-dependent effects on tree survival in an old-growth Douglas fir forest. Journal of Ecology 88(4): 676–688. doi: 10.1046/j.1365-2745.2000.00482.x
38. Gómez-Aparicio L, García-Valdés R, Ruíz-Benito P, Zavala MA (2011) Disentangling the relative importance of climate, size and competition on tree growth in Iberian forests: implications for forest management under global change. Global Change Biology 17: 2400–2414. doi: 10.1111/j.1365-2486.2011.02421.x
39. Ribbens E, Silander JA, Pacala SW (1994) Seedling recruitment in forests- Calibrating models to predict patterns of tree seedling dispersion. Ecology 75(6): 1794–1806. doi: 10.2307/1939638
40. Lambert D (1992) Zero-inflated Poisson regression, with an application to defects in manufacturing. Technometrics 34: 1–14. doi: 10.2307/1269547
41. Hall DB (2000) Zero-inflated Poisson and binomial regression with random effects: A case study. Biometrics 56(4): 1030–1039. doi: 10.1111/j.0006-341x.2000.01030.x
42. Affleck DLR (2006) Poisson mixture models for regression analysis of stand level mortality. Canadian Journal of Forest Research 36: 2994–3006. doi: 10.1139/x06-189
43. Fortin M, DeBlois J (2007) Modeling tree recruitment with zero-inflated models: The example of hardwood stands in southern Quebec, Canada. Forest Science 53(4): 529–539.
44. Gómez-Aparicio L (2008) Spatial patterns of recruitment in Mediterranean plant species: linking the fate of seeds, seedlings and saplings in heterogeneous landscapes at different scales. Journal of Ecology 96(6): 1128–1140. doi: 10.1111/j.1365-2745.2008.01431.x
45. Gómez-Aparicio L, Pérez-Ramos IM, Mendoza I, Matías L, Quero JL, et al. (2008) Oak seedling survival and growth along resource gradients in Mediterranean forests: implications for regeneration in current and future environmental scenarios. Oikos 117(11): 1683–1699. doi: 10.1111/j.1600-0706.2008.16814.x
46. Granda E, Escudero A, de la Cruz M, Valladares F (2012) Juvenile-adult tree associations in a continental Mediterranean ecosystem: no evidence for sustained and general facilitation at increased aridity. Journal of Vegetation Science 23: 164–175. doi: 10.1111/j.1654-1103.2011.01343.x
47. Callaway RM, Walker LR (1997) Competition and facilitation: a synthetic approach to interactions in plant communities. Ecology 78: 1958–1965. doi: 10.2307/2265936
48. Rodríguez-Garcia E, Bravo F, Spies TA (2011) Effects of overstorey canopy, plant-plant interactions and soil properties on Mediterranean maritime pine seedling dynamics. Forest Ecology and Management 262(2): 244–251. doi: 10.1016/j.foreco.2011.03.029

49. Maestre FT, Cortina J (2004) Do positive interactions increase with abiotic stress? - A test from a semi-arid steppe. Proceedings of the Royal Society of London Series B-Biological Sciences 271: S331–S333. doi: 10.1098/rsbl.2004.0181

50. Baraza E, Zamora R, Hódar JA (2006) Conditional outcomes in plant-herbivore interactions: neighbours matters. Oikos 113: 148–156. doi: 10.1111/j.0030-1299.2006.14265.x

51. Davis MA, Wrage KJ, Reich PB, Tjoelker MG, Schaeffer T, et al. (1999) Survival, growth, and photosynthesis of tree seedlings competing with herbaceous vegetation along a water-light-nitrogen gradient. Plant Ecology 145(2): 341–350.

52. Maestre FT, Valladares F, Reynolds JF (2005) Is the change of plant-plant interactions with abiotic stress predictable? A meta-analysis of field results in arid environments. Journal of Ecology 93(4): 748–757. doi: 10.1111/j.1365-2745.2005.01017.x

53. Homgren M, Segura AM, Fuentes ER (2000) Limiting mechanisms in the regeneration of the Chilean matorral. Plant Ecology 147: 49–57.

54. Gónzalez-Rodríguez V, Villar R, Casado R, Suárez-Bonnet E, Luis Quero J, et al. (2011) Spatio-temporal heterogeneity effects on seedling growth and establishment in four *Quercus* species. Annals of Forest Science 68(7): 1217–1232. doi: 10.1007/s13595-011-0069-z

55. Matías L, Zamora R, Castro J (2012) Sporadic rainy events are more critical than increasing of drought intensity for woody species recruitment in a Mediterranean community. Oecologia 169(3): 833–844. doi: 10.1007/s00442-011-2234-3

56. Herrero A (2012) Capacidad de respuesta al estrés ambiental de poblaciones de *Pinus sylvestris* y *Pinus nigra* en el límite sur de distribución: una aproximación multidisciplinar PhD Thesis, University of Granada, Spain.

57. Beland M, Agestam E, Eko PM, Gemmel P, Nilsson U (2000) Scarification and seedfall affects natural regeneration of Scots pine under two shelterwood densities and a clear-cut in southern Sweden. Scandinavian Journal of Forest Research 15: 247–255. doi: 10.1080/028275800750015064

58. Zamora R, Gómez JM, Hódar JA, Castro J, García D (2001) Effect of browsing by ungulates on sapling growth of Scots pine in a Mediterranean environment: consequences for forest regeneration. Forest Ecology and Management 144: 33–42. doi: 10.1016/s0378-1127(00)00362-5

59. Millar CI, Stephenson NL, Stephens SL (2007) Climate change and forests of the future: Managing in the face of uncertainty. Ecological Applications 17(8): 2145–2151. doi: 10.1890/06-1715.1

SECTION III

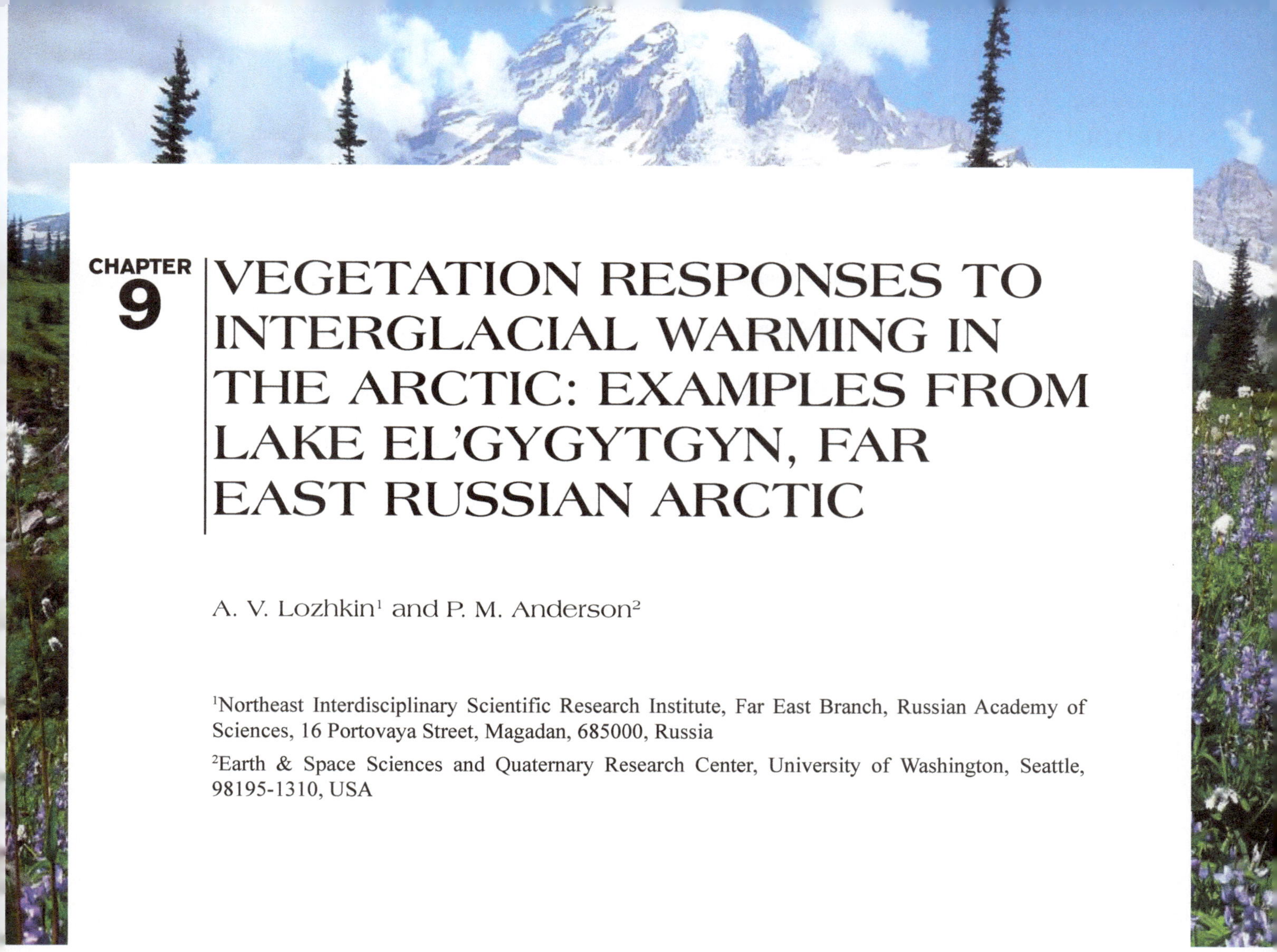

CHAPTER 9

VEGETATION RESPONSES TO INTERGLACIAL WARMING IN THE ARCTIC: EXAMPLES FROM LAKE EL'GYGYTGYN, FAR EAST RUSSIAN ARCTIC

A. V. Lozhkin[1] and P. M. Anderson[2]

[1]Northeast Interdisciplinary Scientific Research Institute, Far East Branch, Russian Academy of Sciences, 16 Portovaya Street, Magadan, 685000, Russia

[2]Earth & Space Sciences and Quaternary Research Center, University of Washington, Seattle, 98195-1310, USA

ABSTRACT

Preliminary analyses of Lake El'gygytgyn sediment indicate a wide range of ecosystem responses to warmer than present climates. While palynological work describing all interglacial vegetation is ongoing, sufficient data exist to compare recent warm events (the postglacial thermal maximum, PGTM, and marine isotope stage, MIS5) with "super" interglaciations (MIS11, MIS31). Palynological assemblages associated with these climatic optima suggest two types of vegetation responses: one dominated by deciduous taxa (PGTM, MIS5) and the second by evergreen conifers (MIS11, MIS31). MIS11 forests show a similarity to modern ***Picea–Larix–Betula–Alnus*** **forests of Siberia. While dark coniferous forest also characterizes MIS31, the pollen taxa show an affinity to the boreal forest of the lower Amur valley (southern Russian Far East). Despite vegetation differences during these thermal maxima, all glacial–interglacial transitions are alike, being dominated by deciduous woody taxa. Initially** ***Betula*** **shrub tundra established and was replaced by tundra with tree-sized shrubs (PGTM),** ***Betula*****woodland (MIS5), or** ***Betula–Larix*** **(MIS11, MIS31) forest. The consistent occurrence of deciduous forest and/or high shrub tundra before the incidence of maximum warmth underscores the importance of this biome for modeling efforts. The El›gygytgyn data also suggest a possible elimination or massive reduction of Arctic plant communities under extreme warm-earth scenarios.**

Citation: Lozhkin, A. V. and Anderson, P. M, Vegetation responses to interglacial warming in the Arctic: examples from Lake El'gygytgyn, Far East Russian Arctic, Clim. Past, 9, 1211-1219, doi:10.5194/cp-9-1211-2013, 2013.

INTRODUCTION

Marine and ice cores contain vital data that document fluctuations in past climate and provide insight into the possible causes of the observed paleoclimatic variability. However, such records cannot provide information on the responses of biotic systems to those changes. Most paleo-biotic records do not have the longevity, continuity, or temporal resolution of ocean and ice cores, which limits the comparison of longterm trends in biological and climatic data. Lake El'gygytgyn (Lake E), whose basin was created following a meteor impact 3.58 ± 0.04 Ma ago (Layer, 2000), presents a rare exception having yielded a continuous sedimentary record dating back ~ 2.8 Ma (Melles et al., 2012). The location of the lake in Arctic Chukotka (Fig. 1) provides a unique look into the response of terrestrial ecosystems at northern high latitudes to a broad range of warm and cold conditions.

Several biological proxies have been examined in cores from Lake E (Brigham-Grette et al., 2007; Melles et al., 2012); we present here interpretations of past vegetation based on palynological data. Previous research has shown that cool, dry intervals are clearly marked in the Lake E record by a dominance of herb pollen (particularly Poaceae) and usually higher values of Selaginella rupestris spores (Lozhkin et al., 2007; Matrosova, 2009). These spectra contrast with interglacial assemblages where arboreal pollen is at a maximum, herbaceous taxa are low, and spores are often dominated by Sphagnum. Continued research on a newly raised core shows that interglacial assemblages display vegetation and climatic variability both within and between isotope stages. For example, marine isotope stage (MIS) 11.3 and MIS31 have been described as "super interglaciations" with maximum summer temperatures ~ 4 to 5 °C warmer than either MIS5.5 or the postglacial thermal maximum (PGTM; Melles et al., 2012).

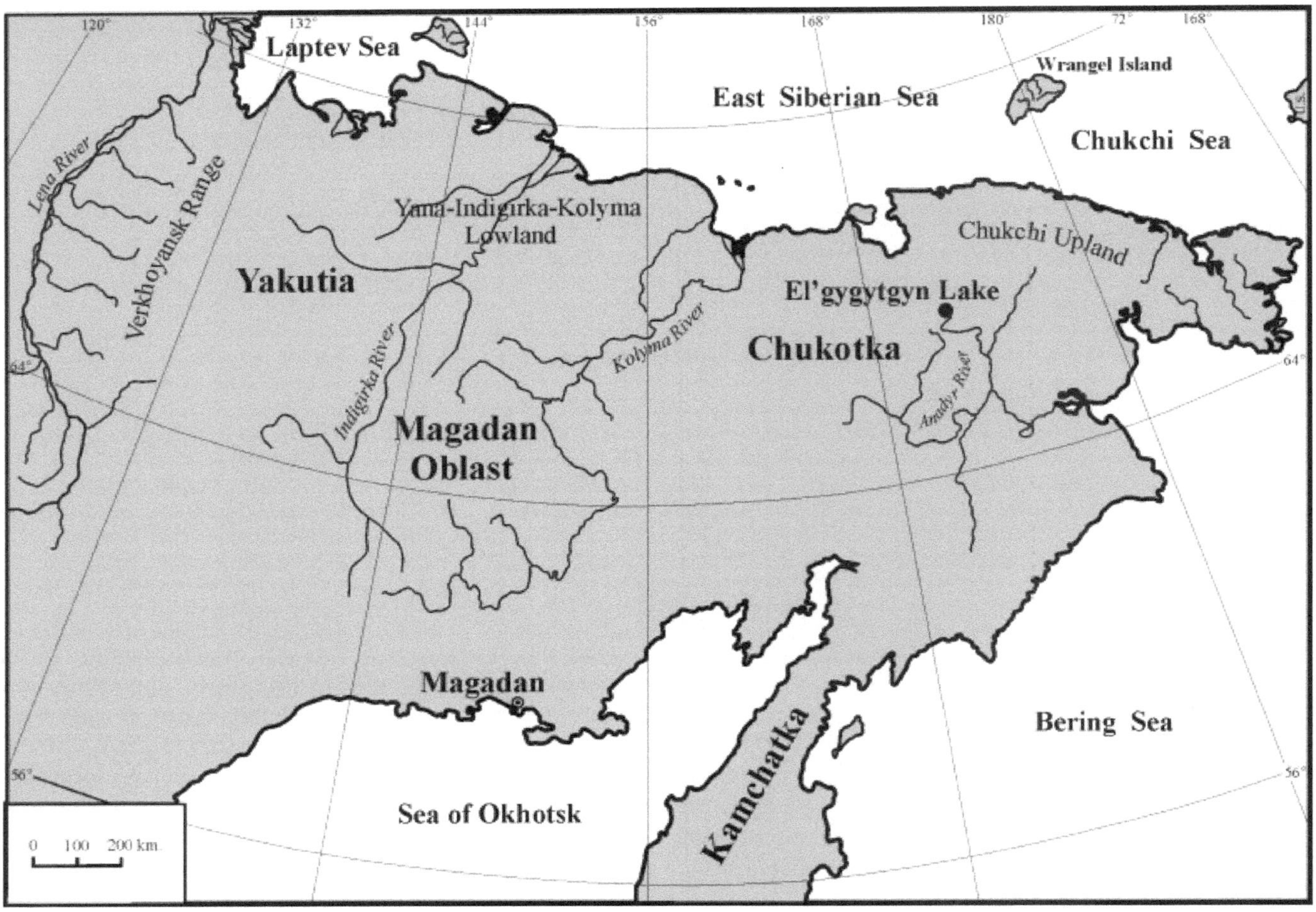

Figure 1. Map of northeast Siberia showing location of Lake El'gygytgyn.

These warmer-than-present intervals are of particular interest for understanding possible responses and feedbacks related to increasing global temperatures (e.g., Harrison et al., 1995; Kaplan et al., 2003). Both conceptual and quantitative models have considered two types of vegetation responses at northern high latitudes: one suggests the expansion of evergreen conifer forest northward across North America and much of Eurasia (e.g., Claussen, 1996; Kaplan et al., 2003), whereas the second postulates a dominance of broadleaf deciduous forest in current areas of tundra (e.g., Bonan et al., 1990; Chapin III and Starfield, 1997). Paleobotanical data are one means by which these alternative scenarios may be evaluated. Although analysis continues on Lake E, sufficient palynological data are available to compare the vegetation responses during the "super interglaciations" and the more moderate warming represented in MIS5.5 and the PGTM.

METHODS

One to two cm^3 sediment samples were processed for

palynological analysis following standard procedures used in preparing organic-poor samples (PALE, 1994). Samples for MIS1 and MIS2 and MIS4 to MIS6 are from core PG-1352 (Lozhkin et al., 2007), whereas core ICDP 5011 provided samples from MIS11 and MIS12, and MIS31 (Melles et al., 2012). Pollen sums for interglacial spectra generally exceed 500 identified pollen grains. Percentages of individual taxa are based on a sum of all arboreal, nonarboreal, and unidentified pollen grains. Subsumed percentages on the left of the diagrams (Figs. 2–4) are calculated from a total of pollen and spores. Pollen zonation was done subjectively using changes in percentages of major taxa. Although many palynologists have differentiated tree and shrub Betula based on morphology and/or size, blind tests in our laboratories of modern reference materials suggest that these criteria are inconsistent and unreliable; thus we do not separate the Betula pollen. Picea obovata pollen is classified within Picea sect. Eupicea, Picea ajanensis within Picea sect. Omorica, Pinus sylvestris within Pinus subg. Diploxylon, and Pinus pumila and Pinus sibirica within Pinus subg. Haploxylon. Following traditional palynological usage, Duschekia fruticosa is referred to as shrub Alnus. All other plant taxonomy follows Czerepanov (1995).

The age model for the El'gygytgyn record is based on paleomagnetic stratigraphy and tuning of additional proxy data to the marine isotope stratigraphy and to variations in regional insolation (see Melles et al., 2012). Because pollen zones do not always correspond to isotope substages, the zones are identified by isotope stage (number) and as a pollen assemblage (letter). Following palynological format the lowest zone is indicated by "a" the next youngest by "b", etc., in contrast to the order of isotope substages.

Modern analogs to the fossil samples were based on a squared chord-distance measure (Overpeck et al., 1985) using palynological surface samples from Beringia (MIS5e and younger; Lozhkin et al., 2007) or a geographically unrestricted modern dataset (Melles et al., 2012; MIS1, MIS5, MIS11, MIS31). Modern climate values assigned to each surface sample were calculated based on interpolated values from a high-resolution surface-climate dataset (see supplement in Melles et al., 2012, and Lozhkin et al., 2007, for more detail).

MODERN SETTING

Lake E (67°300 N, 172°5 0 E; ~ 500 m a.s.l.) has a diameter of 12 km and a maximum depth of 175 m (Melles et al., 2012). The local vegetation is a mix of discontinuous and continuous herb–lichen tundra (Kozhevnikov, 1993). Shrubs (Salix krylovii, S. alaxensis, Betula exilis) are not abundant and limited to protected sites in valleys and saddles within the crater walls and along the outlet valley. The vegetation in the Lake E catchment is sparser than in the surrounding Chukchi Uplands because of unusual soil qualities associated with the meteor impact crater. The uplands are dominated by shrub tundra that includes Betula exilis, Duschekia fruticosa (shrub Alnus), Pinus pumila, Salix, and Ericales. Larix gmelinii tree line is located ~ 150 to the southwest of the lake, but the main body of the forest is found at ~ 300 km distance. Modern pollen studies indicate that the pollen rain in Lake E is more indicative of regional than local vegetation with up to 60 % of the pollen (Alnus, Betula, Pinus pumila) being blown in from the Chukchi Uplands (Lozhkin et al., 2001). The over-representation of upland vegetation, which reflects regional conditions, strengthens the use of the Lake E pollen record for paleoclimatic interpretations. Estimated mean climatic values are as follows: annual temperature −13.5 to −14°C; July temperature 8.8 to 9 °C; January temperature −31 to −32°C; annual precipitation 255 mm (Melles et al., 2012).

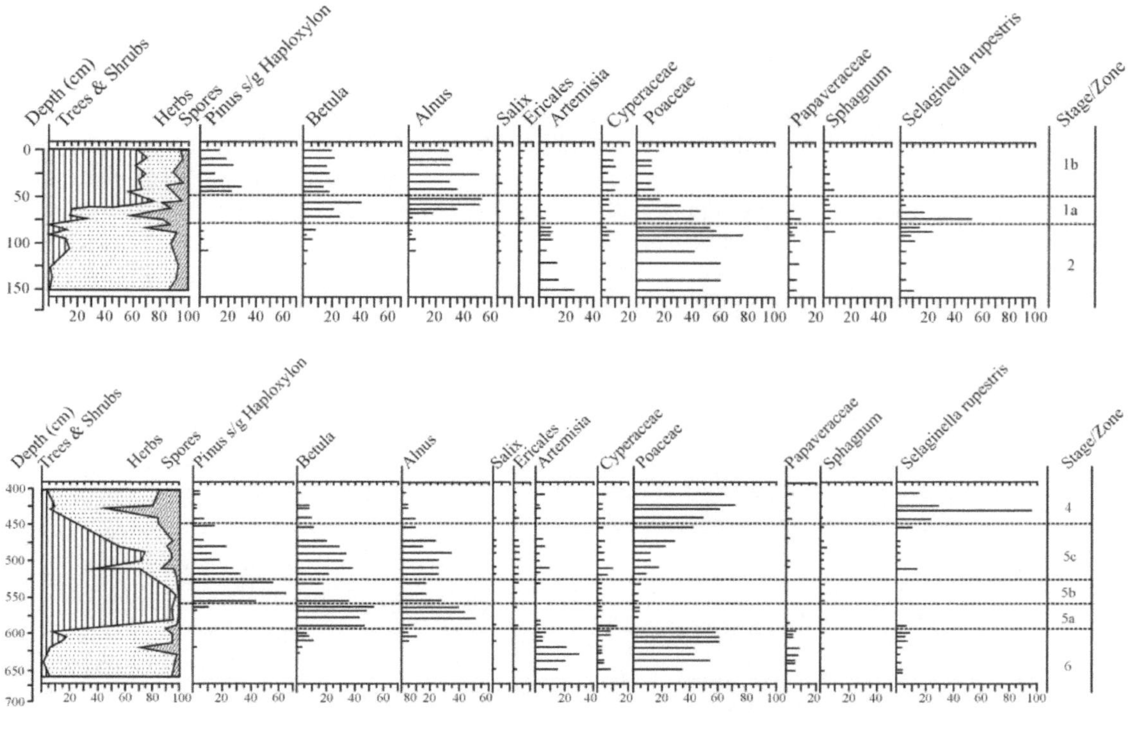

Figure 2. Percentage diagram of select taxa from MIS1 and MIS2 and MIS4 to MIS6. Results are from core PG-1352. Note that labels for the pollen assemblages include both a number, indicating the isotope stage, and a letter, indicative of the pollen zone. Climatic optima are represented by zones 1a and 5a.

INTERGLACIAL VEGETATION OF LAKE E AND THE CHUKCHI UPLANDS

While interglaciations share similarities in boundary conditions (e.g., high summer insolation, warm sea surface temperatures, minimal terrestrial ice cover), variations in the climatic forcings and strengths of feedback mechanisms add a uniqueness to each warm interval (see Melles et al., 2012). The following sections describe the variability of vegetation responses to climates of the PGTM, MIS5, MIS11, and MIS31.

PGTM

High percentages of Betula and Alnus characterize the late glacial and early Holocene pollen assemblages at Lake E (Fig. 2; pollen zone 1a), an assemblage that is common across northern sectors of the Russian Far East (Lozhkin et al., 1993). Statistical comparisons to modern pollen and related climatic parameters reveal mean July temperatures were perhaps 2–4°C warmer than present between ~ 8600 and 10 700 14C yr BP (Lozhkin et al., 2007). Analog analysis also indicates warm January temperatures and increased July precipitation as compared to modern. These conclusions are based on "possible pollen analogs" to southern Beringia shrub tundra (Ibid.). Macrofossil evidence from the Yana– Indigirka–Kolyma lowlands (Fig. 1) and northern Chukotka documents the occurrence of tree or tree-sized shrubs of Salix and Alnus beyond present-day tree line (Lozhkin, 1993). Using both pollen and plant macrofossils to enhance understanding of plant functional types, Edwards et al. (2005) concluded that an unusual biome of deciduous forest and forest-tundra characterized Beringia during the PGTM. In the Chukchi Uplands, vegetation was probably a mosaic of forest and tundra communities including gallery forests of tree Betula and perhaps Salix, and possibly stands of deciduous trees in protected sites within the interfluves. Other areas likely supported high shrub tundra.

MIS5

Because MIS5 vegetation has been described elsewhere (Lozhkin et al., 2007), we focus here on MIS5.5. Like the PGTM, the MIS5 climatic optimum (Fig. 2; pollen zone 5a) is characterized by high percentages of Betula and Alnus pollen. The dominance of these woody taxa suggests that the regional vegetation during MIS5.5 was also a mix of deciduous forest and high shrub tundra. The reduced herb percentages as compared to the PGTM indicate a more extensive and/or denser population of Betula, Alnus, and possibly Salix in the regional vegetation. The spectra may also indicate a greater establishment of trees/ high shrubs within the Lake E catchment as compared to the PGTM. Plant macrofossils of last interglacial age have been found throughout northern sectors of the Far East (Lozhkin and Anderson, 1995). Although chronological control is insuf- ficient to assign sites specifically to MIS5.5, current understanding of the valley stratigraphies suggests they are from MIS5. The reconstructions vary from open Betula forest (Anadyr basin; Fig. 1), to Larix forest with tree Betula, tree Alnus, and Pinus pumila (northern Chukotka, upper Kolyma and Indigirka basins, Yana–Indigirka–Kolyma lowlands) to Larix forest with Picea (upper Kolyma and Indigirka basins, northern lowlands, northern Okhotsk sea coast). Although Picea may have extended almost to the coast of the East Siberian Sea, the Lake E record raises the possibility that these sites may actually be from MIS11 (see below). Nonetheless, the preponderance of woody remains and nature of the MIS5.5 pollen spectra at Lake E strongly indicate the extensive presence of forest in northern areas of the Russian Far East and likely establishment of deciduous forest in the Chukchi Uplands. Analog-based analyses of the pollen data indicate the prevalence of warmer and wetter conditions as compared to modern (Lozhkin et al., 2007; Melles et al., 2012).

MIS11

Three vegetation types characterize MIS11 (Fig. 3), and all types indicate greater summer warmth and annual precipitation as compared to modern (Melles et al., 2012). However, the history of MIS11 diverges dramatically from the vegetation succession seen in MIS5 and MIS1 (i.e., Betula–Alnus– Pinus pumila).

The earliest climatic amelioration in MIS11 is indicated by high percentages of Betula and decreasing percentages of Poaceae pollen (pollen zone 11a). In contrast to the MIS6/MIS5 and MIS2/MIS1 transitions, Betula-dominated vegetation appears to persist for a longer time in MIS11 prior to the arrival of Alnus. Lower zone 11a, representing the earliest effects of warming, was likely characterized by Betula shrub tundra with Salix and Ericales. Possibly shrubs achieved large growth forms as occurred in the PGTM. Betula pollen increases to 45–60 % in the mid-to-upper levels of zone 11a, percentages that generally exceed modern values from shrub tundra environments in the northern Far East (Anderson et al., 2002; Lozhkin et al., 2002). These values more closely approximate modern surface samples from Kamchatka (Lozhkin and Anderson, unpublished data) where forests of Betula ermanii are common. In Alaska, highest Betula percentages (> 40 %) also occur in interior forests, where Betula papyrifera is found in valleys and lower mountain slopes (Anderson and Brubaker, 1993). Inasmuch as we can draw from modern analogs, these data suggest that Betula shrub tundra probably was replaced by Betula woodland. The presence of Larix pollen, even in only a single sample, further suggests the establishment of deciduous woodland. The decreasing percentages of graminoids and Ericales and increases in Betula pollen likely indicate a greater presence of wooded landscapes by upper zone 11a.

Zone 11b, representing the climatic optimum and corresponding to MIS11.3, includes some variation within the pollen assemblage. The pollen spectra between 1855–1905 cm (1) are dominated by Picea sect. Eupicea, Alnus, and Betula; (2) show a consistent appearance of Larix; and (3) have decreased percentages of herb taxa, particularly Poaceae. Pinus subg. Haploxylon pollen appears initially in low amounts but increases in later spectra (1827–1855 cm). This pollen type may represent either Pinus pumila or Pinus sibirica (see below). The increase in Pinus pollen is associated with a slight decline in Betula and Alnus pollen.

The abrupt increase in Picea pollen at the zone 11a–11b boundary suggests a rapid replacement of deciduous woodland by Picea obovata–Larix–Betula–Alnus forest. The expansion of Picea obovata to Lake E represents a range extension northeastward of ~ 2000 km (Kolosova, 1980). Chosenia macrolepis possibly was present in this forest (see below), but its pollen can not be differentiated from shrub Salix. Herb pollen is virtually absent, which is characteristic of modern spectra from closed dark coniferous forest. Although other coniferous taxa (e.g., Picea sect. Omorica, Pinus subg. Diploxylon) and Corylus appear, these grains likely represent long-distance transport, and the pollen is not indicative of a local presence. Shrub communities of Betula, Alnus, and Salix grew along streams and on moist sites near the lake shore. Shrub tundra perhaps occupied higher elevations, but the regional vegetation was dominated by dark coniferous forest.

The inferred vegetation for early MIS11.3 (1855– 1905 cm) is reminiscent of that found today in the middle reaches of the Yama and Maimandzha rivers, which are located ~ 150 km to the east of Magadan (Fig. 1) and thought to be a relict forest from a previous warm interval (Rheutt, 1970). Its plant communities contrast to the regional vegetation of light coniferous forest dominated by Larix gmelinii and Pinus pumila. In the Yama–Maimandzha lowlands, the forest is a mix of Picea obovata, Larix, tree Alnus (Alnus hirsuta), and tree Betula (B. platyphylla, B. lanata). Populus suaveolens and Chosenia macrolepis are limited to floodplains. Picea grows to 20+ m heights, and Larix and Alnus hirsuta have diameters of 30–50 cm (Lozhkin, personal observation). The percentages of Pinus pollen occurring between 1855 and 1905 cm are sufficiently low to suggest Pinus shrubs or trees were likely absent or at best minor components of the ancient vegetation. This pattern is consistent with the modern Yama–Maimandzha Picea forest, where Pinus sibirica is absent and Pinus pumila is restricted to other areas of the drainage

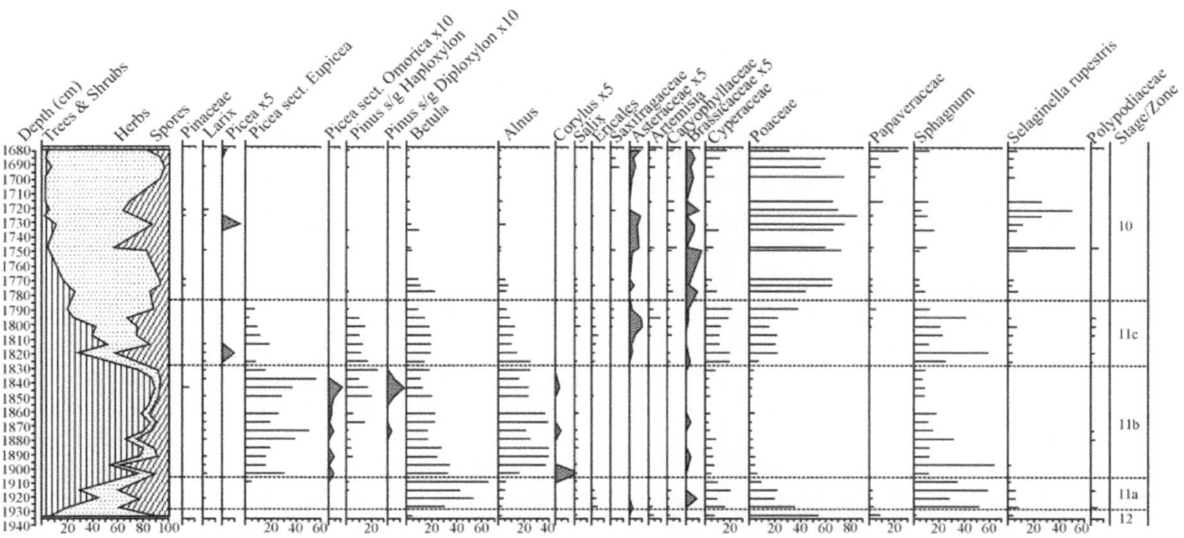

Figure 3. Percentage diagram of select taxa from MIS10 to MIS12. The climatic optimum is represented by zone 11b. Results are from core ICDP 5011.

Dark coniferous forest continues to dominate the landscape during later MIS11.3 (1827–1855 cm). Pollen percentages indicate that Picea obovata, Alnus hirsuta, and Betula remain important elements in the forest communities, although the abundance of the latter two species perhaps decreased from earlier in the zone. The most significant shift in the pollen assemblage of late MIS11.3 is the rise in Pinus subg. Haploxylon pollen. As mentioned above, this pollen type represents both Pinus pumila and Pinus sibirica. Modern range distributions of Picea obovata and Pinus pumila do not overlap in the northern Far East (Pinus pumila only) or to the west of the Lena drainage (Picea obovata only; Kolosova, 1980). However, the two conifers do co-occur in the middle and upper reaches of the Lena catchment, extending southward to the Amur drainage. Here mean July temperatures are usually ≤ 12 ◦C (Mock, 2002). In contrast, Pinus sibirica is found from southeastern Siberia to northern sectors of the Western Siberian Plain, a distribution it shares with Picea obovata. In some of this region, mean July temperatures can be as high as 20 ◦C. The three species ranges overlap in the area of Lake Baikal, where their distribution reflects altitudinal gradients. Larix–Picea–Pinus forest is found along large flat areas between ~ 1100–1900 m, and Pinus pumila is part of a subalpine tundra zone beyond 1900 m (Suslov, 1961). These modern distributions suggest three alternative paleovegetation interpretations. If the pollen is exclusively from Pinus pumila, its increase suggests that the Picea forest became sufficiently open to allow for the establishment of Pinus shrubs in the understory and/or the shrubs formed a mid-elevation shrub zone beyond altitudinal tree line. As compared to early MIS11, this vegetation change would perhaps suggest a slight summer cooling (to open the forest or lower tree line) and/or increased depth/duration of snow fall (required to protect Pinus pumila from winter desiccation; Khokhyrakov, 1985). However, the persistence of low percentages of herb pollen does not suggest a significant opening of the forests or decrease in Picea cover on mountain slopes. If the pollen represents Pinus sibirica, then the Lake E record marks a range extension of ~ 2500 km (Kolosova, 1980) and an extensive forest cover across northern Chukotka. The addition of Pinus sibirica would not require changes in summer conditions as it shares similar requirements with Picea obovata, raising questions as to why Pinus sibirica was absent in early MIS11.3. If both Pinus species are present, then the regional and/or local vegetation probably had a strong altitudinal component with increased snow fall to aid establishment of Pinus pumila.

Quantitative paleoclimatic reconstructions indicate little systematic change in summer temperatures (ranging from ~ 12–16°C) between early and late MIS11.3 (Melles et al., 2012). The absence of Pinus sibirica during early MIS11.3 probably was not caused by insufficient summer temperatures, especially given the establishment of Picea obovata at the time. Possibly delayed migration from an unknown but distant population may account for the tree's absence. Mean annual precipitation is estimated to be at its highest during late MIS11.3 (> 600 mm yr−1 vs. ~ 350 mm yr−1 during early MIS11.3). If this trend reflects greater seasonal and/or annual snow fall, then this shift in climate would provide a reasonable mechanism for the expansion of Pinus pumila. This conclusion, if correct, would indicate that early MIS11.3 experienced drier cool seasons or seasonal transitions as compared to late MIS11.3. This interpretation would also suggest that Pinus pumila was the main Pinus species present during late MIS11.3.

Pollen zone 11c represents the final stages of MIS11 and is marked by the decrease in Picea pollen, a more gradual decline in Alnus pollen, and increases in herb pollen (primarily from Cyperaceae and Poaceae). Areas of graminoid meadows or tundra were more common than previously suggested by the higher percentages of herb pollen, slightly greater amounts of Ericales and Salix pollen, and to a lesser extent the occurrence of higher Sphagnum values. Trees were limited on both local and regional landscapes. The loss of Larix pollen late in the zone suggests that tundra dominated in the latter part of MIS11.

The forest-tundra of zone 11c perhaps was similar to areas found today in eastern Siberia (e.g., lower Lena River valley), where Picea and Larix form latitudinal tree line with isolated stands of Larix scattered northward across the tundra. Pinus pumila may have persisted from zone 11b, having a similar role as seen in the contemporary vegetation. However, paleoclimatic reconstructions show a sharp decline in mean annual precipitation (Melles et al., 2012), a potentially adverse factor in the survival of Pinus pumila. Decreased summer temperatures as indicated by the quantitative climate estimates would have been sufficiently severe to eliminate Pinus sibirica, if it was present on the zone 11b landscapes.

MIS 31

Of all warm periods discussed in this paper, MIS31 presents the most unusual pollen spectra (Fig. 4). While dominated by typical boreal taxa (Picea, Larix, Abies, Betula, Alnus), the MIS31 assemblages also contain small numbers of more temperate deciduous taxa (e.g., Corylus, Carpinus, Acer, Quercus). Not surprisingly given the 1.1 Ma age, these assemblages seem to lack exact analogs in the modern landscapes. Although the pollen diagram is somewhat skeletal, the data are sufficient to determine general characteristics of the MIS31 vegetation.

The MIS32/MIS31 transition (pollen zone 31a) is similar to that of MIS6/MIS5 (pollen zone 5a) with spectra dominated by Betula and Alnus pollen and with very low numbers of herb types. However, differences do exist between the two transitions, including the minor presence of Larix pollen and occasional occurrence of the warm taxa Corylus, Carpinus, and Myrica in zone 31a. As during pollen zone 5a, the regional vegetation during early MIS31 probably was dominated by deciduous trees possibly with tree-sized shrubs including Betula, Alnus, and Salix. Unlike MIS5, Larix was established in the Lake E watershed and likely across the upland. The low percentages of herb pollen suggest that a relatively dense deciduous forest rather than open woodland was established at this time. Although pollen from other deciduous trees is the result of long-distance wind transport, occurrence of these taxa suggests that temperate forests were in closer proximity to north-central Chukotka as compared with today.

The zone 31b assemblage represents the thermal maximum within MIS31 and is characterized by several Picea and Pinus species, Larix, Abies, Betula, Alnus, and the Siberian dark coniferous taiga and Mongolian steppes. In this region, taiga dominates in the mountains and forests between ~ 1000 and 2000 m and includes Larix sibirica, Picea obovata, Abies sibirica, Pinus sibirica, and occasional Pinus sylvestris. A third possible analog is from the western shores of the Sea of Okhotsk (from ~ 51° N southward to the lower Amur valley), where the taiga is a mix of Picea glehnii, Abies nephrolepis, Larix gmelinii, and Betula ermanii. Pinus pumila is limited to protected sites where it can grow in dense thickets achieving over 2 m height (Suslov, 1961). The more temperate taxa Corylus mandshurica, Quercus mongolica, Carpinus cordata, and Acer mandshuricum are present in the Sikhote-Alin Mountains, which border the Amur drainage. Possibly the zone MIS31b vegetation represented a mix of plant communities from the Siberian dark coniferous forest and the Okhotsk taiga, as no modern vegetation type totally corresponds to taxa represented in the pollen assemblage.

The combined percentages of Betula and Alnus pollen, as in zone 31a, dominate the MIS31b spectra and demonstrate that deciduous forest remained an important landscape component. The presence of Corylus in two samples in minor but significant amounts (2–3 %) suggests that the plant was probably present within the region. Today the nearest Corylus population (Corylus mandshurica) is found in the lower reaches of the Amur drainage, and its establishment near Lake E would represent a ~ 2400 km range extension (Kolosova, 1980). Inasmuch as we can draw on the Okhotsk taiga as analog, it is likely that during zone 31b the vegetation in the Lake E crater and neighboring uplands was characterized by a marked altitudinal zonation. Betula–Larix– Alnus forest likely dominated river valleys and lower slopes. Possibly the Salix pollen represents Chosenia macrolepis, a common floodplain tree in modern landscapes. Transitioning to mountain slopes, deciduous forest was replaced by conifer communities of Picea, Pinus, and Abies. Betula ermanii is common today in Picea–Abies forest, sometimes forming pure stands. This Betula species occupies rockier and steeper slopes. Corylus was likely an understory shrub that grew on warmer, south-facing slopes. Other temperate taxa (e.g., Acer, Carpinus), if present, would be restricted to the warmest sites. Pinus sylvestris, if the Pinus subg. Diploxylon pollen is not the result of long-distance transport, was not abundant. As in the case of MIS11, the Pinus subg. Haploxylon pollen may be interpreted in two ways. If from Pinus pumila, the species likely grew in protected, open areas within the forest and possibly formed high shrub tundra beyond altitudinal tree line. As in zone 11b, low percentages of herb taxa suggest closed forests that extended far up mountain slopes, making the occurrence of a Pinus pumila belt less likely. If the pollen represents Pinus sibirica, then the taxon would be a member of the coniferous forest.

Within pollen zones 31c and 31d, the pollen spectra indicate steady vegetation changes implying gradual climatic cooling and increasing aridity. Pollen from evergreen conifers decreases, whereas Betula and Alnus continue as dominant taxa with a consistent appearance of Larix pollen. Deciduous forest probably became the most extensive vegetation type both locally and regionally. The increase in Ericales, Poaceae, and total herb pollen and Sphagnum spores indicates an opening of the forest and establishment of tundra communities by late zone 31c and zone 31d.

DISCUSSION AND CONCLUSIONS

The palynological data presented here and in previous papers (Lozhkin et al., 2007; Matrosova, 2009) are in many senses a preview of the exciting paleoenvironmental information contained within the Lake E archive. For example, a focus on only a small part of the vegetation history has revealed interglaciations with an unexpected magnitude of warming (Melles et al., 2012). The pollen data described herein further indicate that there are at least two types of vegetation responses to interglacial conditions. The vegetation in the first type of thermal optimum is dominated by deciduous tree and shrub species and corresponds to the two most recent warming events (PGTM, MIS5.5): the second type, or "super" interglaciation (MIS11.3, MIS31), indicates climax vegetation dominated by evergreen conifers, particularly Picea. These exceptional climatic events experienced warmer summers (+4.5 ◦C) and greater annual precipitation (~ 300 mm) as compared to the PGTM and MIS5.5 (Melles et al., 2012), which themselves are characterized by summers that are warmer and wetter than present (increases of ~ 2 ◦C and up to 20 mm for mean July temperature and precipitation, respectively; Lozhkin et al., 2007).

Despite the classification into types, the interglaciations share some commonalities. One characteristic of interglacial intervals at Lake E is that they show greater heterogeneity in vegetation types as compared to glaciations. These latter periods display a general uniformity in their pollen assemblages both through time as in the Lake E record (Lozhkin et al., 2007) and also in space, at least in the cases of the latest glaciation when the most paleobotanical data are available (Brigham-Grette et al., 2004). This observation implies that once a "threshold" of coolness and/or dryness was surpassed, the vegetation response during glaciations is limited to herbdominated tundra or steppe tundra. Of course, the establishment of climatic "thresholds" affected interglacial as well as glacial vegetation. Although a level of summer warmth is required for the expansion of boreal tree species, modeling by Kaplan et al. (2003) illustrated the importance of relatively mild winters for the spread of Picea. The Lake E data suggest that once temperature "thresholds" have been achieved, the importance of seasonal precipitation (particularly the duration and volume of snow cover) and/or effective moisture may have been the greatest determinant of interglacial forest composition and distribution.

Another shared trait of the interglaciations is the rapid establishment of plant communities dominated by woody taxa during the glacial to interglacial transition. The earliest interglacial vegetation was Betula shrub tundra (note that this is not evident in early MIS31, but we expect additional pollen counts will show a parallel trend to the other interglaciations) that quickly was replaced by deciduous forest or woodland. For the PGTM and MIS5, the establishment of Betula woodland/high shrub tundra and Betula–Larix forest/high shrub tundra during the PGTM and MIS5, respectively, represents maximum summer warming. However, during the older warm periods, Betula–Larix (MIS11) or Betula–Alnus– Larix (MIS31) forests were transitional as interglacial climates continued to become warmer and wetter. Edwards et al. (2005) noted that the widespread establishment of deciduous forest in Beringia during the PGTM represented a novel biome that differed structurally and functionally from evergreen coniferous forest and tundra that dominate arctoboreal settings today. They also commented that deciduous boreal forest should be included in modeling efforts that explore likely biosphere–atmosphere feedbacks and possible impacts caused by global warming. The new Lake E data underscore the importance of this proposition, as deciduous forest seems to play an essential role in interglacial vegetation history, whether it was as transitional or as climax vegetation.

The different interglacial vs. "super" interglacial modes suggest the need to explore the potential impact of a range of forcing mechanisms and feedbacks involved in "warm" and "warmer" scenarios. Melles et al. (2012), for example, suggested that the melting of the West Antarctic Ice Sheet, with consequent changes in ocean and atmospheric circulation, was possibly a major factor contributing to the magnitude of temperature and precipitation increases reconstructed for the "super" interglaciations of MIS11 and MIS31. The cryosphere and biosphere perhaps exerted additional influences as interactions among sea ice and snow albedo, and forest extent can provide a further positive feedback to climate change (Foley et al., 1994). Moreover, modeling and modern observations indicate the magnitude and direction of change associated with biosphere and atmosphere feedbacks are sensitive to the composition and distribution of plant communities. For example, positive feedbacks to high latitude warming, including such factors as increasing sensible heat flux and reducing albedo (e.g., Foley et al., 1994; Levis et al., 1999; Eugster et al., 2000; Chapin et al., 2000), would be enhanced by the shift from deciduous to evergreen conifer forest. If Lake E is indicative of vegetation responses across northern high latitudes during these "warmer" intervals, then Arctic plant communities during "super" interglaciations constricted to such an extent that they no longer existed as a major biome with consequent feedbacks related to shifts in albedo, sensible heat flux, surface roughness, etc. Such an effect was noted during MIS5 when a 5–10 % decrease in albedo was estimated for northeastern Asia in response to the northern expansion of forest (Velichko, 2009). In summary, the Lake E record provides a set of natural experiments within which scientists might gage the range of vegetation responses, the degree to which the distribution and composition of arcto-boreal communities may alter, and the scope of expansion or contraction of contemporary biomes as means to further improve predictions of possible biotic responses in Arctic regions to projected global warming. The palynological data from Lake E highlight the merits of existing models that predict the northward expansion of either conifer or deciduous forests under warmer-thanpresent climate scenarios. However, the Lake E record also demonstrates that idiosyncrasies of climatic forcings and associated atmosphere–biosphere–cryosphere feedbacks may cause more complex vegetation responses than anticipated by current modeling efforts.

ACKNOWLEDGEMENTS

Funding for investigations at Lake El'gygytgyn was provided by the International Continental Scientific Drilling Program (ICDP), the US National Science Foundation (NSF), the German Federal Ministry of Education and Research (BMBF), Alfred Wegener Institute (AWI) and GeoForschungsZentrum Potsdam (GFZ), the Russian Academy of Sciences Far East Branch (RAS FEB), the Russian Foundation for Basic Research (RFBR), and the Austrian Federal Ministry of Science and Research (BMWF). The Russian GLAD 800 drilling system used for sediment retrieval was developed and operated by DOSECC Inc. The down-hole logging was performed by the ICDP-OSG. Core curation was provided by LacCore, University of Minnesota. The palynological work specifically was supported by the RAS FEB (12-III-A-09-198, 12-II-CO-08-024), RFBR (12-05-00286a), and the Civilian Research and Development Foundation (RUG1-2987-MA-10). We thank E. F. Gorodnichaya, Ju. A. Korzun, T. B. Solomatkina, and E. D. Takmazyan for their great help in the analysis and preparation of the data. Comments by two unnamed referees were greatly appreciated and helped strengthen the original manuscript.

REFERENCES

1. Anderson, P. and Brubaker, L.: Holocene vegetation and climate histories of Alaska, in: Global Climates Since the Last Glacial Maximum, edited by: Wright Jr., H., Kutzbach, J., Webb III, T., Ruddiman, W., Street-Perrot, F., and Bartlein, P., University of Minnesota Press, Minneapolis, 386–400, 1993.
2. Anderson, P., Lozhkin, A., Belaya, B., and Stetsenko, T.: Modern spore-pollen spectra from the mountain regions of the Kolyma and Indigirka rivers from lacustrine sediments, in: Quaternary Paleogeography of Beringia, Simakov, K., NEISRI FEB RAS, Magadan, Russia, 28–39, 2002 (in Russian).
3. Bonan, G., Shugart, H., and Urban, D.: The sensitivity of some high-latitude boreal forests to climatic parameters, Clim. Change, 16, 9–29, 1990.
4. Brigham-Grette, J., Lozhkin, A., Anderson, P., and Glushkova, O.: Paleoenvironmental conditions in western Beringia before and during the last glacial maximum, in: Entering America Northeast Asia and Beringia before the Last Glacial Maximum, edited by: Madsen, D., The University of Utah Press, Salt Lake City, UT, USA, 29–62, 2004.
5. Brigham-Grette, J., Melles, M., and Minyuk, P. (Eds.): Arctic Record of the Last 250 ka from El'gygytgyn Crater Lake, NE Russia, Special Issue, J. Paleolimnol., 37, 162 pp., 2007.
6. Chapin III, F. and Starfield, A.: Time lags and novel ecosystems in response to transient climatic change in arctic Alaska, Clim. Change, 35, 449–461, 1997.
7. Chapin III, F., McGuire, A., Randerson, J., Pielke, S., Baldocchi, D., Hobbie, S., Roulet, N., Eugster, W., Kasischke, E., Rastetter, E., Zimov, S., and Running, S.: Arctic and boreal ecosystems of western North America as components of the climate system, Global Change Biol., 6 (Supplement 1), 211–223, 2000.
8. Claussen, M.: Variability of global biome patterns as a function of initial and boundary conditions in a climate model, Clim. Dynam., 12, 371–379, 1996.
9. Czerepanov, S.: Vascular Plants of Russia and Adjacent States (the Former USSR), Cambridge University Press, Cambridge, UK, 516 pp., 1995.
10. Edwards, M., Brubaker, L., Lozhkin, A., and Anderson, P.: Structurally novel biomes: a response to past warming in Beringia, Ecology, 86, 1696–1703, 2005.
11. Eugster, W., Rouse, W., Pielke, R., McFadden, J., Baldocchi, D., Kittel, T., Chapin, F., Liston, G., Vidale, P., Vaganov, E., and Chambers, S.: Land-atmosphere energy exchange in arctic tundra and boreal forest: available data and feedbacks to climate, Global Change Biol., 6, Supplement 1, 84–115, 2000.
12. Foley, J., Kutzbach, J., Coe, M., and Levis, S.: Feedbacks between climate and boreal forests during the Holocene epoch, Nature, 371, 52–52, 1994.
13. Harrison, S., Kutzbach, J., Prentice, I., Behling, P., and Sykes, M.: The response of northern hemisphere extratropical climate and vegetation to orbitally induced changes in insolation during the last interglaciation, Quaternary Res., 43, 174–184, 1995.
14. Kaplan, J., Bigelow, N., Prentice, I., Harrison, S., Bartlein, P., Christensen, T., Cramer, W., Matveyeva, N., McGuire, A., Murray, D., Razzhivin, V., Smth, B., Walker, D., Anderson, P., Andreev, A., Brubaker, L., Edwards, M., and Lozhkin, A.: Climate change and arctic ecosystems: 2. Modeling, paleodatamodel comparisons, and future projections, J. Geophys. Res., 108, 8171, doi:10.1029/2002JD002559, 2003.
15. Khokhyrakov, A.: Flora of Magadan Oblast, IBPB, FEB, USSR Academy of Sciences, Vladivostok, Russia, 48 pp., 1985 (in Russian).
16. Kolosova, L. (Ed.): Geographic Atlas, Central Office of Geodesy and Cartography of the Soviet Ministry of the USSR, 4th Edn., Moscow, Russia, 1980 (in Russian).
17. Kozhevnikov, Y.: Vascular plants near El'gygytgyn Lake, in, Natural Depression El'gygytgyn Lake (Problems of Study and Protection), NEISRI, FEB, RAS, Magadan, Russia, 62–82, 1993 (in Russian).
18. Layer, P.: Argon-40/argon-39 age of the El'gygytgyn impact event, Chukotka, Russia, Meteroitics and Planetary Science, 35, 591– 599, 2000.
19. Levis, S., Foley, J., and Pollard, D.: Potential high-latitude vegetation feedbacks on CO2-induced climate change, Geophys. Res. Lett., 26, 747–750, 1999.
20. Lozhkin, A.: Geochronology of late Quaternary events in northeastern Russia, Radiocarbon, 35, 429–433, 1993.
21. Lozhkin, A. and Anderson, P.: The last interglaciation in northeast Siberia, Quaternary Res., 43, 147–158, 1995.
22. Lozhkin, A., Anderson, P., Eisner, W., Ravako, L., Hopkins, D., Brubaker, L., Colinvaux, P., and Miller, M.: Late Quaternary lacustrine pollen records from southwestern Beringia, Quaternary Res., 39, 314–324, 1993.
23. Lozhkin, A., Anderson, P., Vartanyan, S., Brown, T., Belaya, B., and Kotov, A.: Reconstructions of late

Quaternary paleoenvironments and modern pollen data from Wrangel Island (northern Chukotka), Quaternary Sci. Rev., 20, 217–233, 2001.

24. Lozhkin, A., Anderson, P., Belaya, B., and Stetsekno, T.: Reflections on modern pollen rain of Chukotka from bottom lake sediments, in: Quaternary Paleogeography of Beringia, edited by: Simakov, K., NEISRI FEB RAS, Magadan, Russia, 40–50, 2002 (in Russian).

25. Lozhkin, A., Anderson, P., Matrosova, T., and Minyuk, P.: The pollen record from El'gygytgyn Lake: implications for vegetation and climate histories of northern Chukotka since the late middle Pleistocene, J. Paleolimnol., 37, 135–153, 2007.

26. Matrosova, T.: Reconstruction of vegetation and climate in northern Chukotka during the last 350 thousand years (according to palynological evidence from El'gygytgyn Lake), Vesnik, 1, 23–30, 2009 (in Russian).

27. Melles, M., Brigham-Grette, J., Minyuk, P., Nowaczyk, N., Wennirch, V., DeConto, R., Anderson, P., Andreev, A., Coletti, A., Cook, T., Haltia-Hovi, E., Kukkonen, M., Lozhkin, A., Rosen, P., Tarasov, P., Vogel, H., and Wagner, B.: 2.8 million years of arctic climate change from Lake El'gygytgyn, NE Russia, Science, 337, 315–320, 2012.

28. Mock, C. J.: Regional climates of Russia, in: Late Quaternary Vegetation and Climate of Siberia and the Russian Far East (Palynological and Radiocarbon Database), edited by: Anderson, P. and Lozhkin, A., NOAA and Russian Academy of Science, Far East Branch, Magadan, 17–26, 2002.

29. PALE Steering Committee: Research Protocols for PALE: Paleoclimatic of Arctic Lakes and Estuaries, PAGES Workshop Report Series 94-1, PAGES Core Project Office, Bern, Switzerland, 53 pp., 1994.

30. Overpeck, J., Prentice, I., and Webb III, T.: Quantitative interpretation of fossil pollen spectra: dissimilarity coefficients and the method of modern analogs, Quaternary Res., 23, 87–108, 1985.

31. Reutt, A.: Vegetation. In, North of the Far East, edited by: Shilo, N., Nauka, Moscow, Russia, 257–299, 1970 (in Russian). Suslov, S.: Physical Geography of Asiatic Russia, W.H. Freeman, London, UK, 594 pp., 1961.

32. Velichko, A.: Paleoclimates and Paleoenvironments of ExtraTropical Area of the Northern Hemisphere Late PleistoceneHolocene, GEOS, Moscow, Russia, 120 pp., 2009.

CHAPTER 10

GRASSLAND VEGETATION CHANGES AND NOCTURNAL GLOBAL WARMING

Richard D. Alward[1], James K. Detling[2] and Daniel G. Milchunas[3]

[1]University of Nebraska–Lincoln

[2]Colorado State University - Fort Collins

[3]Colorado State University - Fort Collins

ABSTRACT

Global minimum temperatures (T_{MIN}) are increasing faster than maximum temperatures, but the ecological consequences of this are largely unexplored. Long-term data sets from the shortgrass steppe were used to identify correlations between T_{MIN} and several vegetation variables. This ecosystem is potentially sensitive to increases in T_{MIN}. Most notably, increased spring T_{MIN} was correlated with decreased net primary production by the dominant C4 grass (Bouteloua gracilis) and with increased abundance and production by exotic and native C3 forbs. Reductions in B. gracilis may make this system more vulnerable to invasion by exotic species and less tolerant of drought and grazing.

There is general consensus that there is an anthropogenic warming signal in the long-term climate record (1). Over land, this is primarily due to average annual minimum temperatures (T_{MIN}) having increased at twice the rate of maximum temperatures (T_{MAX}) (1, 2). At the global scale, these increases in T_{MIN} are related to increases in global cloudiness (1, 3). Experiments with agricultural plants and insect pests suggest important roles for T_{MIN} in influencing plant and insect development (4, 5). However, there has been little research on the consequences of elevated T_{MIN} for natural ecosystems (6, 7). If elevated T_{MIN} leads to longer growing seasons, net primary production and carbon sequestration may increase as a consequence (8). However, the opposite may occur if elevated T_{MIN} leads to increased plant and microbial nocturnal respiration rates without a compensatory increase in photosynthesis. Additionally, elevated T_{MIN} could shift competitive interactions among C3 (cool-season) and C4 (warm-season) plants.

It is important to identify features of ecosystems that are sensitive to changes in T_{MIN}. To date, most modeling efforts and experimental manipulations investigating ecosystem responses to climate change have assumed that future warming will occur primarily during the day or uniformly over the diurnal cycle. This assumption clearly is not valid on a global level nor at most regional scales (2). Furthermore, there is no a priori reason to assume that ecosystems will respond similarly to changes in T_{MIN} and T_{MAX}. To investigate potential ecological consequences of elevated TMIN, we examined a 23-year data set for correlations between temperature [T_{MIN}, T_{MAX}, and mean

Citation: Alward, Richard D.; Detling, James K.; and Milchunas, Daniel G., "Grassland Vegetation Changes and Nocturnal Global Warming" (1999). Faculty Publications in the Biological Sciences. Paper 150. http://digitalcommons.unl.edu/bioscifacpub/150.

annual temperature (T_{AVE}) ($T_{AVE} = (T_{MIN} + T_{MAX})/2$)] and both the abundance and aboveground net primary productivity (ANPP) of several key plant species and functional groups found at the Central Plains Experimental Range (9) in northeastern Colorado.

We identified seasonal and annual trends in T_{MIN} and T_{MAX} to determine whether asymmetric diurnal temperature increases held true for this site (10). The densities of most species were determined by counting all individuals within permanently marked quadrats (11). Harvests at time of peak standing crop were used as estimates of ANPP (12, 13). Plants in the short grass steppe are commonly water limited, and variation in precipitation could obscure plant responses to gradually changing temperatures (9, 14). Therefore, we included annual and seasonal precipitation totals, in addition to annual and seasonal minimum and maximum temperatures, as variables for stepwise regression model selection (15). We constructed linear models to evaluate significant correlations between these variables and ANPP or plant species density (16).

Mean annual temperatures (T_{AVE}) have increased by an average of 0.12°C year-1 at this site since 1964 ($P = 0.0001$, $R^2 = 0.52$). During this period, T_{MAX} increased 0.085°C year-1 (Figure 1A), whereas T_{MIN} increased 0.15°C year-1 (Figure 1B). We limited further analyses of temperature to the period beginning in 1970, when standardized monitoring of vegetation density was initiated. Since 1970, T_{AVE} has risen over 1.3°C, largely due to a significant increase in T_{MIN} of 0.12°C year-1 ($P = 0.003$; $R^2 = 0.44$). However, there was no significant trend for T_{MAX} ($P = 0.49$). Averages of seasonal minimum temperatures since 1970 also exhibited significant warming, with similar trends in winter (0.17°C year-1, $P = 0.0013$, $R^2 = 0.40$), spring (0.16°C year-1, $P = 0.0007$, $R^2 = 0.43$), and summer TMIN (0.12°C year-1, $P = 0.004$, $R^2 = 0.33$). No significant trends were detected in fall T_{MIN} ($P = 0.64$, $R^2 = 0.01$). Annual precipitation (Figure 1C) varied from 230 to 480 mm and has also exhibited a significant linear increase since 1970 (6 mm year-1, $P = 0.007$, $R^2 = 0.30$). However, there were no significant correlations between annual or seasonal T_{MIN} and annual or seasonal precipitation ($P > 0.1$). Since 1983 (12), ANPP of Bouteloua gracilis, the dominant C4 grass of the shortgrass steppe, declined over time (-12.2 g m-2 year-1; $P = 0.002$; $R^2 = 0.78$), and was negatively correlated with average spring T_{MIN} (Figure 2A). ANPP of the most abundant C3 forb, Sphaeralcea coccinea, was negatively correlated with winter T_{MIN} (Figure 2B). In contrast, ANPP of both the C3 sedge Carex eleocharis (Figure 2C) and of all C3 forbs combined (Figure 2D) was positively correlated with fall and summer T_{MIN}, respectively. Plant density was also correlated with T_{MIN}. Exotic forb density was positively correlated with spring TMIN (Figure 2E), whereas the density of the C3 grass Sitanion hystrix was positively correlated with winter T_{MIN} (Figure 2F).

The relationships between T_{MIN} and vegetation revealed by these analyses highlight potential effects of climate change on natural ecosystems. This shortgrass steppe site has experienced increases in TMIN over the past few decades that are similar to trends found by others at larger spatial and temporal scales (2). For each 1°C increase in average spring T_{MIN}, ANPP of the dominant grass decreased by nearly one-third (Figure 2A). Because this one species (B. gracilis) represents an average of 66% of total ANPP and nearly 90% of the total basal cover (9), this result has serious implications for both the structure and function of the shortgrass steppe, if its productivity is causally related to T_{MIN}. Bouteloua gracilis is a drought- and grazing-tolerant species that makes up as much as 40% of the diet of cattle on the shortgrass steppe (17). A major reduction in this species could have substantial consequences for livestock production if it were not replaced by other palatable species. Also of concern is the increase in exotic forb density, because invasive exotic plants are already recognized as a threat to the structure and function of numerous natural ecosystems (18); increasing TMIN may exacerbate this threat.

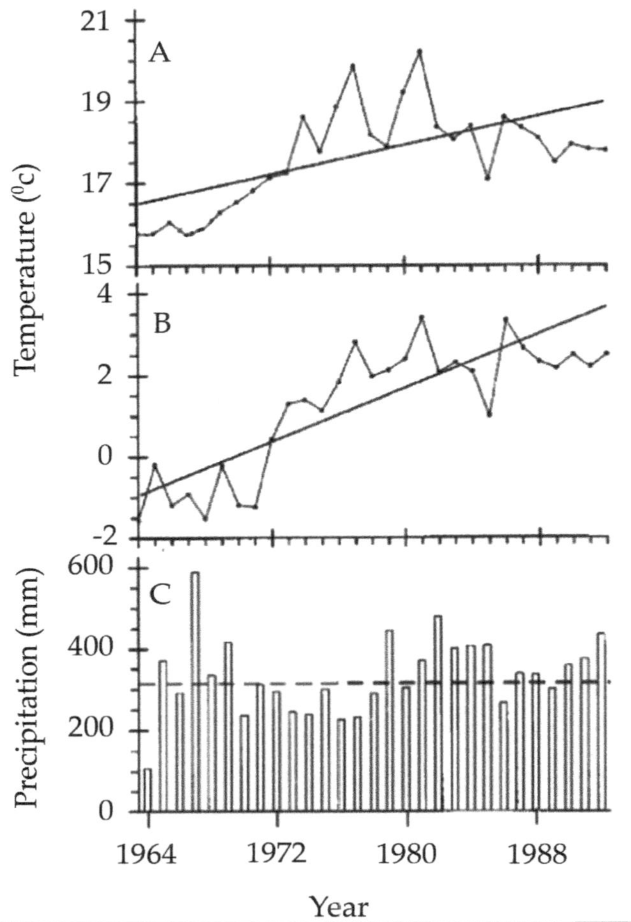

Figure 1. Summary of climate data for the Central Plains Experimental Range site. (A) Average annual T_{MAX}. The heavy line is the significant linear trend in T_{MAX} [$T_{MAX} = -150 + 0.085$ (year); $P = 0.001$; $R^2 = 0.36$]. (B) Average annual T_{MIN}. The heavy line is the significant linear trend in T_{MIN} [$T_{MIN} = -299 + 0.15$ (year); $P = 3.3 \times 10\text{-}8$; $R^2 = 0.68$]. (C) Total annual precipitation. The horizontal dashed line identifies the average annual precipitation (323 mm) at this site since 1939.

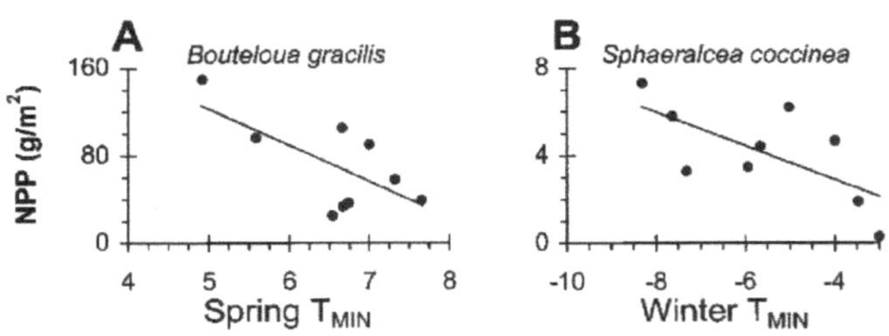

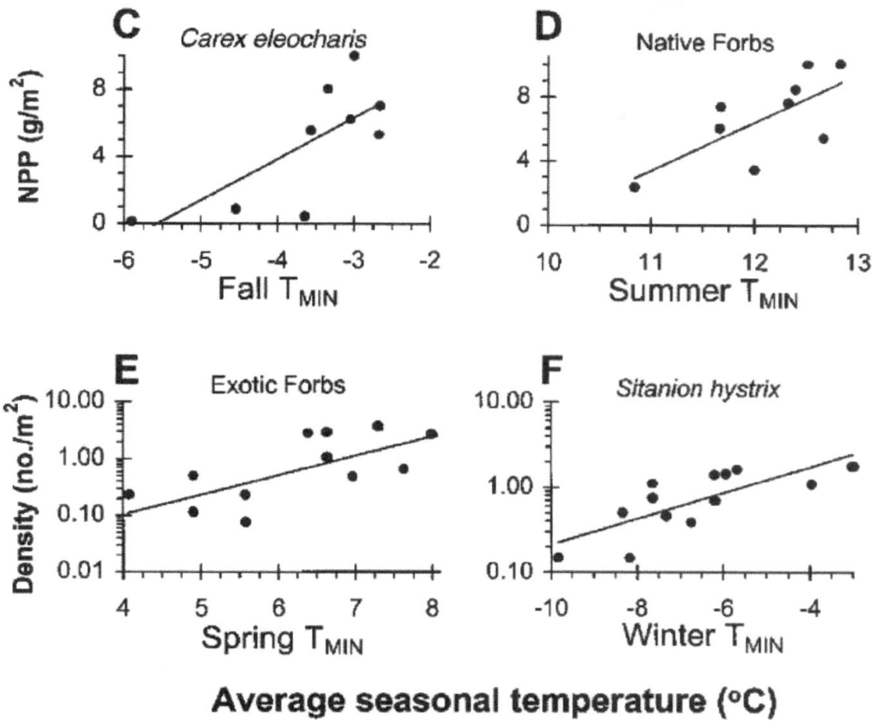

Figure 2. Vegetation correlations with seasonal average T_{MIN}. (A) Bouteloua gracilis and spring T_{MIN} [ANPP = 288 - 33.1 (T_{MIN}); P = 0.039; R^2 = 0.48]. (B) Sphaeralcea coccinea and winter T_{MIN} [ANPP = 0.149 - 0.77 (T_{MIN}); P = 0.038; R^2 = 0.48]. (C) Carex eleocharis and fall T_{MIN} [ANPP = 13.6 + 2.44 (T_{MIN}); P = 0.019; R^2 = 0.56]. (D) Native forb (herbaceous dicots) and summer T_{MIN} [ANPP = -29.5 + 2.99 (T_{MIN}); P = 0.028; R^2 = 0.52]. (E) Exotic (nonnative) forb density and spring T_{MIN} [density = 0.008e0.71 (T_{MIN}) ; P = 0.014; R^2 = 0.46]. (F) Sitanion hystrix density and winter T_{MIN} [density = 6.4e0.33 (T_{MIN}) ; P = 0.002; R^2 = 0.57]. Methods for obtaining density and ANPP data are described in (11, 12).

Elevated T_{MIN} may have direct, but counterbalancing, effects on ANPP and the abundance of plants through mechanisms such as increased rates of carbon assimilation due to warmer mornings, accelerated carbon loss through increased rates of respiration due to warmer nights, and differential effects on C3- versus C4-photosynthesizing plants. Positive correlations between T_{MIN} and both forb ANPP and exotic plant densities (Figure 2, D and E) support the hypothesis that increased production will be observed in some plants, and the negative correlation between T_{MIN} and B. gracilis ANPP (Figure 2A) is consistent with the increased respiration hypothesis. In addition to direct effects on rates of plant physiological processes, increases in T_{MIN} could affect plant growth indirectly through changes in the length of growing seasons through increased duration of the frost-free period or changes in the availability of soil water. An increase in season duration would be expected to primarily benefit cool-season plants that are growing most rapidly, and preemptively consuming resources, early and late in the growing season. The positive correlation between spring TMIN and exotic C3 forbs (Figure 2E) and between fall T_{MIN} and C. eleocharis (Figure 2C) is consistent with this hypothesis.

Some of the correlations could be the result of effects of elevated T_{MIN} on biotic interactions. Some plants may be increasing (Figure 2, C through F) in response to the decrease in B. gracilis ANPP and the consequent increase in availability of space, nutrients, or water, rather than because of any direct effects of elevated TMIN on these plants. Alternatively, if increased T_{MIN} benefits the growth of C3 plants (Figure 2, C through F), this could subsequently result in negative effects on the C4 B. gracilis (Figure 2A). Such a scenario might occur if cool-season plants were able to reduce available soil moisture before the period of rapid growth of warm-season plants. Intertrophic interactions might also be affected. If increased developmental and consumption rates by insects in response to elevated T_{MIN} (4) are common, increased herbivory could alter plantresponses to climate change.

Without a clear causal link, there is no compelling evidence to eliminate factors other than increased T_{MIN} as causes of observed changes in ANPP and plant densities. Unfortunately, most experiments and models designed to investigate climate change effects have focused on manipulating T_{MAX} or have assumed equal contributions by T_{MIN} and T_{MAX} toward achieving an increase in T_{AVE}. The outcomes of such experiments may not realistically predict the future structure and dynamics of ecosystems if climate change continues to be manifested primarily as increases in T_{MIN}. There is a need for experiments that define the relationship between T_{MIN}, plant abundance, and ANPP and that identify mechanisms behind the relationship.

REFERENCES AND NOTES

1. Intergovernmental Panel on Climate Change (IPCC), Climate Change 1995: The Science of Climate Change, J. T. Houghton et al., Eds. (Cambridge Univ. Press, Cambridge, 1995).
2. G. Kukla, T. Karl, M. Riches, Eds., Asymmetric Change of Daily Temperature Range: Proceedings of the International MINIMAX WORKSHOP, College Park, MD, 27 to 30 September 1993 [U.S. Department of

Energy (DOE), Washington, DC, 1994]; T. R. Karl, D. R. Easterling, R. W. Knight, P. Y. Hughes, in Trends '93: A Compendium of Data on Global Change, T. A. Boden, D. P. Kaiser, R. J. Sepanski, F. W. Stoss, Eds. (ORNL/CDIAC-65, Carbon DIoxide Information Analysis Center, Oak Ridge National Laboratory, Oak Ridge, TN, 1994), pp. 686-736; D. R. Easterling, et al., Science 277, 364 (1997) .

3. A. Henderson-Sellars, GeoJournal 27, 255 (1992); A. Dai, A . D. Del Genio and I. Y. Fung, Nature 386, 665 (1997).

4. Y. Yang and N. E. Stamp, Oecologia 104, 225 (1995).

5. M. Seddigh, G. D. Jolliff, J. H. Orf, Crop Sci. 29, 400 (1989); C. A. Zeiher, P. W. Brown, J. C. Silvertooth, N. Matumba, N. Milton, in Cotton: A College of Agriculture Report, J. Silvertooth, Ed. (University of Arizona, Tucson, AZ, 1994), pp. 89-96; P. Manunta and M. B. Kirkham, J. Agron. Crop Sci. 176, 267 (1996) .

6. H. A. Mooney, G. W. Koch, C. B. Field, in Asymmetric Change of Daily Temperature Range, G. Kukla, T. Karl, M. Riches, Eds. (DOE, Washington, DC, 1994), pp. 467-484.

7. R. A. Kerr, Science 255, 682 (1992) ; ibid. 268, 1567 (1995); ibid. 270, 1565 (1995).

8. M. B. Coughenour and D.-X. Chen, Ecol. Appl. 7, 802 (1997); R. B. Myneni, C. D. Keeling, C. J. Tucker, G. Asrar, R. R. Nemani, Nature 386, 698 (1997) .

9. For site descriptions, see D. G. Milchunas, W. K. Lauenroth, P. L. Chapman, M. K. Kazempour, Vegetatio 80, 11 (1989); W. K. Lauenroth and D. G. Milchunas, in Ecosystems of the World, R. T. Coupland, Ed. (Elsevier, Amsterdam, 1991), vol. 8A, chap. 11.

10. Daily TMIN and T_{MAX} were used to calculate monthly averages; monthly data were used to calculate seasonal averages identified as winter (JanuaryMarch), spring (April-June), summer (July-September) and the previous fall (October-December).

11. D. G. Milchunas, W. K. Lauenroth, P. L. Chapman, M. K. Kazempour, J. Veg. Sci. 1, 375 (1990); D. G. Milchunas and W. K. Lauenroth, Ecology 73, 593 (1992); Ecol. Appl. 5, 452 (1995).

12. In August or September of each year, all vegetation within 15 0.25-m2 circular quadrats was harvested, sorted, and dried. We restricted analyses of ANPP to data collected in 1983-1992, because the method used during these years was different from that used before 1983.

13. The following species and functional groups were represented in the analyses: Artemesia frigida, Aristida longiseta, Bouteloua gracilis, Carex eleocharis, Kochia scoparia, Opuntia polyacantha, Sitanion hystrix, and Sphaeralcea coccinea, warm-season grasses, cool-season grasses and sedges, annual grasses, all grasses and sedges combined, forbs (native herbaceous dicots), exotic (nonnative) forbs, dwarf shrubs, and total biomass from all species. Species richness was also analyzed.

14. O. E. Sala, W. J. Parton, L. A. Joyce, W. K. Lauenroth, Ecology 69, 40 (1988); D. G. Milchunas, J. R. Forwood, W. K. Lauenroth, J. Range Manage. 47, 133 (1994) .

15. The timing of vegetation measurements determined appropriate precipitation variables. Plant density was measured in early July. Although in July, August, and September potential evapotranspiration exceeds precipitation, these dry months would not affect July plant density. Thus, in addition to total annual (January-December) precipitation, "seasonal" precipitation totals included "useful" annual precipitation (OctoberJune) and winter (October-March) and spring (April-June) precipitation. However, ANPP was measured in September; thus the seasonal precipitation totals used in the ANPP analyses included total annual (January-December), useful annual (October-September), winter (October-March), spring (AprilJune), and summer (July-September) precipitation.

16. Significant correlations were identified with a stepwise model-building routine [SAS, version 6.11 (SAS Institute, Cary, NC, 1996); PROC REG with STEPWISE method; enter and exit = 0.1]. Plant density values were log-transformed. We report the single factor that explained the most variation in each species' responses.

17. R. W. Rice and M. Vavra, US IBP Grassland Biome Technical Report No. 103 (Colorado State University, Fort Collins, CO, 1971).

18. D. A. Wedin and D. Tilman, Science 274, 1720 (1996) ; A. N. Cohen and J. T. Carlton, ibid. 279, 555 (1998); T. J. Stohlgren et al., Ecol. Monogr., in press.

19. We thank T. J. Minnick for bringing the data sets to our attention and H. W. Polley for comments that improved the manuscript. Support included a NASA Graduate Student Fellowship in Global Change Research to R.D.A., the Shortgrass Steppe Long Term Ecological Research Project (NSF grants DEB- 9632852 and BSR-8114822), a National Park Service Agreement (1268-2-9004, CEGR-R92-0043,174), and NSF grant DEB-9708596. The Central Plains Experimental Range is administered by the U.S. Department of Agriculture's Agricultural Research Service.

SECTION IV

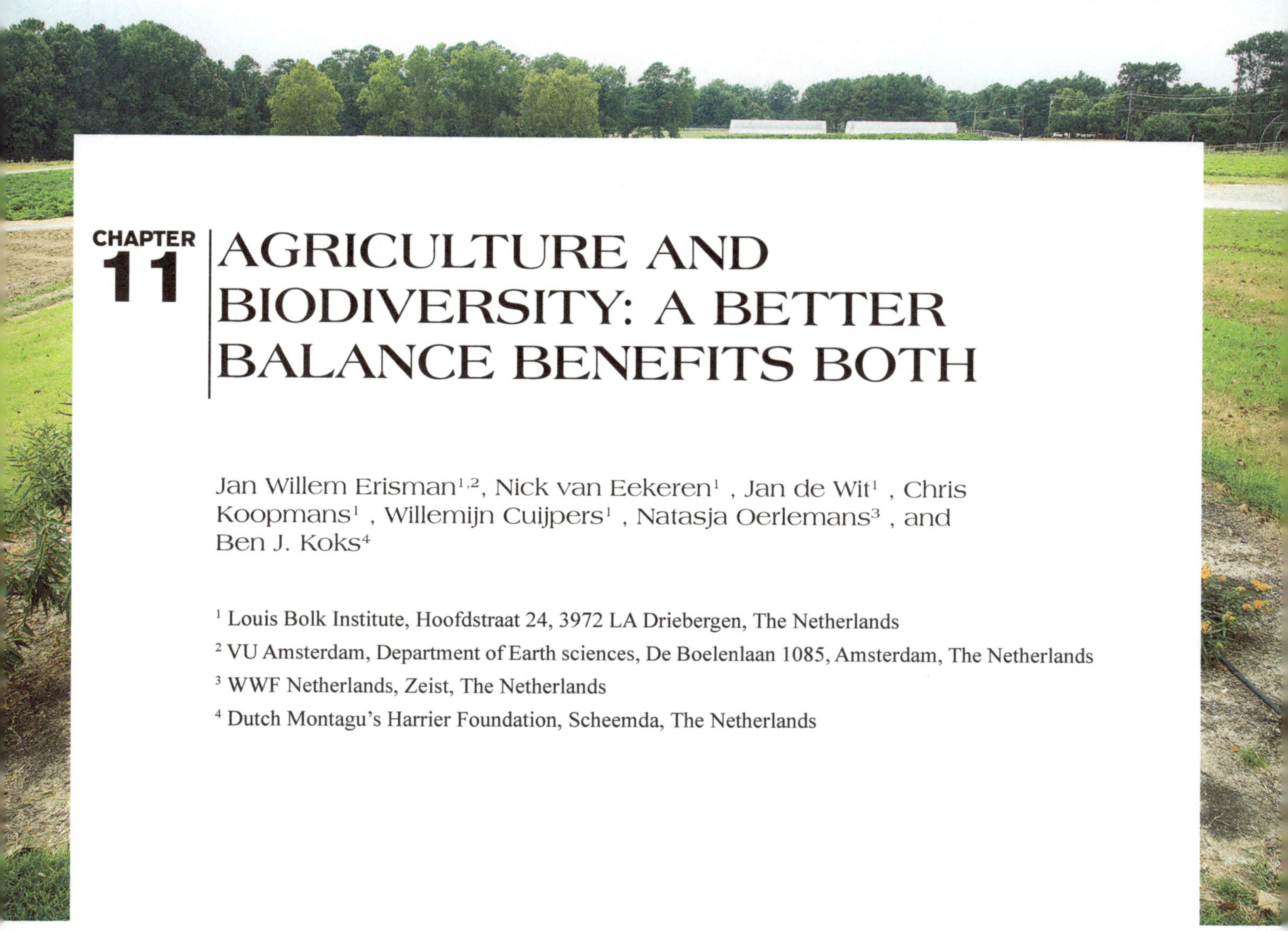

CHAPTER 11

AGRICULTURE AND BIODIVERSITY: A BETTER BALANCE BENEFITS BOTH

Jan Willem Erisman[1,2], Nick van Eekeren[1] , Jan de Wit[1] , Chris Koopmans[1] , Willemijn Cuijpers[1] , Natasja Oerlemans[3] , and Ben J. Koks[4]

[1] Louis Bolk Institute, Hoofdstraat 24, 3972 LA Driebergen, The Netherlands

[2] VU Amsterdam, Department of Earth sciences, De Boelenlaan 1085, Amsterdam, The Netherlands

[3] WWF Netherlands, Zeist, The Netherlands

[4] Dutch Montagu's Harrier Foundation, Scheemda, The Netherlands

ABSTRACT

Sustainable agriculture is an important component of many of the 17 Sustainable Development Goals agreed upon by the UN in 2015 (https://sustainabledevelopment.un.org/sdgs). However, the trend in agriculture is moving in the opposite, non-sustainable direction. Agriculture is one of the major drivers of biodiversity loss. Next to biodiversity loss due to habitat destruction by conversion of natural lands into agriculture, intensification of agriculture has led to a strong decline of specific farmland biodiversity. Furthermore, many agricultural landscapes face pollution by pesticides and fertilizers, and encounter depleted soils and erosion due to unsustainable farming practices. This is threatening not only biodiversity but also complete ecosystems and the ecosystem services on which agriculture itself depends. Moreover, the pressure of feeding an increasing number of people in combination with a change in diets towards more animal protein puts a lot of additional pressure on the current available agricultural lands and nature areas.

We propose a holistic approach that contributes to the development and implementation of sustainable agricultural practices that both make use and support biodiversity and ecosystem services both in agricultural and in semi-natural areas. An agricultural system based on the full potential of (functional agro) biodiversity provides opportunities to create a resilient system in which both food production and nature can thrive.

Citation: Jan Willem Erisman, Nick van Eekeren, Jan de Wit, Chris Koopmans, Willemijn Cuijpers, Natasja Oerlemans, and Ben J. Koks, "Agriculture and biodiversity: a better balance benefits both," AIMS Agriculture and Food, 1(2): 157-174 DOI: 10.3934/agrfood.2016.2.157.

INTRODUCTION

Biodiversity can be described as the richness and diversity of all life on earth. Biodiversity is not just about the individual species, but also about the diversity of ecosystems, species and genes, and the relationship between them [1]. Biodiversity is not only relevant in/for (semi-)natural areas, but also for agricultural areas, which often have specific biodiversity which contributes to ecosystem services. Agriculture in turn can contribute to the increase and conservation of biodiversity, for example by smarter management of marginal land, but also by the management of fertile areas.

Agriculture has different functions of which production of food, feed and fibers and sustaining socio-economic structures and management of ecosystem services are the most important. In doing so, agriculture often makes use of and contributes to the services provided by ecosystems such as healthy soils. About 40% of the world's terrestrial surface has been converted into agriculture [2], Thus, in many parts of the world, humankind has intervened tremendously into the natural succession of ecosystems. In a resilient agricultural system, farming practices provide a good balance between the exploitation and use of biodiversity, ecosystems services and the natural surroundings. In these systems, the challenge is to optimize food production while at the same time minimizing impacts on the environment and the ecosystem.

The notion that agriculture depends on biodiversity and that many specific species of animals and plants depend on sustainable agricultural landscapes is key in the approach of resilient agricultural systems. Both agriculture and nature can benefit from a holistic approach towards resilient systems. This approach focuses on an optimal use of agro-biodiversity and a reduction of long-term (economic and natural) risks by using ecosystem services rather than external inputs.

Agriculture Benefits from Biodiversity

According to Odum (1969) [3] succession is the "increased control of, or homeostasis with, the physical environment in the sense of achieving maximum protection from its perturbations. "... the strategy of maximum protection (that is, trying to achieve maximum support of complex biomass structure) often conflicts with man's goal of maximum production (trying to obtain the highest possible yield)." Agriculture is in ecological sense an early-successional system which is continually out of (homeostatic) balance and therefore makes stability hard to achieve. In a resilient agricultural system there should be a good balance between protection and production. Holling (1973) [4] discusses the difference between resilience and stability in ecological systems. Resilience is defined as the propensity of a system to retain its organizational structure and functions such as productivity following a perturbation. Stability, on the other hand, is the ability of a system to return to an equilibrium state after a temporary disturbance. The more rapidly it returns, and with the least fluctuation, the more stable it is. In this definition stability is the property of the system and the degree of fluctuation around specific states. The goal of producing a maximum sustained yield may result in a more stable system of reduced resilience [4]. Nowadays Holling's definition of resilience is defined as "resistance" and "stability" defined as "resilience". Here, resilience is thus defined as "the ability of a system to return to an equilibrium state after a temporary disturbance". But we go a step further and add to this definition

Agriculture hampers ecosystem succession, as with each cropping cycle, the system is brought back to an early-successional stage. At the same time, agriculture stimulates productivity and manages biodiversity. By converting natural ecosystems into agriculture a new environment was created that allowed for a different biodiversity. A good example is the strong increase in the black-tailed godwit (Limosa limosa) population in the Netherlands when forests and peatbogs where converted into meadows and agricultural grasslands [5].

However, today's agriculture generally aims to produce large quantities of food, against the lowest economic costs, in the short term. These short-term goals often lead to conflict with the conservation and management of biodiversity and other long-term ecosystem services (e.g. [6-8]). Moreover, a short term focus on maximizing productivity can endanger the ecosystem services in such a way that dependence on external inputs such as fertilizers and pesticides increases in order to maintain productivity [9-12]. When out of balance, these systems inevitably collapse or need more technical interventions to remain productive, all at great economic, social and ecological costs (e.g. [11,13,14]).

The importance of agricultural biodiversity (agrobiodiversity) is thus large: living organisms are involved in all natural processes used in agriculture to produce food. If only considering soil, many types of bacteria, fungi, nematodes, springtails, earthworms, etc. are active in the soil to provide ecosystem services e.g. digesting manure, crop residues and roots. This digestion process releases nutrients for crops, and at the same time contributes to the carrying capacity of soils, to disease suppression, water retention and water-supplying capacity of the soil. But also on the level of plant and grassland biodiversity, the diversity of farm animals, the agricultural gene pools, or the on-farm habitat diversity, agriculture basically depends on biodiversity to function well (e.g. [15-18]). Recently more attention has been paid to these services in agriculture (e.g. [19,20]).

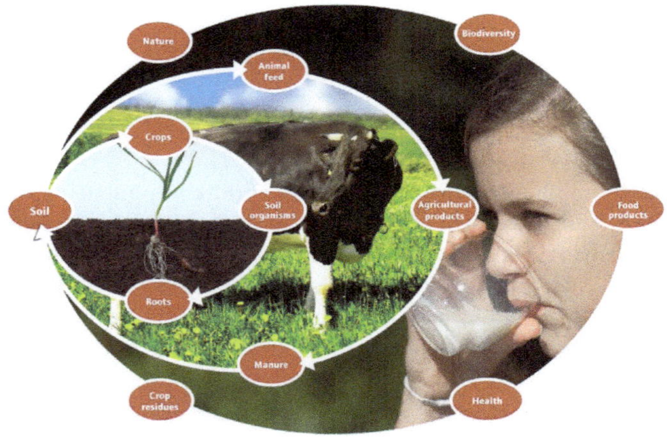

Figure 1. Relationship between sustainable agriculture, nutrition and health. Sustainable agriculture, nutrition and health begins with a healthy soil. The outer cycle represents the main drivers of the agriculture system.

Research shows that the resilience of the (eco)system increases with a larger (functional) biodiversity [21-24]. This also holds for biodiversity in (agro)ecosystems [25,26]. Crops grown in genetically uniform monocultures show a higher vulnerability against diseases, pests and nutrient deficiencies and increase the pressure on environmental resources [27]. When growing mixtures of different crops or varieties, the system becomes more resilient to fungal diseases and pests, yields are stabilized and the sustainability of the system as a whole increases (e.g. [28-30]). Biodiversity increases productivity and biodiversity increases stability of ecosystems. Biodiversity is therefore strongly linked to the resilience of systems and is essential for food security. Maintaining or improving biodiversity means obtaining a true system view and approach. The interaction between the production of food, nature and health is shown in Figure 1: Sustainable agriculture, nutrition and health begins with a healthy soil.

Biodiversity Can Benefit From Agriculture

Besides the direct function of biodiversity in agriculture, (bio) diversity also provides other values. Due to the presence of hedges, ditches, field margins, hedgerows, etc. the cultural aspects of landscape design is preserved, but these elements also form the specific habitat for insects, birds, plants and other animals. This biodiversity has a high cultural and nature value, but can also support agricultural production e.g. nutrition, animal health (as leaves of shrubs and trees contain health-promoting substances) and welfare of livestock (animal behavior and shade), or the provisioning of insects for pollination or biological plague reduction. When aiming for a durable and robust farming system and thus for sustainable agriculture, it is essential to preserve, support, use and promote biodiversity.

There are many examples where agriculture has a positive (and negative) effect of biodiversity. Here we use examples of grassland management. Recent outbreaks of common voles (Microtus arvalis) in the North of the Netherlands (Friesland) caused extensive damage to the grassland and therewith milk production. It was shown that land use changes, management and lowering the groundwater table was the main cause of the outbreaks. Increasing the water table led to better control of the vole population but at the same time increased grassland productivity and the meadow bird population. These unpublished results are confirmed by extensive research on the factors associated with the colonization of agricultural areas by common voles in NW Spain [31]. Increased predator densities over the years have been suggested as an important cause of the trend towards lower nest survival in agricultural grasslands (e.g. [32]). Kentie et al. [33] showed that high-quality habitat in the form of herb-rich meadows provides a degree of protection against predators with higher nest survival on herbrich meadows compared to monocultures. Therefore, increased predator densities are an increased threat during the egg stage only if habitat quality is low. Several studies show that agriculture landscape heterogeneity contribute to the highest diversity of birds [7,34,35].

At present, we propose to implement this principle of biodiversity-based agriculture in dairy systems in the Netherlands as an example to develop a conceptual framework for biodiversity as the basis for sustainable agriculture. This framework in essence can be applied to other agricultural sectors (e.g. arable cropping, horticulture or mixed farming). Here we will discuss the state of biodiversity in Dutch agriculture and characterize the agricultural model. We propose a resilient model as the basis for the conceptual framework for biodiversity and ecosystem services in agriculture. After explanation of the framework we will provide development paths and clusters of management options to successfully improve biodiversity.

AGRICULTURE THREAT TO BIODIVERSITY

Occupying 70% of the land area of the Netherlands, the agricultural landscape is the largest habitat for plants and animals. There are more than 45,000 plants, fungi, insects, animals and other organisms known in the Netherlands[36]. A large number of these species depend on the agricultural landscape as their prime habitat. However, in recent years species for which the Netherlands is extremely important, like the black-tailed godwit (Limosa limosa), lapwing (Vanellus vanellus), partridge (Perdix perdix) and the skylark (Alauda arvensis), have been in decline. Strikingly, the most significant cause of the decline of meadow birds lies in their breeding grounds in the Netherlands, rather than elsewhere along their migratory route [6,8,37-39]. The dairy sector uses 40% of the terrestrial area of the Netherlands. As such the sector has a large impact on the biodiversity in the agricultural landscape. Moreover, the dairy sector is one of the largest contributors to nitrogen deposition in nature areas, which is considered one of the main causes of ongoing biodiversity loss in open nature areas [8,40].

The average number of species of breeding birds and vascular plants (such as flowers and grasses) has increased in semi-natural areas between 1990–2005 compared to 1975–1989 (see Figure 2, left). In contrast, in the same period the figure shows a continuous decline of all species in the agricultural zone (see Figure 2, right). Butterflies are declining in both areas, but at a faster rate in agricultural areas compared to semi-natural areas [8,36].

Not only species on farmland are in decline; agriculture is an important cause of the stagnant recovery of Dutch nature [38,39]. For example, the critical nitrogen load is exceeded in two thirds of the semi-natural ecosystems. The quality of these areas is thus under pressure through agriculture, which is next to burning of fossil fuels the main source of nitrogen deposition [39,41]. Moreover, both nature and agriculture profits from species that live in both areas, such as pollinators, butterflies, etc. Therefore, decline of these species in nature affects agriculture.

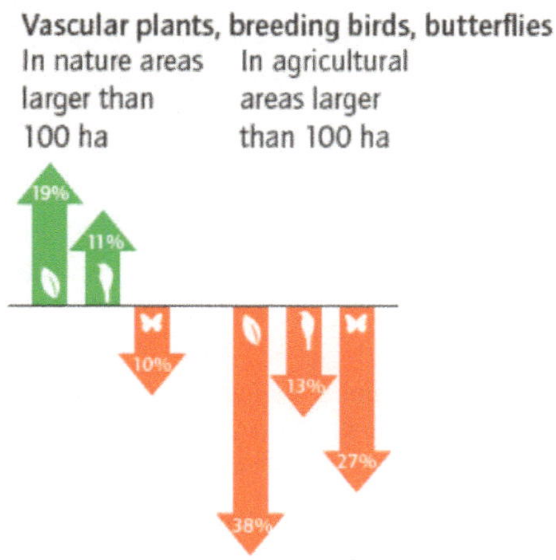

Figure 2. Changes (%) in vascular plants, breeding birds and butterflies in seminatural areas and in agricultural areas as compared between the periods 1990–2005 and 1975–1989 [8,36].

RISK MANAGEMENT MODEL VERSUS RESILIENCE MODEL

In the Netherlands, the loss of biodiversity is mainly due to intensification of agriculture these past decades. The main characteristic of the intensification of agriculture has been the focus on as much production per hectare or per animal as possible resulting in a simplification of ecosystems: mixed farms have switched to specialization, fields have been stripped from hedges and trees, and ditches were straightened as the result of land consolidation policies [42]. Furthermore, the mowing of grassland was intensified and groundwater levels lowered [14,37]. This was made possible by the availability of techniques and tools, such as machinery, fertilizers, pesticides and irrigation. These services that were previously provided by the ecosystem were gradually unconsciously replaced by these tools.

We call this the so-called "control" or risk management model, where the balance is heavily dependent on external inputs such as seed companies and suppliers, processors, supermarkets, price control, etc. and the focus is on short-term revenues and control actions (Figure 3). In this model, natural processes and associated biodiversity have become under severe pressure and are not optimally utilized. As the plant depends on readily available fertilizers, the soil processes for nutrient supply play only a marginal role. Weeds are controlled with herbicides and no use is made of natural pest control. There is a high risk that soil quality decreases in the long term. Because the natural functions are not used, there is a high degree of dependence on external inputs. Price fluctuations make the system very vulnerable when insufficient buffer is built. This can lead to risks such as the long-term loss of production, and thus additional costs. The "control" to keep the system in balance and minimize risk of production loss leads to increased dependency on external resources [41,43,44].

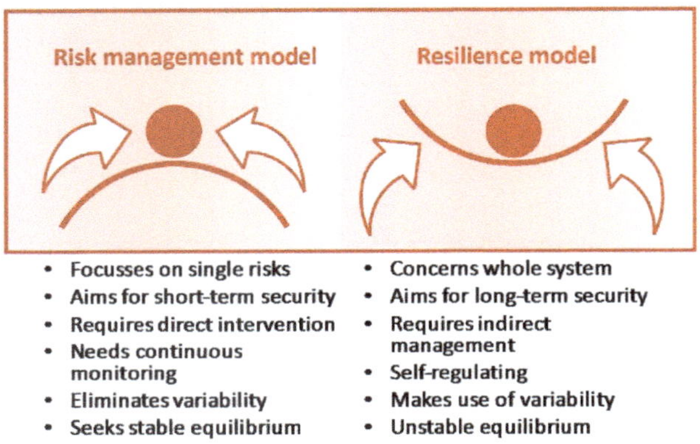

Figure 3. Outline of the current agricultural system that fits into a Risk management model and the characteristics of a resilience model. Robust, resilient systems are less vulnerable [43,44].

This contrasts with the resilience model which optimally uses biodiversity and natural processes in order to increase the resilience and reduce risks [43,45]. By supporting and strengthening the natural processes in agriculture, productive systems can be realized to increase the productivity and stability. Short-term productivity may be equal or lower, but long-term productivity is warranted, through higher resilience and a better ability to cope with disturbances (e.g. [45-47]).

This is a system approach and focuses on the stable equilibrium and usage of the variability as a basis for resilience. As an example, promoting the root system (e.g. by breeding, [48]) increases the resilience of the grassland system. Roots play an important role in reconciling production and environmental goals. By increasing the rooting system, nutrient uptake and yield are improved, while environmental losses (e.g. nitrate leaching) are decreased. It also reduces the risk of the effects of climate change through better water absorption, better soil structure and less erosion. Due to the buildup of organic material more carbon is sequestered, contributing to the mitigation of climate change (e.g. [49]). A resilient system focuses on strengthening the interactions within the system.

The Risk management model has led to a large increase in food production per hectare of farmland, creating a more productive system. Side effects, such as environmental impacts and reduced animal welfare are now increasingly recognized which has led to more focus on sustainability and "precision" agriculture. These side effects have led to additional costs for society of which the estimates by Van Grinsven et al. [50] for nitrogen pollution in Europe are a clear example. The intensification of agriculture has negative effects on soil quality and on the above- and belowground biodiversity. For sustainable production good management of soil and biodiversity is essential. Another characteristic of the risk management model is that it is focused on short term economic revenues and a large dependency of individual farmers on large companies that mainly profit from the farmer's activities. These companies include banks, retail, insurance companies, machinery producers, seed companies, etc., who all have limited incentives to change this system from their own economic interest.

The resilience model is put into practice through e.g. organic farming permaculture, agroforestry, agro-ecology, etc. [7,51]. These are proven systems, but do not deliver the same production levels as conventional farming systems. The difference is on average between 20 to 30% less yield as measured on the short-term [52,53], but there are also production systems that achieve similar levels of production, see e.g., [54]. Furthermore, at the long-term, especially when climate extremes are experienced, these systems appear to be much more resilient. Yield losses are expected to be much smaller than in case of conventional, low-resilience farming systems (e.g. [55-57]). In the resilience model soil quality and biodiversity are essential. A healthy soil can much better cope with external stress such as drought and heavy rainfall, disease, plagues, etc. In this model, therefore the long-term risks are much lower than in the Risk management model. Eventually, on the long term, the lower costs will compensate for the lower production level, yielding a better margin for the farmer and less social (external) costs for society due to lower pollution levels, more carbon sequestration and higher levels of biodiversity.

THE VALUE OF LANDSCAPE ELEMENTS

Landscape elements are essential to support the functional (agro)biodiversity on the farm [7,35,58-60]. If these habitat elements are not present, the overall effect of (agro) biodiversity at farm level remains limited, regardless whether the management focused on the functional biodiversity, according to a recent international study [61]. The importance of these elements for biodiversity and species is illustrated by Kretschmer et al. [62] and Hoffmann et al. [63] who used observations to show that species richness increases with an increasing proportion of landscape elements. Benton et al. [64] show the effect of the disappearance of mosaic structures on the decrease of farmland biodiversity.

There is a large amount of different landscape elements: roadsides, groves, springs, ditches, hedges, sunken roads, orchards, land vegetation, ditches, thickets, ponds, and livestock drinking pools and watercourses [60]. Landscape elements are part of functional agro-biodiversity and may consist of species or ecological communities (e.g. trees, hedges, ditches or land vegetation) that are also part of biodiversity. Landscape elements often also have a cultural and historical value. A large proportion of landscape elements, however, is no guarantee of high biodiversity value. Their quality and the connections made between individual elements, are of great importance. By fragmentation of the area some species cannot thrive.

BIODIVERSITY VERSUS INTENSITY OF A FARM

The intensity of a farm is interpreted in different ways and is not a unique term. Intensity, for example can be related to land-based systems, defined as the amount of food that can be produced on agricultural land in relation to the production of milk/meat without affecting soil quality and within the environmental standards. The intensity also determines how the nutrient cycles are closed at the farm. The intensity is higher when nutrients (or feed) need to be imported from elsewhere.

Allan et al. [65] characterized intensity as a measure composed of: the level of fertilization, the grazing and mowing regime. They show higher loss of farm scale biodiversity at higher levels of fertilization, less grazing and higher cutting regime (intensive agriculture). Fertilization has a big impact on biodiversity, as has been shown by changes in plant species composition (e.g. [40,66]), but also plagues of Common Vole Microtis arvalis are associated with increased fertilization [67]. At higher levels of fertilization the aboveground biodiversity decreases rapidly. Also probably the function of the underground biodiversity is reduced at higher levels of fertilization, although there seems to be an optimum in functionality at a given level of fertilization [68]. The positive effect of earthworms disappears when fertilization increases: the higher the fertilization, the lower the efficiency of nitrogen. Van Groeningen et al. [69] showed that the positive effect of earthworms can be demonstrated and leads to 25% more production especially when organic material is returned to the soil. The positive effect of earthworms disappears when fertilization increases: the higher the fertilization, the lower the efficiency of earthworms to efficiently use nitrogen.

Geiger et al. [10,70] examined 13 variables that have an effect on biodiversity in European agriculture. They discovered that fungicides had the greatest influence. In practice on dairy farms the parameters affecting biodiversity are perhaps more complex and next to fertilization, grazing and mowing regime also turning grassland into continuous maize production, tillage, use of pesticides and antibiotics, concentrates and indirect the percentage of grass in the feed ration play a significant role.

It is important to derive an unequivocal indicator of the intensity of a farm, because within classes of intensity biodiversity varies considerably and that gives opportunities to achieve improvement in biodiversity at farm level. Besides an indicator of intensity also indicators for biodiversity are essential. This is illustrated for fertilization as one of the factors of intensity. Much is known about the relationship between nitrogen and yields and (above ground) biodiversity. While above ground biodiversity rapidly decreases with increasing intensity, resilience decreases and the risks increase. Not only short-term, direct risks for the farmer, e.g. the risk of damage due to climate or weather conditions such as drought or excessive rainfall; but also social risks, e.g. the probability of increasing the social costs by increasing the environmental pressures; and finally farmers are confronted with long-term risks of higher production costs and loss of production, due to soil degradation and the loss of the functioning of soil life. Recently, the environmental and social costs of loss of nitrogen to the environment were quantified on the basis of the "willingness to pay method". The method estimates what people are willing to pay, or the cost of actions they are willing to take, to avoid the adverse effects of nitrogen pollution [50,71]. The costs due to nitrogen losses to the environment are quantified by PBL for Europe and EPA for the US and increase as the fertilization and associated losses increase to the environment [50,71,72]. These costs are indirectly paid by society.

Intensification has been historically linked with "uniformisation" and "simplification" of agricultural systems. The negative correlation between intensification and biodiversity reflects the loss of diversity within and among agro-ecosystems If (bio)diversity would be restored and/ or properly managed it could be possible to have intensive agricultural systems with relatively high biodiversity [73,74]. Furthermore, it has been proven that management of biodiversity creates benefits for agriculture in terms of productivity (e.g. [44,45,73,75]) and overall sustainability (e.g. [51,55,56,73,74]). Even intensive agriculture can benefit from biodiversity management to increase species and at the same time maintain or increase productivity [70]. Genetic diversity may be deliberately "engineered" to increase crop yield further, which has come to a standstill with the breeding for uniformity in e.g. wheat.

CONCEPTUAL FRAMEWORK AS DEVELOPED FOR DAIRY FARMING

To promote resilience (as described in the resilience model) on dairy farms an integrated approach is required [9]. The starting point of the conceptual framework is the desire to reverse the decline of biodiversity, and on the other hand to better use and enhance biodiversity at the farm. This at the same time increases the efficiency in the cycles at the farm and reduced stress factors causing biodiversity loss in (semi)natural areas (Figure 4).

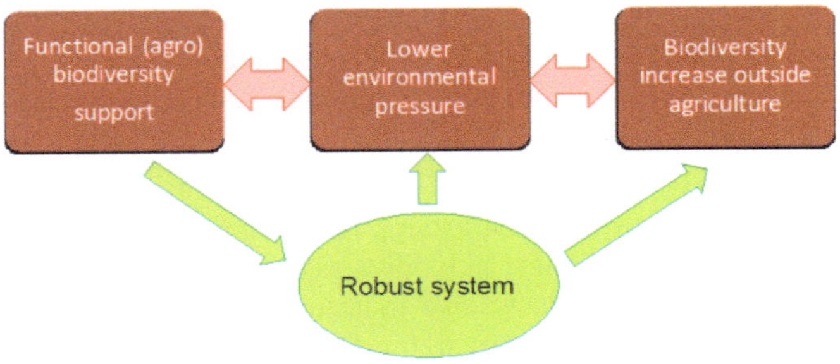

Figure 4. Functional agrobiodiversity strengthens a robust system, lowering environmental pressure and therewith increasing biodiversity outside the agricultural system.

Considering the importance and function (promoting resilience and reducing risks) of biodiversity on dairy farms it is important to enhance the functional (agro)biodiversity. Functional (agro)biodiversity enhancement however is not enough and should be supported by landscape elements and diversity and the connections of biodiversity source areas in an area including several other farms and land uses. Therefore, we distinguish four interconnected pillars for biodiversity (Figure 5):

- Functional (agro)biodiversity at the farm. This encompasses management of soil biodiversity, including rooting systems, grass- and cropland biodiversity, and the diversity of farm animals,
- and the cycles of nutrients, water and energy at the farm (soil, crop, cow, business); optimized by using the functional agro-biodiversity and to serve as a basis for underground and aboveground biodiversity, water management, carbon sequestration, nutrient use, etc. The intensity of a farm largely determines whether cycles are closed at the farm level. 2. Landscape diversity at the farm: influence of the landscape elements (hedges, ditches, flower zones, trees and forests, etc.) to support the functional agro-biodiversity.
- Specific species management (mowing, fertilization timing, water management, etc.) at the farm for maintaining and increasing specific species (e.g. farmland/meadow birds).
- Source areas and connection zones (landscape): management within an area (ecological corridors, exchange and connection of dry and wet zones, regional biodiversity, etc.).

Biodiversity is enhanced when clusters of measures are taken within these four levels of biodiversity.

These four interconnected pillars form the basis of the conceptual model and helps the farmer to manage the farm sustainably. It builds on the force of nature, which is determined by the potential of the land, region and climatic circumstances by focusing on soil, roots, grass and mineral cycles (pillar 1), supported by the region specific landscape elements (pillar 2) in connection with biodiversity sources areas, hydrological and landscape features (pillar 4). Except for the third pillar, the farmer benefits from focusing on the improvement of the other 3. The third pillar is mainly for promoting the biodiversity value of agriculture and to preserve vulnerable species which depend on the agricultural land for breeding and food.

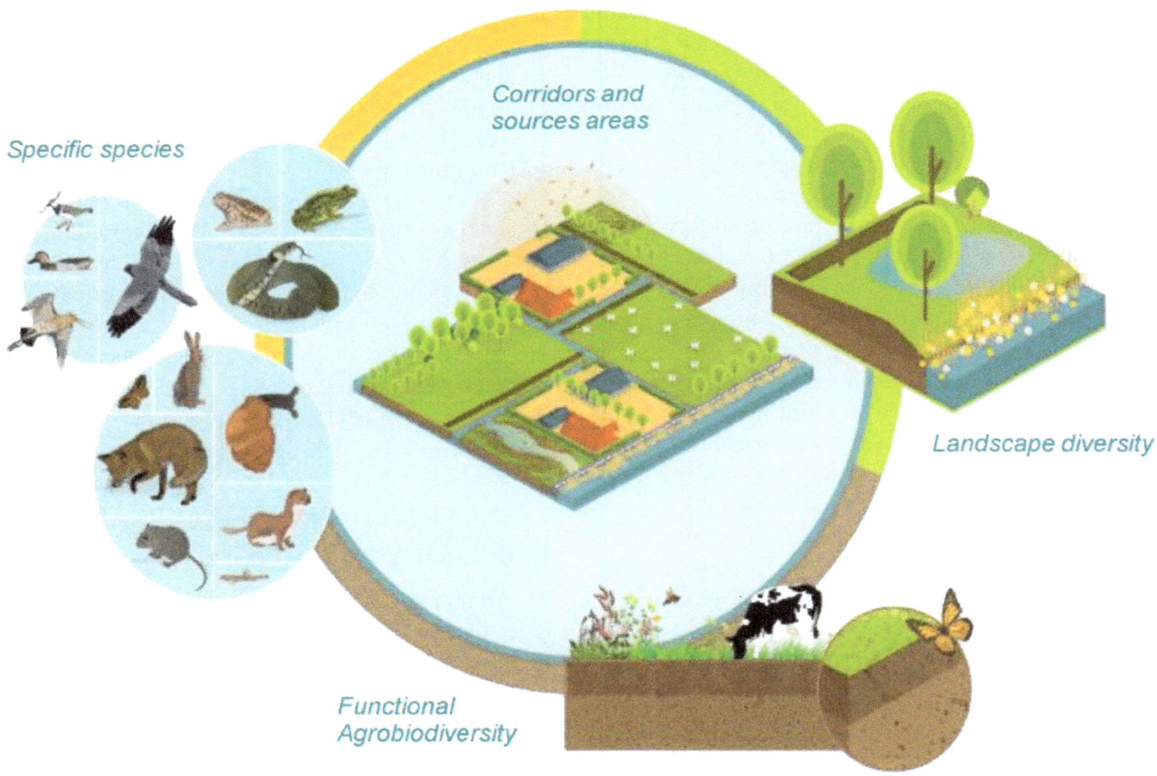

Figure 5. Four interconnected pillars for biodiversity in and around agriculture. It starts with optimizing the functional Agrobiodiversity (pillar 1), supported by landscape diversity (pillar 2) and corridors and source areas for biodiversity (pillar 4). When desired, measures for maintaining specific species can be taken (pillar 3).

CLUSTERS OF MEASURES DEPENDING ON THE DAIRY FARMER'S AMBITION

Within the four pillars of biodiversity, it is possible to identify levels of ambition that lead to an in- or decrease and an optimal use of functional (agro)biodiversity. Within the levels of ambitions clusters of integrated measures contribute to biodiversity enhancement and functioning. The clusters form a logical sequence of actions: they start with soil and grassland management, and then expand towards measures in the area of grass and crop diversity, animal health and welfare, and farm level measures. The exact content of the action is open and depends on the farmer's ambition. The below mentioned levels of ambition serve as examples. The four proposed ambition levels are shown in Figure 6 and explained below. Figure 6 shows a score of farms (the individual dots) in relation to intensity and biodiversity score. The maximum biodiversity per intensity class is determined by connecting the best farms in terms of biodiversity for each intensity class.

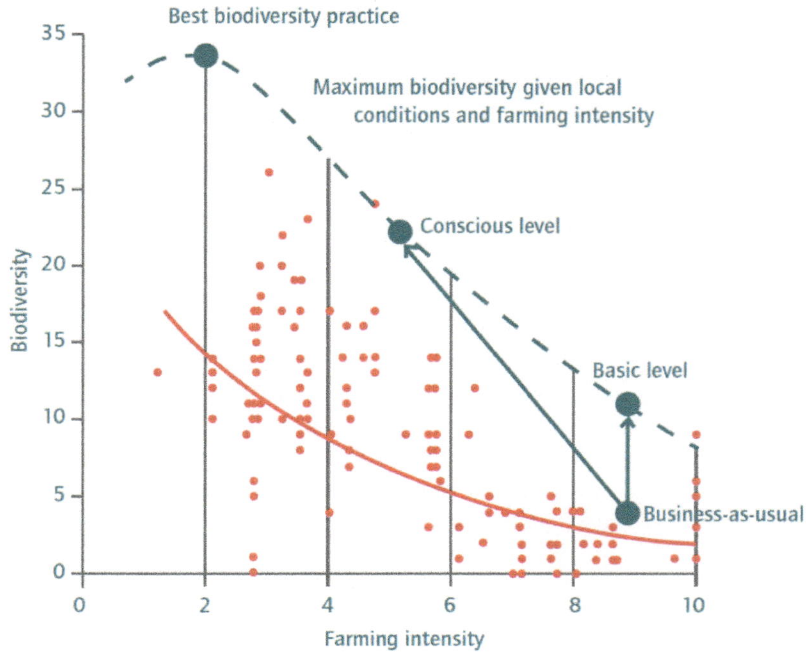

Figure 6. Different farms scored on intensity level and biodiversity adopted from [10,65]. The maximum biodiversity line is the line connecting the farms scoring the highest biodiversity per intensity. The four ambition levels are explained in the text.

Business as Usual

The business as usual level makes little or no use of stimulating functional biodiversity on the farm. The focus is mainly on optimizing or maximizing the various production levels and their control. In addition, the production is highly dependent on external inputs, which both ensure a constant quality (concentrate, genetic material, fertilizers, pesticides), but on the other hand also entail risks regarding the (international) market. In addition, a lack of basic care of biodiversity (starting with management of the soil) in the long run undermine the production system, and to weaken the resilience that is needed to cope with changes in climate, for example.

Basis

The basic level is current practice of agricultural nature conservation and essentially only focusses on pillar 3: no changes are made regarding the on-farm intensity and management strategy of the farm, but specific biodiversity is stimulated on the farm itself. This can be done by setting aside a piece of (marginal) land or using considerate mowing. This ambition is not enough to sustain agriculture on the long term and a more conscious ambition is needed.

Conscious

The next level of ambition goes towards an adaptive, resilient system. The basis of soil and grassland management has been laid by a focus on careful management of the most important production factor: the soil. Attention is paid to the organic matter supply to the soil, feeding the soil, and there is a good base of grassland care, which includes a greater diversity in the grassland (herbs, clover), minimization of (land-based) arable land (and feed import), the use of cereals in rotation with maize, and less externally imported concentrates. Not only the level of soil, grass and crop management is optimal, but also the animal breeding strategy can play a role as an important aspect of resilience. At this level there is much focus on the landscape elements to support the functional biodiversity at the farm. The elements that are chosen depend on the specific location of the farm in a region.

Best of Biodiversity Practice

The last level of ambition essentially is a switch from a control system to an adaptive or resilient system, taking the full advantage of the conceptual framework. It makes maximum use of functional biodiversity at farm level, and a transition has been made to a resilient system. At all levels: from soil to grass and crop, animal and plant, the landscape elements and the integration within the region, the farmer makes maximum use of the functional aspects of biodiversity. This ensures less dependence on external inputs, greater internal resilience and robustness. However, these measures sometimes lead to production reduction and / or increase the cost price of products. The advantages of a greater resilience and more stable production levels become visible only in the long term (4–6 years).

INDICATORS

The conceptual framework focusses on the improvement of agrobiodiversity on different levels and by doing that it implicitly assumes that this leads to an increase in resilience and biodiversity elsewhere (Figure 4). This is supported by literature, but the extent of improvement has yet to be determined. This requires the development of simple indicators of biodiversity and of farm performance (for production and economics) for the four pillars above. There are several studies on biodiversity indicators, or even wider sustainability in agriculture. A survey of the literature indicates 500 indicators for sustainability in agriculture that occur one or more times in different systems (certification, etc.; [9]). This indicates that as yet there is no single set of indicators available that can easily be used in practice. Our conceptual framework might provide a basis to derive verifiable indicators at the four pillars and therewith form a basis for measures of biodiversity enhancement and building farm resilience. It should be realized that these are region specific because regional circumstances, environments and ecosystems determine the potential biodiversity.

The conceptual framework requires a limited set of integral key performance indicators (KPIs) that can be used for management purposes and for monitoring progress. In a recent assessment [76] a set of 17 KPI's were identified for dairy farming which should all be followed and then could be used to integrally optimize the system and to monitor progress and results. We propose to reduce these to a limited set of primary KPI's which are needed to integrally manage the system in such a way that it induces measures that lead to improvement of different other parameters. An example is the % protein from and % manure to own land. If the aim is to set this % very high (from currently 60 to 90%), the farmer has to optimize the nutrient cycles, manage the grasslands in such a way that it yields highest productivity (type and age of grassland, rooting, water management, etc.), resulting in improved soil, organic matter content, biodiversity, water quality and emissions. These KPI's are now tested in practice.

HOW TO APPLY AND USE THE CONCEPTUAL FRAMEWORK?

How a concrete development plan for (functional) biodiversity looks at an individual farm depends on several factors? First, it is important to clarify what is already realized in terms of management and where there is room for improvement for the farmer and for biodiversity. The personal ambition and inspiration of the dairy farmer plays a major role. Additionally, a checklist can make clear where current utilization of functional biodiversity is strong or weak (see e.g. www.biodiversiteitsadvies.nl).

The Development plan for biodiversity can be prepared by the farmer with other interested parties. In the Development plan, the farmer determines the concrete measures leading to improved biodiversity. The choice of measures depends on the eventual fees or other incentives available for the preservation and promotion of biodiversity.

It is important to know where the farm is located geographically, which provides the maximum possible level of biodiversity given the region, soil type, source areas, etc. Then the farmer determines his or her level of ambition (partly dependent on financial capacity and incentives or rewards for biodiversity promotion). This level of ambition must be in agreement with the goals of other stakeholders in the region (water managers, ecological zones, tourism/recreation, etc. [77]. The level of ambition then determines which clusters of measures are applicable.

Management of biodiversity in agriculture is all about the knowledge and experience of the farmer [78]. A shift back from sensor, monitoring tools and indicator management (control) to knowledge intensive management together with technological support is essential (e.g. [77-79]). Farmers have to regain the knowledge about the ecological processes and drivers that enable production of enough and healthy food. This can be done in two ways, both based on agronomic knowledge: reducing the external input (and the dependency on these inputs) to as low as possible requirements for the system to optimal function and be resilient (precision agriculture). The second option is to start optimizing the system without external inputs and gradually increasing the production using agronomic knowledge and necessary external inputs. When the optimum is reached it can be decided to add external inputs to further increase its productivity searching for the resilient limits

CONCLUSION

A paradigm shift is needed in agriculture to stop the large-scale loss of biodiversity in the agricultural landscape and soil, but even more than that to rethink the use of the role of soil life, landscape elements and biodiversity in sustainable agriculture. In Europe, where long-term data are available it is shown that most species are in decline. For some species as birds, butterflies and insects the speed and extend of decline is alarming. In hotspot intensive agricultural areas such as the Netherlands for some years policies have been implemented that should help improve the situation (manure policy, nitrate directive, air quality, nature protection...). While in the semi-natural areas this has led to some success stories and species decline is stopped or reversed, this is not the case in agricultural areas. Here a steady decline is still observed. The agricultural intensification which focuses on increasing productivity per hectare or per unit input (such as feed) is the main cause of the decline. Apparently we reached a tipping point where rethinking our agricultural practices is needed in the face of sustainability and production of food. There are different successful initiatives, but a large scale change is not observed.

At present, the risks in intensive agriculture are managed by a so-called (risk) control model based on externalities, with important side effects such as risks of social costs and decreased function of natural processes. In a resilient system risks are lower, reducing costs and increasing biodiversity. Adaptive risk management is most successful when a portfolio of measures is taken. This can be achieved by focusing on four interconnected levels of biodiversity for sustainable management of ecosystem services (Functional (agro)biodiversity; Landscape diversity; Specific species management; and Source areas and connection zones).

The conceptual framework provides a basis to derive strong indicators at the four levels and therewith forms a basis for measures of biodiversity enhancement and building farm resilience. These levels provide also the basis for a system of reward or other incentives. Such incentives could be based on the relative size between maximum achievable and current biodiversity in an intensity class of the farm and taking into account the current level of biodiversity.

The framework focusses on the use and management of biodiversity to improve productivity in agricultural system along with sustainability and therewith on lowering the long term risk of production loss. By managing (bio)diversity in agriculture, sustainable agriculture can contribute to reaching several Sustainable Development Goals

ACKNOWLEDGEMENT

We are very grateful for the fruitful discussions with Guus van Laarhoven, Richard Piechocki and Frank Verhoeven.

REFERENCES

1. Convention on Biological Diversity, United Nations. 1992. Available from: https://www.cbd.int/doc/legal/cbd-en.pdf
2. Fahrig L, Baudry J, Brotons L, et al. (2011) Functional landscape heterogeneity and animal biodiversity in agricultural landscapes. Ecol Lett 14: 101-112.
3. Odum EP (1969) The strategy of ecosystem development. Science 164: 262-270
4. Holling CS (1973) Resilience and stability of ecological systems. Annu Rev Ecol Syst 4: 1-23.
5. Beintema AJ (1986) Man-made Polders in the Netherlands: a Traditional Habitat for shorebirds. Colon Waterbirds 9: 196-202
6. EEA (2015) European ecosystem assessment — concept, data, and implementation. Contribution to Target 2 Action 5 Mapping and Assessment of Ecosystems and their Services (MAES) of the EU Biodiversity Strategy to 2020. EEA Technical Report no.6/2015, EEA, Copenhagen, Denmark.
7. Henckel L, Borger L, Meiss H, et al. (2015) Organic fields sustain weed metacommunity dynamics in farmland landscapes. Proc R Soc B 282: 20150002.
8. Wereld Natuur Fonds (2015) Living Planet Report. Natuur in Nederland. WNF, Zeist, The Netherlands.
9. Buckwell A, Uhre AN, Williams A, et al. (2012) Sustainable Intensification of European Agriculture A review sponsored by the RISE Foundation. www.risefoundation.eu
10. Geiger F, Bengtsson J, Berendse F, et al. (2010) Persistent negative effects of pesticides and biological control potential on European farmland. Basic Appl Ecol 11: 97-105.
11. Foley JA, DeFries R, Asner GP, et al. (2005) Global consequences of land use. Science 309: 570-574.
12. Lechenet M, Bretagnolle V, Bockstaller C, et al. (2014) Reconciling pesticide reduction with economic and environmental sustainability in arable farming. PLoS ONE 9: e97922.
13. FAO, WFP and IFAD (2012) The State of Food Insecurity in the World 2012. Economic growth is necessary but not sufficient to accelerate reduction of hunger and malnutrition. Rome, FAO.
14. Tittonell P (2013) Farming Systems Ecology. Towards ecological intensification of world agriculture. Inaugural lecture upon taking up the position of Chair in Farming Systems Ecology at Wageningen University on 16 May 2013. http://www.wageningenur.nl/en/show/Feeding-theworld-population-sustainably-and- efficiently-with-ecologically-intensive-agriculture.htm.
15. Tomich TP, Brodt S, Ferris H, et al. (2011) Agroecology: A Review from a Global-Change Perspective. Annu Rev Environ Resour 36: 193-222.
16. Davis AS, Hill JD, Chase CA, et al. (2012) Increasing Cropping System Diversity Balances Productivity, Profitability and Environmental Health. PLoS ONE 7: e47149.
17. Tilman D, Wedin D, Knops J (1996) Productivity and sustainability influenced by biodiversity in grassland ecosystems. Nature 379: 718-720.
18. Tilman D, Reich P, Isbell F (2012) Biodiversity impacts ecosystem productivity as much as resources, disturbance or herbivory. Proc Natl Acad Sci 109: 10394-10397.
19. FAO (2010) Sustainable Diets and Biodiversity. Directions and Solutions for Policy, Research and Action. Proceedings of the International Scientific Symposium 'Biodiversity and Sustainable Diets United Against Hunger' on 3-5 November 2010 Food and Agriculture Organization of the United Nations, Rome.
20. Cardinale BJ, Duffy JE, Gonzalez A, et al. (2012) Biodiversity loss and its impact on humanity. Nature 486: 59-67.
21. Fischer J, Lindenmayer DB, Manning AD (2006) Biodiversity, ecosystem function, and resilience: ten guiding principles for commodity production landscapes. Front Ecol Environ 4: 80-86.
22. Huyghe C, Litrico I, Surault F (2012) Agronomic value and provisioning services of multispecies swards. Grassl Sci Eur 17: 35-46.
23. Schmid B (2002) The species richness-productivity controversy. Trends Ecol Evol 17: 113-114.
24. Zuppinger-Dingley D, Schmid B, Petermann JS, et al. (2014) Selection for niche differentiation in plant communities increases biodiversity effects. Nature 515: 108-111.
25. Vandermeer J, van Noordwijk M, Anderson J, et al. (1998) Global change and multi-species agroecosystems: Concepts and issues. Agric Ecosyst Environ 67: 1-22.
26. Walker B (1995) Conserving biological diversity through ecosystem resilience. Conserv Biol 9: 747-752.
27. Jacobsen SE, Sørensen M, Pedersen S, et al. (2013) Feeding the world: genetically modified crops versus agricultural biodiversity. Agron Sustain Dev 33: 651-

662.

28. Altieri MA (2009) Agroecology, small farms and food sovereignty. Mon Rev 61: 102.

29. Mundt CC (2002) Use of multiline cultivars and cultivar mixtures for disease management. Annu Rev Phytopathol 40: 381-410.

30. Wolfe MS (2000) Crop strength through diversity. Nature 406: 681-682.

31. Jareño D, Viñuela J, Luque-Larena JJ, et al. (2015) Factors associated with the colonization of agricultural areas by common voles Microtus arvalis in NW Spain. Biol Invasions 17: 2315-2327.

32. Evans KL (2004) The potential for interactions between predation and habitat change to cause population declines of farmland birds. Ibis 146: 1-13.

33. Kentie R, Both C, Hooijmeijer JCE, et al. (2015) Management of modern agricultural landscapes increases nest predation rates in Black-tailed Godwits Limosa limosa. Ibis 157: 614-625.

34. Bretagnolle V, Gauffre B, Meiss H, et al. (2011) The Role of Grassland Areas within Arable Cropping Systems for the Conservation of Biodiversity at the Regional Level. In: Grassland Productivity and Ecosystem Services, ed. Lemaire, G. et al. 251-60. USA: CABI

35. Pywell RF, Heard MS, Woodcock BA, et al. (2015) Wildlife friendly farming increases crop yield: evidence for ecological intensification. Proc R Soc B 282: 20151740.

36. Compendium, 2015. Available from: https://www.Compendiumvoordeleefomgeving.nl

37. Donald PF, Sanderson FJ, Burfield IJ, et al. (2006) Further evidence of continent-wide impacts of agricultural intensification on European farmland birds, 1990-2000. Agric Ecosyst Environ 116: 189-196.

38. Ministry of Economic Affairs (2014) Convention on Biological Diversity. Fifth National Report of the Kingdom of the Netherlands. Den Haag, The Netherlands.

39. PBL (2014) Balans van de Leefomgeving 2014. De toekomst is nú, Den Haag: Planbureau voor de Leefomgeving. The Netherlands.

40. Dise NB, Ashmore M, Belyazid S, et al. (2011) Nitrogen as a threat to European terrestrial biodiversity. In: The European Nitrogen Assessment, ed. Sutton MA, Howard CM, Erisman JW, et al. Cambridge University Press.

41. Erisman JW, Galloway JN, Dise NB, et al. (2015) Nitrogen: Too much of a vital resource. WWF Netherlands, Zeist, the Netherlands. 27.

42. Van den Noort PC (1987) Land consolidation in the Netherlands. Land Use Policy 4: 11-13.

43. Napel J, Bianchi F, Bestman MWP (2006) Utilising intrinsic robustness in agricultural production systems. TransForum Working Papers. 32-53. (available at www.louisbolk.org)

44. Erisman JW, Brasseur G, Ciais P, et al. (2015) Global change: Put people at the centre of global risk management. Nature 519: 151-153.

45. Altieri MA (1999) The ecological role of biodiversity in agroecosystems. Agric Ecosyst Environ 74: 19-31.

46. Rodale Institute (2011) The farming systems trial. Celebrating 30 years. The Rodale Institute, USA. Available from: http://rodaleinstitute.org/assets/FSTbooklet.pdf

47. De Schutter O (2010) Report submitted by the Special Rapporteur on the right to food, Olivier De Schutter. United Nations, General Assembly. Human Rights Council, Sixteenth session, Agenda item 3: Promotion and protection of all human rights, civil, political, economic, social and cultural rights, including the right to development. Available from: www2.ohchr.org/english/issues/food/docs/A-HRC-16-49. pdf). August 17, 2011.

48. Deru J, Schilder H, Van der Schoot JR, et al. (2014) Genetic differences in root mass of Lolium perenne varieties under eld conditions. Euphytica 199: 223-232.

49. Gewin V (2010) An underground revolution. Nature 466: 552-553.

50. Van Grinsven HJM, Holland M, Jacobsen BH, et al. (2013) Costs and Benefits of Nitrogen for Europe and Implications for Mitigation. Environ Sci Technol 47: 3571-3579.

51. Lin BB (2011) Resilience in Agriculture through Crop Diversification: Adaptive Management for Environmental Change. BioScience 61: 183-193.

52. De Ponti T, Rijk B, Van Ittersum MK (2012) The crop yield gap between organic and conventional agriculture. Agric Syst 108: 1-9.

53. Seufert V, Ramankutty N, Foley JA (2012) Comparing the yields of organic and conventional agriculture. Nature 485: 229-232

54. Ponisio LC, M'Gonigle LK, Mace KC, et al. (2015) Diversification practices reduce organic to conventional yield gap. Proc R Soc B 282: 20141396.

55. Altieri MA, Nicholls CI, Henao A, et al. (2015) Agroecology and the design of climate changeresilient farming systems. Agron Sustain Dev 35: 869-890

56. Isbell F, Craven D, Connolly J, et al. (2015) Biodiversity increases the resistance of ecosystem productivity to climate extremes. Nature 526: 574-577.

57. Nemes N (2009) Comparative analysis of organic and non-organic farming systems: a critical assessment of farm profitability. FAO, Rome, Italy.

58. Gonthier DJ, Ennis KJ, Farinas S, et al. (2014) Biodiversity conservation in agriculture requires a multi-scale approach. Proc R Soc B 281: 20141358.

59. Smith HG, Dänhardt J, Lindström A, et al. (2010) Consequences of organic farming and landscape heterogeneity for species richness and abundance of farmland birds. Oecologia 162: 1071-1079

60. Smits MJW, van Alebeek FAN (2007) Biodiversiteit en kleine landschapselementen in de biologische landbouw; Een literatuurstudie. Wageningen, Wettelijke Onderzoekstaken Natuur & Milieu, WOT-rapport 39. 84.

61. Schneider MK, Lüscher G, Jeanneret P, et al. (2014) Gains to species diversity in organically farmed fields are not propagated at the farm level. Nat Commun 5: 1-9.

62. Kretschmer H, Pfeffer H, Hoffmann J, et al. (1995) Strukturelemente in Agrarlandschaften Ostdeutschlands - Bedeutung für den Biotop- und Artenschutz. ZALF-Bericht 19, Selbstverlag ZALF, Müncheberg: 164 S.
63. Hoffmann J, Kretschmer H, Pfeffer H (1999) Effects of Current Landscape Patterns on Biodiversity. -In: Tenhunen, J.D., R. Lenz und R.E. Hantschel (Hrsg.): Ecosystem properties and landscape function in Central Europe. Ecological Studies. Springer.
64. Benton TG, Vickery JA, Wilson JD (2003) Farmland biodiversity: is habitat heterogeneity the key? Trends Ecol Evol 18: 182-188.
65. Allan E, Bossdorf O, Dormann CF, et al. (2014) Interannual variation in land-use intensity enhances grassland multidiversity. Proc Natl Acad Sci 111: 308-313.
66. Payne RJ, Thompson AM, Standen V, et al. (2012) Impact of simulated nitrogen pollution on heathland microfauna, mesofauna and plants. Eur J Soil Biol 49: 73-79.
67. Wymenga E, Latour J, Beemster N, et al. (2015) Terugkerende muizenplagen in Nederland. Inventarisatie, sturende factoren en beheersing. A&W-rapport 2123. Altenburg & Wymenga bv, Alterra Wageningen UR, Livestock Research Wageningen, Wetterskip Fryslân, Stichting Werkgroep Grauwe Kiekendief. Feanwâlden.
68. Van Eekeren N, Heeres E, Smeding F (2003) Leven onder de graszode - Discussiestuk over het beoordelen en beïnvloeden van bodemleven in de biologische melkveehouderij. Louis Bolk publikatie LV 52. Louis Bolk Instituut Driebergen, The Netherlands.
69. Van Groenigen JW, Lubbers IM, Vos HMJ, et al. (2014) Earthworms increase plant production: a meta-analysis. Sci Reports 4: 6365.
70. Geiger F (2011) Agricultural intensification and farmland birds. Thesis Wageningen University, The Netherlands.
71. Brink C, van Grinsven H, Jacobsen BH, et al. (2011) Costs and benefits of nitrogen in the environment. In: Sutton MA, Howard CM, Erisman JW, et al. (Eds). The European Nitrogen Assessment. Cambridge, UK, Cambridge University Press, Chapter 19, pp. 434-462.
72. Compton JE, Harrison JA, Dennis RL, et al. (2011) Ecosystem services altered by human changes in the nitrogen cycle: a new perspective for US decision making. Ecol Lett 14: 804-815.
73. Frison EA, Cherfas J, Hodgkin T (2011) Agricultural biodiversity is essential for a sustainable improvement in food and nutrition security. Sustainability 3: 238-253.
74. Lemaire G, Franzluebbers A, Carvalho PCF, et al. (2014) Integrated crop-livestock systems: Strategies to achieve synergy between agricultural production and environmental quality. Agric Ecosyst Environ 190: 4-8.
75. Grace JB, Anderson TM, Seabloom EW, et al. (2016) Integrative modelling reveals mechanisms linking productivity and plant species richness. Nature 529: 390-393.
76. Van Eekeren N, Verhoeven F, Erisman JW (2015) Verkenning Kritische Prestatie Indicatoren voor stimulering van een biodiverse melkveehouderij. Louis Bolk, Driebergen, The Netherlands.
77. Groot JCJ, Rossing WAH, Jellema A, et al. (2007) Exploring multi-scale trade-offs between nature conservation, agricultural profits and landscape quality - a methodology to support discussions on land-use perspectives. Agric Ecosyst Environ 120: 58-69.
78. De Snoo GR, Herzon I, Staats H, et al. (2012) Toward effective nature conservation on farmland: making farmers matter. Conserv Lett 6: 66-72.
79. Duru M, Therond O, Martin G (2015) How to implement biodiversity-based agriculture to enhance ecosystem services: A review. Agron Sustain Dev 35: 1259-1281.

CHAPTER 12

RELATIONSHIPS BETWEEN SUBSTRATE, SURFACE CHARACTERISTICS, AND VEGETATION IN AN INITIAL ECOSYSTEM

P. Biber[1], S. Seifert[1], M. K. Zaplata[2], W. Schaaf[3], H. Pretzsch[1], and A. Fischer[2]

[1]Chair for Forest Growth and Yield, Department Ecology and Ecosystem Management, Technische Universität München, Freising, Germany

[2]Geobotany, Department Ecology and Ecosystem Management, Technische Universität München, Freising, Germany

[3]Soil Protection and Recultivation, Brandenburg University of Technology, Cottbus, Germany

ABSTRACT

We investigated surface and vegetation dynamics in the artificial initial ecosystem "Chicken Creek" (Lusatia, Germany) in the years 2006–2011 across a wide spectrum of empirical data. We scrutinized three overarching hypotheses concerning (1) the relations between initial geomorphological and substrate characteristics with surface structure and terrain properties, (2) the effects of the latter on the occurrence of grouped plant species, and (3) vegetation density effects on terrain surface change. Our data comprise and conflate annual vegetation monitoring results, biennial terrestrial laser scans (starting in 2008), annual groundwater levels, and initially measured soil characteristics. The empirical evidence mostly confirms the hypotheses, revealing statistically significant relations for several goal variables: (1) the surface structure properties, local rill density, local relief energy and terrain surface height change; (2) the cover of different plant groups (annual, herbaceous, grass-like, woody, *Fabaceae*), and local vegetation height; and (3) terrain surface height change showed significant time-dependent relations with a variable that proxies local plant biomass. Additionally, period specific effects (like a calendar-year optimum effect for the occurrence of *Fabaceae*) were proven. Further and beyond the hypotheses, our findings on the spatiotemporal dynamics during the system's early development grasp processes which generally mark the transition from a geo-hydro-system towards a bio-geo-hydro system (weakening geomorphology effects on substrate surface dynamics, while vegetation effects intensify with time), where pure geomorphology or substrate feedbacks are changing into vegetation–substrate feedback processes.

Citation: Biber, P., Seifert, S., Zaplata, M. K., Schaaf, W., Pretzsch, H., and Fischer, A.: Relationships between substrate, surface characteristics, and vegetation in an initial ecosystem, Biogeosciences, 10, 8283-8303, doi:10.5194/bg-10-8283-2013, 2013.

ABSTRACT

We investigated surface and vegetation dynamics in the artificial initial ecosystem "Chicken Creek" (Lusatia, Germany) in the years 2006–2011 across a wide spectrum of empirical data. We scrutinized three overarching hypotheses concerning (1) the relations between initial geomorphological and substrate characteristics with surface structure and terrain properties,

(2) the effects of the latter on the occurrence of grouped plant species, and (3) vegetation density effects on terrain surface change.

Our data comprise and conflate annual vegetation monitoring results, biennial terrestrial laser scans (starting in 2008), annual groundwater levels, and initially measured soil characteristics. The empirical evidence mostly confirms the hypotheses, revealing statistically significant relations for several goal variables: (1) the surface structure properties, local rill density, local relief energy and terrain surface height change; (2) the cover of different plant groups (annual, herbaceous, grass-like, woody, Fabaceae), and local vegetation height; and (3) terrain surface height change showed significant time-dependent relations with a variable that proxies local plant biomass. Additionally, period specific effects (like a calendar-year optimum effect for the occurrence of Fabaceae) were proven.

Further and beyond the hypotheses, our findings on the spatiotemporal dynamics during the system's early development grasp processes which generally mark the transition from a geo-hydro-system towards a bio-geo-hydro system (weakening geomorphology effects on substrate surface dynamics, while vegetation effects intensify with time), where pure geomorphology or substrate feedbacks are changing into vegetation–substrate feedback processes.

INTRODUCTION

While a lot of studies on ecosystem development have been conducted in mature ecosystems (e.g. Campbell et al., 2007; Ellenberg et al., 1986; Fränzle et al., 2008; Pennisi, 2010) our information about initial systems is comparably weak. This is remarkable because initial ecosystems are usually less complex compared to mature systems. Therefore, the study of the development of initial ecosystems could be very helpful to achieve a better understanding of the complex relationships and feedback mechanisms which typically arise over time (Jørgensen et al., 2000). Tracing the development of young ecosystems and observing how new relationships and feedbacks emerge with increasing complexity would help to get a better insight on key processes and a basic understanding of their interactions.

Initial ecosystems are also important from a very practical point of view. Worldwide, human activities regularly leave behind huge bare areas (Walker and Willig, 1999; Bradshaw, 1983; Hüttl and Weber, 2001; Schaaf, 2001; Zikeli et al., 2002), which create serious environmental challenges due to their ecological instability and low productivity. Here, further knowledge is urgently needed for finding optimal ways to transform such landscapes into landscapes that can be used in a sustainable way by society.

The land surface is an interface where geomorphological and biological processes connect. It can also be seen as an interface where pedosphere, biosphere, hydrosphere and atmosphere are strongly interlinked (Brantley et al., 2007). During early developmental stages, the evolution of a rill system which channels surface runoff and material transport probably is the most visible outcome of such interactions. At the very beginning of ecosystem development, starting with bare ground, these interactions are expected to be especially important (Schaaf et al., 2012).

Scientifically investigated areas, where the first interactions between atmosphere, pedosphere and biosphere are just starting to develop, are rare. Among them are areas created by volcanic activity (Bishop, 2002; Dahlgren et al., 1999; del Moral and Wood, 1993; Müller-Dombois and Fosberg, 1998; Friðriksson, 2005) or areas that become exposed after glacier retreat (Cooper, 1923; Matthews, 1992). However, information about the initial conditions at "point zero" is generally incomplete.

The creation and examination of artificial areas is one approach to close the existing knowledge gap on processes determining early ecosystem development. Ideally, such areas are complete water catchments with a well-defined and if possible homogeneous substrate, and with negligible traces of life and therefore without any successional history. Their value for understanding the emerging interactions between atmosphere, pedosphere, hydrosphere and biosphere cannot be overestimated (Schaaf et al., 2011). The 6 ha artificial catchment "Chicken Creek", established in 2004/05 in the open-cast mining area Welzow-Süd near Cottbus, eastern Germany, was designed to represent the theoretical ideal conditions as close as possible. Here, over a period of seven years, so far, a broad range of key properties of geomorphology, vegetation succession and ecosystem development have been studied with high spatio-temporal resolution.

Combining interdisciplinary data from the Chicken-Creek catchment, our goal within this study is to capture the developing complexity in an initial ecosystem by scrutinizing the following overarching hypotheses.

- Initial substrate characteristics determine the terrain surface's structure (rills, local relief energy) and its change. The degree of determination of these processes weakens with the ecosystem's development (H1).
- Surface and substrate characteristics determine the plant species groups to be found initially (H2).
- Increasing density of vegetation reduces the erosion and therefore the degree of change of terrain surface structure (H3).

More specifically, in the context of H1 we expect to find the occurrence of rills, the local relief energy, and the terrain surface height change from one year to another being time dependently correlated to the annual groundwater level, the near-surface substrate grain size, and other substrate characteristics as measured after catchment construction.

From the perspective of H2, vascular plant species are grouped by ecological and morphological criteria into annual, herbaceous, grass-like, and woody, and affiliation to the N-fixing Fabaceae family. The occurrence of a given plant species group is expected to depend on the same variables as mentioned above. Moreover, H2 comprises total plant cover, overall vegetation height and density as goal variables which are assumed to be related to the above-mentioned explanatory variables in terms of a progressive development as well. In contrast to H1, where only groundwater and substrate characteristics are related to the change of surface height, H3 includes a measure of vegetation density as an explanatory variable.

MATERIAL AND METHODS

The Artificial Catchment Chicken Creek

The artificial catchment, Chicken Creek, with an area of 6 ha was constructed in the open-cast mining area of Lusatia, Germany (51.6049°N, 14.2667°E) in 2004–2005. It is a 2–4 m layer of post-glacial sandy to loamy sediments overlying a 1–2 m layer of Tertiary clay which forms a shallow pan and seals the whole catchment at the base. No further measures of restoration like planting, amelioration or fertilization were carried out; natural succession and undisturbed development is allowed. Due to the artificial construction, subsoil boundary conditions of this site are clearly defined including well documented inner structures as compared to natural catchments.

The region around the catchment is characterized by a temperate climate with a sub-continental character and comparatively low precipitation. During the last five decades, the mean annual temperature amounted to, on average, 8.9°C (January mean: −0.8°C, July mean: 18.4°C), the annual precipitation to 563 mm (Gerwin et al., 2009).

Up to now, the set of plant species that has been observed during succession is very similar to what field botanists can find in the region (e.g. Felinks, 2000). Propagules of course are brought in by seed rain, but there has also been a small but undeniable seed bank (Zaplata et al., 2010, 2011a).

The floristic starting situation is characterized by an almost monodominant stand of the Canadian horseweed, Conyza canadensis (L.) Cronquist, a prolific spreader (Zaplata et al., 2011a). Soon the study site became quite abundantly inhabited by a herb-rich pioneer population also with Cirsium arvense (L.) Scop. and the grasses Bromus hordeaceus L., Echinochloa crus-galli (L.) P. Beauv., and Hordeum juba tum L. Meanwhile, the abundances of these species decreased and the recently predominant vegetation types might be classified as species-rich sandy xeric grasslands (e.g. with Achillea millefolium subsp. pannonica (Scheele) Hayek, Arenaria serpyllifolia L. s.l., Helichrysum arenarium (L.) Moench, Lotus corniculatus L., Petrorhagia prolifera (L.) P. W. Ball and Heywood, Rumex acetosella var. tenuifolius Wallr., Senecio vernalis Waldst. and Kit., the grass Calamagrostis epigejos (L.) Roth, and tall herb communities (e.g. with Artemisia vulgaris L., Centaurea stoebe L. s.l., Daucus carota L., and the grasses Agrostis capillaris L. and Poa compressa L.). Beyond, some woodland-associated species are present already, e.g. Betula pendula Roth, Pinus sylvestris L., Populus tremula L., the herbaceous Hieracium umbellatum L. and Hypochaeris radicata L., and the gramineous species Carex hirta L. and Holcus mollis L.

In the initial situation, organic substance was not completely absent from the soil, concentrations however, were very low (about 2 mg g−1 substrate). Radio-carbon dating yielded ages between 3000 and 16 000 which indicate the organic substance being mostly of fossil origin (Gerwin et al., 2009; Risse-Buhl et al., 2013).

For more details on the construction process, initial site conditions as well as the monitoring program carried out since 2005 see Gerwin et al. (2009, 2010, 2011) and Schaaf et al. (2011, 2012).

Data Used in this Study

A 20 m × 20 m grid for sampling and orientation purposes was established across the catchment area (Fig. 1). The data used in this study cover the sub-area between the grid-point lines A and Q as shown in Fig. 1 which amounts to approximately 5 ha. The other part, the area around the pond in the lower part of the catchment, was excluded from our study since it is influenced by a high, constant water table and evolved into a semi-aquatic ecosystem with different site conditions.

For the statistical scrutiny of our hypotheses a combined data set comprising terrain surface information from terrestrial laser scanning (TLS), vegetation information from field records, soil and groundwater information was used. The data sources have different spatial and temporal resolutions. Soil samples were taken only once at the beginning of the study, vegetation records were made once every year in summertime and TLS was done at seven different times between 2008 and 2011.

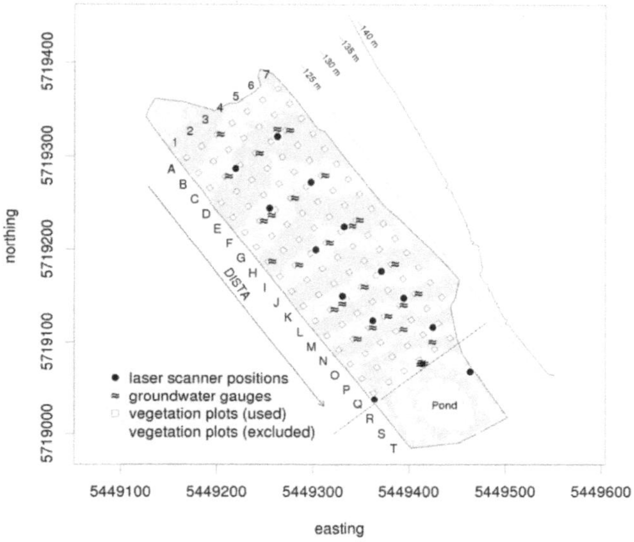

Figure 1. Map of the Chicken Creek area (with Gauss–Krüger zone 5 coordinates) showing locations of the data sources (vegetation plots, laser scanner positions, groundwater gauge positions) relevant for this study. Vegetation plots are identified by "row" (A–T) and "column" (1–7). In this work only rows A–Q were used. The variable DISTA as shown in the figure indicates the distance of a point from row A. Ground heights a.s.l. are given as mean values from the laser scan in September 2008 around the center points of vegetation plots A6, F6, L6 and Q6. The height profile along row 4 is given (top right) in the map. It shows the ground elevation as obtained from the laser scans in September 2008.

The above-mentioned 20 m × 20 m grid points form the basic units of each analysis carried out. A spatial data fusion was thus done in order to assign the different data sources to the grid (see below). Temporally, vegetation record data were assigned to the same year's laser scan data. Thus, dependent on the analysis of interest, the same vegetation record data were assigned to both laser scan measurements of a year. Groundwater measurements were assigned as annual means to each grid point. As there was only one survey of soil properties, a specific temporal assignment was not possible. Thus, for interpretation purposes they represent initial conditions and not a temporal development.

As the catchment is sloped downwards from row A to row Q, the distance of any grid point to row A is indicated as DISTA. The time since January 2005 is encoded in the variable TIME with month as yearly fractions (e.g. 6.42 for May 2011) and CALYEAR as the calendar year. In the following text, the resulting time component of the variables used in this study is identified by the index j and the spatial location by the index i. In Table 1 we list all variables and indexes together with their definitions, while in Table 2 we present their summary statistics (more details below). All response variables were chosen and calculated a priori, based on considerations about ecological interpretability and plausibility.

Table 1. Names, data source and description of the variables used in this study

Source	Variable	Description	Unit
TLS	$RILLR_{ij}$	Proportion of laser scan cells classified as belonging to a rill related to the total amount of laser scan cells per vegetation plot	
TLS	$RELEN_{ij}$	Mean value of local relief energy (0.8 m radius neighbourship)	m
TLS	$DHEIGHT_{ij}$	Mean ground surface elevation changes between September 2008 and May 2009, May 2009 and April 2010, and April 2010 and February 2011. Positive values of DHEIGHT indicate an increase of the surface height indicating a net gain of substrate material in the vegetation square, while negative values indicate a net substrate material export (erosion) from the respective square	m
TLS	$VEGHEIGHT_{ij}$	Mean vegetation height	m
TLS	$VEGDENS_{ij}$	Mean vegetation density surrogate	
TLS	$VEGDH_{ij}$	Product of $VEGHEIGHT_{ij}$ and $VEGDENS_{ij}$	
VEG	$PROPANN_{ij}$	Proportion of annual plant cover related to total plant cover	
VEG	$PROPHERB_{ij}$	Proportion of herbaceous plant cover related to total plant cover	
VEG	$PROPGRASS_{ij}$	Proportion of grass like plants cover related to total plant cover	
VEG	$PROPWOOD_{ij}$	Proportion of woody plants cover related to total plant cover	
VEG	$PROPFAB_{ij}$	Proportion of cover of plants belonging to the *Fabaceae* family related to total plant cover	
VEG	$COVTOT_{ij}$	Total degree of plant coverage	
SOIL	$CORGP_i$	Percentage of organic carbon in the soil	
SOIL	$SKELCONT_i$	Percentage of skeletal content in the soil	
SOIL	$MSAND_i$	Percentage of medium sand (0.2–0.63 mm) in the soil	
AQ	$GAUGE_{ij}$	Mean annual groundwater level below surface	m
	$DISTA_i$	Distance from the row of the grid point to the row A on the catchment (see Fig. 1)	m
	$TIME_j$	Time in years since January 2005, months included as fractions of years	a
	$CALYEAR_j$	Calendar year	a

TLS: derived from laser scans, available for each laser scan grid cell, aggregated for each vegetation plot. VEG: derived from the vegetation monitoring, summer of each year. SOIL: values from soil samples in 2005 at the centre of each vegetation plot. AQ: values assigned from the nearest of the groundwater gauges, yearly averages. Indices: *i* values for a specific location (grid point), constant over time. *j* values for points in time, spatially independent. *ij* values for a specific point in time, at a specific location.

Vegetation

Altogether 119 permanent plots for long-term vegetation monitoring were established at the above-mentioned 20 m × 20 m grid points (Zaplata et al., 2010; Fig. 1). Each plot had a square shape of 5 m × 5 m with the respective grid point in its center. At each plot all vascular plant species were recorded including cover estimates for each species in 30 distinct percentage classes (Zaplata et al., 2013). Bryophyte covers were estimated in the same way, distinguishing mosses and liverworts. For this study, a time series with 6 yr of annual monitoring (2006–2011) is available. The plant species were grouped by (i) lifespan according to Rothmaler (2000), but with only the two categories "annual" and "perennial"; (ii) life forms "grass-like", "herb" or "wood" according to Rothmaler (2000), species of the genus Rubus which in general form woody stems were thus labelled as "wood"; and (iii) affiliation to the Fabaceae family ("Fabaceae" versus "no Fabaceae"). Fabaceae species are of special interest, because they are the only N-fixing vascular plants found on the catchment. Other N-fixing organisms are cyanobacteria, which are part of biological soil crusts, also found at the study site (Fischer et al., 2012). However, in this context Fabaceae is the most important group by far. See Table A.1 in the Appendix for each vascular plant species' group affiliation. In Table 3 we summarize the covers separately for the above-mentioned groups and for all plants together. These group-wise and overall covers were obtained by summing up the single species' covers as recorded in the field. The total plant cover increases from an average of 2.0 % in 2006 up to 105.7 % in 2011. The latter value, being greater than 100 %, indicates an overlap of the areas covered by different species. This dramatic rise occurs rather homogeneously across the area, albeit with a slightly higher cover on its central longitudinal axis. While from 2006 to 2008 the vascular plants (annuals and perennials) almost exclusively contribute to the total cover, the further increase of total cover is carried by a massive increase of the bryophytes between 2010 and 2011 (Elmer et al., 2013). The vascular plants' cover, however, seems to stabilize between 30 and 40 % during 2009–2011.

The annual plants reach a distinct maximum cover of more than 30 % in 2009 which considerably decreases in the following years. A similar development, but with a less steep decline, can be seen for the herbaceous plants because this group largely overlaps with the annuals. Since 2008, Fabaceae occur with very similar cover values as the annuals do. This is due to the annual species Trifolium arvense L. (Fabaceae) which is the most prominent annual species from 2008 to 2011 (cf. Zaplata et al., 2011b).

Table 2. Summary statistics of the variables used in this study (see Table 1 for variable explanations). DHEIGHT, indicating a change between subsequent years, is given always at the start year. The soil-related variables CORGP, GRAVCONT, and MSAND were only measured once in 2005. DISTA does not change with time and is presented for the time when the catchment's establishment was completed. The variable COVTOT is shown in Table 3 together with other vegetation properties

		2005			2006			2007			2008			2009			2010			2011		
variable	**unit**	**min**	**mean**	**max**	**min**	**mean**	**max**	**min**	**mean**	**max**	**min**	**mean**	**max**	**min**	**mean**	**max**	**min**	**mean**	**max**	**min**	**mean**	**max**
RILLR											0.00	**0.10**	1.00	0.00	**0.09**	1.00	0.00	**0.10**	1.00	0.00	**0.16**	1.00
RELEN	m										0.01	**0.13**	1.08	0.00	**0.12**	1.27	0.01	**0.13**	1.70	0.01	**0.22**	2.03
DHEIGHT	m										−0.77	**−0.03**	0.39	−1.51	**0.00**	0.51	−0.55	**0.04**	0.97			
VEGHEIGHT	m										0.00	**0.35**	2.16	0.00	**0.40**	3.63	0.01	**0.51**	4.49	0.00	**0.80**	5.60
VEGDENS											0.01	**0.38**	1.00	0.01	**0.37**	1.00	0.01	**0.38**	1.00	0.00	**0.37**	1.00
VEGDH	m										0.00	**0.13**	1.30	0.00	**0.15**	1.83	0.01	**0.20**	3.45	0.00	**0.30**	4.39
PROPANN					0.00	**0.71**	1.00	0.07	**0.63**	0.99	0.08	**0.59**	0.88	0.05	**0.64**	0.96	0.03	**0.29**	0.74	0.01	**0.21**	0.56
PROPHERB					0.32	**0.81**	1.00	0.34	**0.80**	1.00	0.18	**0.69**	0.94	0.13	**0.70**	0.94	0.15	**0.42**	0.84	0.09	**0.29**	0.67
PROPGRASS					0.00	**0.19**	0.68	0.00	**0.19**	0.66	0.06	**0.28**	0.59	0.01	**0.13**	0.86	0.03	**0.13**	0.44	0.01	**0.07**	0.57
PROPWOOD					0.00	**0.00**	0.08	0.00	**0.00**	0.05	0.00	**0.01**	0.15	0.00	**0.01**	0.34	0.00	**0.02**	0.61	0.00	**0.02**	0.50
PROPFAB					0.00	**0.00**	0.06	0.00	**0.11**	0.78	0.00	**0.33**	0.85	0.01	**0.57**	0.91	0.01	**0.24**	0.64	0.01	**0.20**	0.57
CORGP	%	0.00	**0.16**	0.57																		
GRAVCONT	%	0.24	**15.21**	25.30																		
MSAND	%	9.65	**46.91**	62.43																		
GAUGE	m				0.39	**1.78**	2.79	0.27	**1.46**	2.40	0.17	**1.30**	2.15	0.10	**1.12**	2.05	0.11	**0.75**	1.45	0.16	**0.90**	1.62
DISTA	m	0.00	**154.21**	320.00																		

Table 3. Minimum, mean and maximum covers in % per plant species group and year of survey. The allocation of the single plant species to the groups is shown in Table 14. COVTOT is the total degree of coverage by plants (see Table 1). This variable has larger values than the cover sum of annual and perennial, or grass-like, herbaceous, and woody plants, because it also includes the bryophytes

	2006			2007			2008			2009			2010			2011		
variable	**min**	**mean**	**max**	**min**	**mean**	**max**	**min**	**mean**	**max**	**min**	**mean**	**max**	**min**	**mean**	**max**	**min**	**mean**	**max**
annual	0.0	**1.5**	19.5	0.5	**5.1**	33.4	1.5	**9.1**	46.6	2.3	**31.5**	81.7	1.4	**14.7**	63.0	1.3	**22.0**	55.6
perennial	0.0	**0.6**	5.6	0.1	**2.9**	26.9	0.8	**5.3**	41.7	1.5	**7.7**	42.7	3.6	**14.3**	39.2	4.3	**14.8**	50.9
grass-like	0.0	**0.5**	10.5	0.0	**1.5**	9.9	0.6	**3.7**	18.5	1.0	**4.5**	17.6	2.0	**6.0**	13.9	1.3	**5.6**	21.0
herbaceous	0.1	**1.6**	14.5	0.9	**6.5**	35.2	1.9	**10.6**	48.6	1.5	**34.3**	84.5	4.1	**21.9**	66.5	7.8	**29.4**	79.7
woody	0.0	**0.0**	0.1	0.0	**0.0**	0.7	0.0	**0.1**	1.6	0.0	**0.4**	6.5	0.0	**1.1**	20.7	0.0	**1.7**	40.1
Fabaceae	0.0	**0.0**	0.1	0.0	**1.4**	30.0	0.0	**6.1**	45.1	0.1	**29.8**	82.0	0.7	**12.6**	60.1	0.3	**21.5**	54.6
COVTOT	0.1	**2.0**	25.1	1.1	**8.3**	68.8	2.5	**15.2**	112.0	5.9	**48.0**	147.8	10.3	**59.6**	134.8	13.2	**105.7**	166.3

The perennial plants exhibit a steady area-wide increase of cover during the whole observation period, with however still considerably lower covers than for the annuals in 2011. On a lower level, the grass-like plants show a mostly parallel development, seeming to stabilize in 2010 and 2011. Compared to the other plant groups, the covers of woody plants seem negligible so far. However, their average cover increased exponentially and their presence on the whole area might be a precursor of their future dominance.

In order to capture the different groups' relative dominance and to eliminate the time trend which is inherent in the absolute cover values, we related their covers for each grid point and each observation to the respective total plant cover COVTOT for further analysis. This resulted in the proportions of cover for annual plants (PROPANN), grasslike plants (PROPGRASS), woody plants (PROPWOOD) and Fabaceae family (PROPFAB) (cf. Tables 1, 2). As Table 2 shows, the herbaceous plants are the dominating vascular plants throughout the observation period, however with a pronounced decrease of their share from between 0.7 and 0.8 down to about 0.4 and 0.3 in 2010 and 2011. Annual plants follow the same pattern, albeit on a slightly lower level. Grass-like plants increase their share from 0.2 to almost 0.3 in 2008. After that, they also drop down, but by far not as dramatically as observed for the annuals and herbaceous plants. The Fabaceae, almost not present in 2006, extend their share, starting from the area's western part, up to nearly 0.6 in 2009 and drop – in line with the annuals – back to 0.2 in 2011. The low shares of all these groups in 2010 and 2011 come from the above-mentioned vast increase of bryophyte covers. Only the woody plants' share of the total cover keeps steadily increasing, albeit on a very low level so far.

Soil and Groundwater

After completion of the catchment construction, soil samples were taken at each grid point at a depth of 0–30 cm between October 2005 and April 2006 (see Fig. 1). The samples were air-dried, passed through a 2 mm mesh and analysed for pH (water extract, Beckmann pH34 glass electrode and WTW pH537), electrical conductivity (EC, Hanna HI 8733 and WTW LF537), texture (sieving and sedimentation procedure with Köhn pipette method), total content of carbon (CT), nitrogen (NT) and sulfur (ST, elementar analyser Vario EL III), inorganic (carbonate, Scheibler calcimeter) and or ganic carbon (Corg, calculated as difference CT–$CaCO_3$-C) content (Sparks, 1996; Dane and Topp, 2002).

We assumed the soil data from a grid point being valid for the whole vegetation square. Variables that turned out to be promising for this study from a theoretical point of view as well as in visual exploratory data analyses were the percentage of Corg(CORGP), the percentage of gravel content (GRAVCONT), and MSAND, the percentage of medium sand (see Tables 1, 2). Hereby, the gravel fraction of a soil is defined by grain sizes of more than 2 mm. Corg – mostly from fossil sources – was present in the initial situation, albeit with low concentrations only (Risse-Buhl et al., 2013). These generally low concentrations of CORGP show considerable variation between 0 and almost 0.6 % (Table 2). Spatially, CORGP is not evenly distributed, higher concentrations are found in the south-west half of the area. Mean gravel content (GRAVCONT) is about 15 % (Table 2) with a slight tendency towards higher values again in the south-west zone. At an average of 47 % MSAND shows a homogeneous spatial distribution across the catchment (Table 2).

A total of 21 groundwater gauges were initially installed across the catchment. Nine of them were equipped with pressure transducers to register groundwater levels automatically. The levels at the other gauges were measured manually every two weeks. During the following years, five additional groundwater gauges were installed. The positions of all 26 gauges can be taken from Fig. 1. We simply attributed the mean annual distance of the groundwater level from the surface (variable GAUGE, see Table 1) from the nearest measurement shaft to each grid point, with an average distance of 19 m and maximum distance of 53 m from a square midpoint. We used that simple method because no complex assumptions are made as is the case for any interpolation method.

Since 2006, GAUGE shows a declining trend from on average 1.8 down to 0.9 m in 2011, indicating rising groundwater levels over time. Their spatial distribution is homogeneous in the beginning with higher levels only in the lowest part (south-east) with a slight tendency towards drier conditions at the south-west edge developing during the observation period.

Terrestrial Laser Scans

We used a terrestrial laser scanner (mod. Riegl LMS Z420i) in last target mode to measure 3-D ground surface and vegetation height and density simultaneously. In order to keep the impact on the catchment at a minimum, we only measured from 13 permanently fixed scan positions (Fig. 1), each geo-referenced with 30 DGPS-fixed points (DGPS, differential global positioning system), which allowed us to keep the standard deviation of the geo-referencing error below 2.5 cm.

The scan positions were spatially arranged in a way that allowed us to maintain a horizontal measurement resolution of at least 10 cm × 10 cm at a hypothetical horizontal ground surface. Given the limited number of scan positions, this was only possible by mounting the laser scanner on a portable 6 m tower. This height is sufficient for achieving the desired minimum resolution but not too high in order to hit existing vegetation mainly from more lateral and not so much from vertical directions. This is important, because vertically oriented vegetation like grasses is more likely detected from the side than from above and the side view is better suited to detect vegetation layers under emerging woody plants. See the method comparison by Schneider et al. (2012) for more technical details.

We conducted seven area-wide laser scan campaigns between 2008 and 2011 (September 2008, Mai 2009, August 2009, April 2010, October 2010, February 2011, and May 2011). Although it was not easy to find appropriate time windows (for such a campaign several days with clear weather and low wind are required) the measurements were timed so that at least in the years 2009 and 2010 roughly the catchment's state at the beginning and the end of the growing season could be covered. With our scans we covered the whole catchment amounting to 6 ha altogether. The raw laser scan data are three-dimensional point clouds, each point indicating the position where the laser beam hit an object. In order to separate vegetation from ground surface we divided the covered area into about 200 000 0.5 m × 0.5 m grid cells. On each cell we extracted the minimum vertical coordinate from the point cloud and assumed it to best represent the ground surface level. Our estimate for vegetation height on such a cell is the vertical range of point coordinates, i.e. maximum minus minimum vertical coordinate. As a proxy for vegetation density on each cell, we took the median vertical coordinate (with the estimated ground level being zero) divided by the estimated vegetation height. This results in a number between 0 and 1. Values near 1 would indicate a high density, because the laser signal will be more often reflected in higher regions if vegetation is dense, resulting in a median nearer the maximum height and for less dense vegetation more reflections are measured at the ground, shifting the median lower. Values near zero result from a small median compared to vegetation height, thus indicating the obvious high penetrability when density is low. We also calculated the product of vegetation height and vegetation density, which could serve as a rough proxy for biomass.

In order to detect rills we used a regression-based approach. For each 0.5 m × 0.5 m laser scan cell we took all neighbor cells whose midpoints were inside a radius of 6 m around the midpoint of the central cell. With the horizontal coordinates (x, y) of the cell midpoints as the independent variables and the vertical coordinate (z) of the neighboring cells' ground level as the dependent variable we fitted a simple ordinary least squares (OLS) regression model of the following form:

$$z = a \cdot x + b \cdot y + c + \varepsilon, \quad (1)$$

with a, b, c being parameters to be estimated and ε representing the error term. Equation (1) thus represents a plane that after fitting can be interpreted as a smoothed description of the terrain surface in the cell of interest's proximity defined by the 6 m radius as explained above. If the actual ground level z of the central cell is more than 5 cm lower than the one estimated from Eq. (1), then this cell is assumed to belong to a rill. Visual comparisons with aerial photographs from the catchment show a good agreement with this rill detection method.

In addition to this 6 m neighbor analysis, we used the same method with a 0.8 m radius around the cell of interest and calculated the difference between the greatest and the smallest vertical deviation from the regression plane. This can be interpreted as a description of relief energy in the nearest neighborhood. This potential energy is used as an indicator if the cell is located on a jagged or smooth surface and thus how large the influence of surface runoff events might be or how likely a successful seed deposition may happen. Changes of the ground surface level inform about erosion and deposition processes. For each cell we calculated the difference between the surface level at a given and a previous survey, requesting a survey-to-survey time distance closest to one year. Negative numbers, i.e. decreasing surface levels, indicate erosion while positive numbers indicate deposition. For our analysis the changes were always attributed to the first of the two surveys, in other words to the status before the change.

To each 20 m × 20 m grid point we attributed every laser scan grid cell with its midpoint inside the corresponding 5 m × 5 m vegetation square and used only those cells thereafter. This resulted in about 100 laser scan cells per vegetation square and implied two options for data evaluation. First is the extraction and evaluation of sum and mean values from the laser scan data, one number per vegetation square, such as surface roughness or percentage of rill cells. Second is an analysis on the level of single laser scan grid cells. However, preliminary analyses showed that the latter tended to be less revealing than the former while consuming disproportional computing time. Thus, we focused on the former, more aggregated option.

For further analysis, the following variables were calculated from the laser scan data, resulting in one number per grid point and survey (cf. Table 1): VEGHEIGHT and VEGDENS are the mean vegetation height and the mean vegetation density of all laser scan grid cells on a vegetation square. VEGHD is their product. RILLR is the proportion of rill cells as defined above on a given vegetation square while RELEN is the average relief energy of all laser scan cells on one vegetation square. DHEIGHT represents the average surface height change of all laser scan cells belonging to a given vegetation square.

On average VEGHEIGHT more than doubles from 2008 to 2011 (0.35 to 0.80 m, Table 2). The highest vegetation concentrates in the southern and south-west part of the catchment in 2011. The vegetation density, VEGDENS, on average remains constant during the observation period, albeit with changing spatial distribution (Table 2). Its overall constancy despite increasing heights confirms its adequacy for expressing vegetation density independent from plant size. Accordingly VEGDH, the product of both, doubles from 2008 to 2011 in a similar way as VEGHEIGHT, being greatest in the southern corner and the east side of the area, where both, height and density are high (Table 2,).

The average proportion of rill cells per vegetation square, RILLR, does only slightly increase from 0.10 to 0.16 from 2008 to 2011 (Table 2). Being concentrated on the northern, western and southern edges of the investigated area from 2008 to 2010 a redistribution towards the south-eastern part of the catchment seems to have occurred between 2010 and 2011. Similar to RILLR, the local relief energy, RELEN, stays nearly constant between 0.12 and 0.13 from 2008 to 2010 but almost doubles from 2010 to 2011 accompanied with a spatial redistribution of the highest values towards the eastern edge (Table 2). The surface height change, DHEIGHT, undergoes a reversion during the observation time span. In average, one vegetation square lost 3 cm of terrain height between 2008 and 2009, from 2009 to 2010 there was on average no change, and from 2010 to 2011 we observe an average surface height increase of 4 cm (Table 2). From 2010 to 2011 deposition dominates the whole area, being strongest in the north-western part.

Statistical Evaluation

We chose generalized linear models (GLMs) and generalized additive models (GAMs) as the basic method for our statistical evaluations, because they allow us to deal with different distribution families for the response variable. In addition, GAMs offer a convenient way to incorporate nonlinear effects into the analyses. However, as we take a vegetation square as our basic observation unit, several subsequent observations are available per vegetation square. We covered the resulting autocorrelation by introducing a random effect on vegetation square level. This enlarges GLMs to generalized linear mixed models (GLMMs) and GAMs to generalized additive mixed models (GAMM's). See Zuur et al. (2009) for a detailed description of these model types and Pinheiro and Bates (2000) for further information about mixed models. Preliminary tests showed that including spatial correlation between vegetation squares did not improve our models.

The core of a GLMM in the context of this study is a linear predictor function η of the following form:

$$\eta(X_{1\,ij},\ldots,X_{q\,ij}) = \alpha + \beta_1 \cdot X_{1\,ij} + \ldots + \beta_q \cdot X_{q\,ij} + a_i$$
$$a_i \sim N(0,\sigma_a^2), \tag{2}$$

where $X_1 \ldots X_q$, is a set of q explanatory variables. The indices i and j denote the j th observation on the ith vegetation square. α and β are regression parameters and a_i is a vegetation-square specific random effect. In case of a GAMM, the predictor function enlarges to

$$\eta\left(X_{1\,ij},\ldots,X_{q\,ij},K_{1\,ij},\ldots,K_{n\,ij}\right) = \alpha + \beta_1 \cdot X_{1\,ij} + \ldots + \beta_q \cdot X_{q\,ij} + f_1\left(K_{1\,ij}\right) + \ldots + f_n\left(K_{n\,ij}\right) + a_i, \tag{3}$$

where, in addition, $K_1 \ldots K_n$ is a set of n explanatory variables, and $f_1 \ldots f_n$ is a set of nonparametric smoother functions (cf. Zuur et al., 2009).

The way, that η is transformed into an expected value μij for the actual response variable Y_{ij} is defined by the link function g.

$$g\left(\mu_{ij}\right) = \eta\left(X_{1\,ij},\ldots,X_{q\,ij},K_{1\,ij},\ldots,K_{n\,ij}\right) \tag{4}$$

If the response variable Y_{ij} is assumed to be normally distributed ($Y_{ij} \sim N(\mu_{ij}, \sigma^2)$) then the usual link function is identity:

$$\mu_{ij} = \eta\left(X_{1\,ij},\ldots,X_{q\,ij},K_{1\,ij},\ldots,K_{n\,ij}.\right) \tag{5}$$

This is the first of three options we used in this study.

In cases where Y_{ij} was proportional data with a tendency towards over-dispersion, a quasi-binomial distribution for Y_{ij} with $var(Y_{ij}) = \tau \cdot \mu_{ij} \cdot (1-\mu_{ij})$ and the dispersion parameter τ not being fixed to 1 and a logit link

$$\ln\left(\frac{\mu_{ij}}{1-\mu_{ij}}\right) = \eta\left(X_{1\,ij},\ldots,X_{q\,ij},K_{1\,ij},\ldots,K_{n\,ij}\right) \tag{6}$$

was used. When $Y_{ij} > 0$, the assumption of a gamma distribution for Y_{ij} with $var(Y_{ij}) = \frac{\mu_{ij}^2}{\nu}$ and ν being the dispersion parameter turned out to be useful in combination with a logarithmic link:

$$\ln\left(\mu_{ij}\right) = \eta(X_{1\,ij},\ldots,X_{q\,ij},K_{1\,ij},\ldots,K_{n\,ij}). \tag{7}$$

Promising explanatory variables for each single regression analysis were pre-selected by theoretical considerations and, as proposed by Zuur et al. (2009), based on descriptive exploratory data analyses such as pair-plots and variance inflation factors. If plausibility, visual data inspections or residual plots suggested (partly) nonlinear relationships, a GAMM was chosen instead of a GLMM. In each analysis, we started with the full set of pre-selected explanatory variables, leaving them out stepwise, by AIC comparison and re- fitting each time (Zuur et al., 2009). Model assumptions were confirmed by graphical displays.

Presenting the most important results of a model fit is straightforward for a GLMM (Eq. 2). Presenting the linear parameters' β_1, ..., β_q estimates together with their standard errors and significance levels is standard. In the case of a GAMM, this can be done in exactly the same way for the linear components of the model. However, the components which are expressed by nonparametric smoother functions ($f_1(K_{1ij})$, ..., $f_n(K_{nij})$ in Eq. 3) can only be displayed visually. Thus, for a GAMM, a table is needed for presenting the linear components and a diagram is needed for each component that is represented by a smoother function.

In order to scrutinize H1, "do the initial geomorphological and substrate characteristics determine the surface's structure and properties of the terrain?", we investigated several response variables (see Table 1):

- RILLR: the proportion of rill cells per vegetation square and measurement. As RILLR represents proportional data, the quasibinomial distribution with logit link was used.
- RELEN: the mean value of the local relief energy of all laser scan cells belonging to one vegetation square. Assuming a gamma distribution combined with the logarithmic link matched the data well from a theoretical point of view and yielded good model fits for this response variable.
- DHEIGHT: the ground's height change (see Table 1 for details) was investigated with a regression model that assumes normal distribution of the response variable and uses the identity link.

Due to the definition of H1 we did not include any vegetation properties as explanatory variables in these analyses.

In the context of H2, asking if surface and substrate characteristics determine the plant species groups to be found initially, the following set of vegetation properties was chosen as response variables (see Table 1):

- PROPANN: the portion of annual plants related to the total coverage per vegetation square and survey.
- PROPHERB: same for the portion of herbaceous plants.
- PROPGRASS: same for grass including sedges.
- PROPFAB: same for plants belonging to the nitrogen-fixing Fabaceae family.
- PROPWOOD: same for woody plant species.

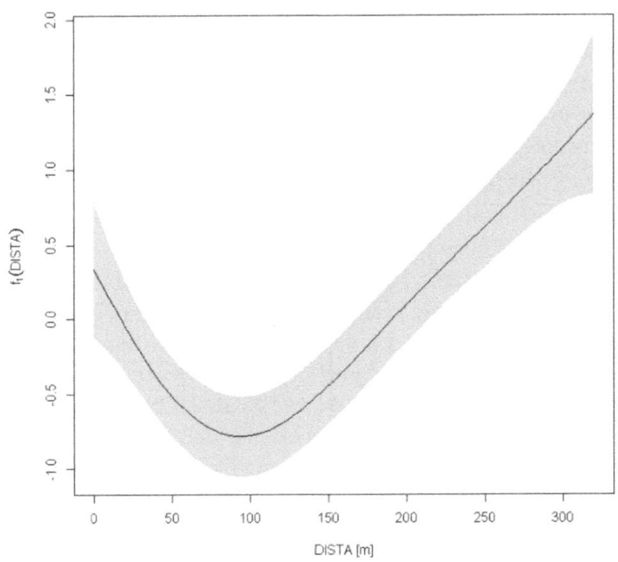

Figure 2. Figure 2 Nonlinear effect of a vegetation square's distance from gridrow A (DISTA) on the portion of rill cells (RILLR) (see Eq. 8). Shaded: 95 %-confidence area.

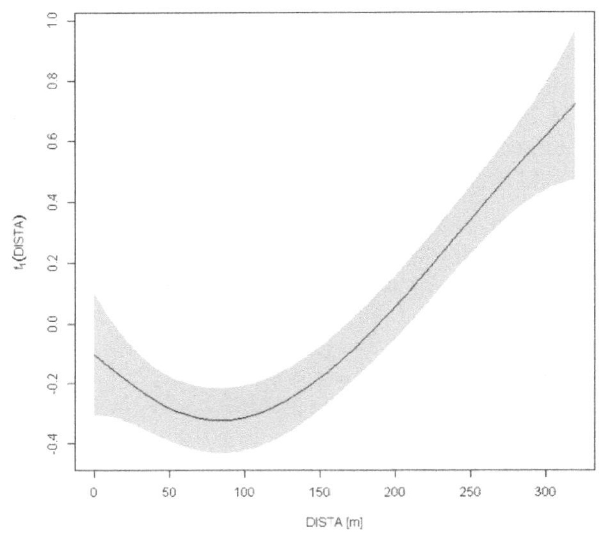

Figure 3. Nonlinear effect of a vegetation square's distance from gridrow A (DISTA) on the local relief energy (RELEN) (see Eq. 9). Shaded: 95 %-confidence area.

For all these proportions, assuming quasi-binomial distribution and using a logit link turned out to be the best option. In addition, we also used two laser-scanned vegetation properties as response variables:

- VEGHEIGHT and VEGDENS: the mean vegetation height and vegetation density per vegetation square and survey.

The former was investigated assuming normal distribution and using the identity link while for the latter quasi-binomial distribution with logit link was applied. For testing H3, if increasing density of vegetation reduces the erosion, we used DHEIGHT: the laser-scanned surface height changes as a response variable.

We did so in the same way as for H1, but tested several ways of including the laser-scanned vegetation properties VEGHEIGHT and VEGDENS as additional explanatory variables. The other explanatory variables were those that showed significance in the context of H1.

The overall important components of temporality and spatiality are included in the variables TIME and DISTA. The final models with the final set of explanatory variables are shown in the results section. Table 1 explains all variables used in this study, and Table 2 presents their means, minima and maxima for each observation year.

For all statistical evaluations we used the free software R version 2.15.1 (R Core Team, 2012), namely the package mgcv (Wood, 2006).

RESULTS

Influence of Geomorphological and Substrate Characteristics on Surface Structure (H1)

The final model for describing the dependency of the portion of rill cells, RILLR, was a GAMM with logit link (cf. Sect. 2.3, Eqs. 3, 6):

$$\ln\left(\frac{E(\mathrm{RILLR}_{ij})}{1-E(\mathrm{RILLR}_{ij})}\right) = \alpha + \beta_1 \cdot \mathrm{CORGP}_i + \beta_2 \cdot \mathrm{GAUGE}_{ij} + \beta_3 \cdot \mathrm{TIME}_j + \beta_4 \cdot \mathrm{CORGP}_i \cdot \mathrm{TIME}_j + f_1(\mathrm{DISTA}_i) + a_i. \quad (8)$$

The indices i and j , and the random effect a have the same meaning as in Eq. (2), and E(RILLR) represents the expected value of RILLR. CORGP is the percentage of organic carbon in the soil, GAUGE is the mean annual groundwater level in m below surface, and f_1(DISTA) is a smoother function that represents the effect of DISTA (see Table 1). All explanatory variables show highly significant ($p < 0.001$) effects.

Figure 2 shows the effect of DISTA. Clearly there is a general trend of more likely finding rills the more one moves further downslope, however with a minimum between 50 and 150 m from row A. Table 4 lists the parameter estimates for α and β_1, ..., β_4. The parameters for the variables CORGP, GAUGE, and TIME are greater than 0, indicating that the probability of encountering rills increases with the percentage of organic carbon (as measured in 2005), a lower annual groundwater level (higher GAUGE) and with time. However, there is also a significant interaction that indicates the influence of CORGP weakens with time.

Table 4. Parameter estimates and significances for the model estimating the portion of rill cells (RILLR) as shown in Eq. (8). Significance levels: ***: $p < 0.001$, **: $p < 0.01$, *: $p < 0.05$. The variance of the random effect a in Eq. (8) is 0.5456. Trend: qualitative illustration of linear predictor variables' significant influences. ↑, ↓: RILLR increases (decreases) with increasing values of the respective predictor variable

Variable	Trend	Parameter	Estimate	Std. Error	Significance
		α	−8.1121	0.5397	***
CORGP	↑	β_1	10.4752	2.0661	***
GAUGE	↑	β_2	1.0374	0.1753	***
TIME	↑	β_3	0.8624	0.0820	***
CORGP*TIME	↓	β_4	−1.7634	0.3555	***
f_1(DISTA)		Nonparametric smoother	See Fig. 2		***

Table 5. Parameter estimates and significances for the model estimating the local relief energy (RELEN) as shown in Eq. (9). Significance levels: ***: $p < 0.001$, **: $p < 0.01$, *: $p < 0.05$. The variance of the random effect a in Eq. (9) is 0.1119. Trend: qualitative illustration of linear predictor variables' significant influences. ↑, ↓: RELEN increases (decreases) with increasing values of the respective predictor variable.

Variable	Trend	Parameter	Estimate	Std. Error	Significance
		α	−5.5417	0.2758	***
CORGP	↑	β_1	5.7311	1.1252	***
GAUGE	↑	β_2	0.5085	0.0894	***
TIME	↑	β_3	0.5726	0.0434	***
CORGP*TIME	↓	β_4	−1.0869	0.2031	***
f_1(DISTA)		Nonparametric smoother	See Fig. 3		***

Table 6. Parameter estimates and significances for the model estimating surface height change (DHEIGHT) as shown in Eq. (10). Significance levels: ***: $p < 0.001$, **: $p < 0.01$, *: $p < 0.05$. The variance of the random effect a in Eq. (10) is 0.0002. Trend: qualitative illustration of linear predictor variables' significant influences. ↑, ↓: DHEIGHT increases (decreases) with increasing values of the respective predictor variable. Thus, ↓ indicates a tendency towards erosion while ↑ indicates trends that counteract erosion

Variable	Trend	Parameter	Estimate	Std. Error	Significance
		α	−0.0414	0.0825	
MSAND	↓	β_1	−0.0031	0.0015	*
GAUGE	↓	β_2	−0.0286	0.0058	***
RILLR	↓	β_3	−0.2559	0.0970	**
DISTA	↑	β_4	0.0006	0.0001	***
TIME		β_5	0.0226	0.0169	
MSAND*TIME	↑	β_6	0.0007	0.0003	*
RILLR*TIME	↑	β_7	0.0429	0.0217	*
DISTA*TIME	↓	β_8	−0.0002	0.0000	***

Testing the local relief energy, RELEN, the right hand side of the final GAMM model was the same as in Eq. (8), however, due to the logarithmic link function the left-hand side is different:

$$\ln\left(E\left(\mathrm{RELEN}_{ij}\right)\right) = \alpha + \beta_1 \cdot \mathrm{CORGP}_i + \beta_2 \cdot \mathrm{GAUGE}_{ij} + \beta_3 \cdot \mathrm{TIME}_j + \beta_4 \cdot \mathrm{CORGP}_i \cdot \mathrm{TIME}_j + f_1\left(\mathrm{DISTA}_i\right) + a_i. \quad (9)$$

The results as presented in Table 5 and Fig. 3 show a strong affinity between the local relief energy and the rill probability. Again, relief energy generally increases with increasing DISTA, however with a minimum at about 100 m from row A (Fig. 3). Relief energy also increases with CORGP (Table 5), lower groundwater levels (higher GAUGE) and, expectedly, increases with TIME (parameters $\beta_1, \ldots, \beta_3 > 0$), however the negative value obtained for $\beta4$ shows a weakening influence of CORGP with time.

The change of the surface height, DHEIGHT, in relation to non-vegetation variables only, yielded a final model of the following GLMM structure:

$$E\left(\mathrm{DHEIGHT}_{ij}\right) = \alpha + \beta_1 \cdot \mathrm{MSAND}_i + \beta_2 \cdot \mathrm{GAUGE}_{ij} + \beta_3 \cdot \mathrm{RILLR}_{ij} + \beta_4 \cdot \mathrm{DISTA}_i + \beta_5 \cdot \mathrm{TIME}_j + \beta_6 \cdot \mathrm{MSAND}_i \cdot \mathrm{TIME}_j + \beta_7 \cdot \mathrm{RILLR}_{ij} \cdot \mathrm{TIME}_j + \beta_8 \cdot \mathrm{DISTA}_i \cdot \mathrm{TIME}_j + a_i. \quad (4)$$

In addition to the afore-mentioned symbols, MSAND means the percentage of medium sand (grain size 0.2–0.63 mm) in the substrate (as measured in 2005). The parameter estimates and significances are listed in Table 6. Significances are partly weaker than in the models shown above. The effect of MSAND ($\beta_1 < 0$) means that more material (volume) erodes at places where more medium sand was found in 2005. Lower groundwater levels are connected with greater material export ($\beta_2 < 0$), the more rills we find, the more material is exported ($\beta_3 < 0$). There is a weak, but highly significant tendency towards a surface height increase – indicating deposition effects – with increasing DISTA ($\beta_4 > 0$) while time has no significant isolated effect, however the analysis reveals significant time effects in interaction with other variables. Parameter $\beta6$, being significantly greater than 0 indicates that the above-mentioned effect of MSAND weakens with time, and the same is true for the effect of RILLR ($\beta_7 > 0$). A reverse effect is observed for DISTA in interaction with time ($\beta_8 < 0$), which describes well the spatial change of DHEIGHT as visualized.

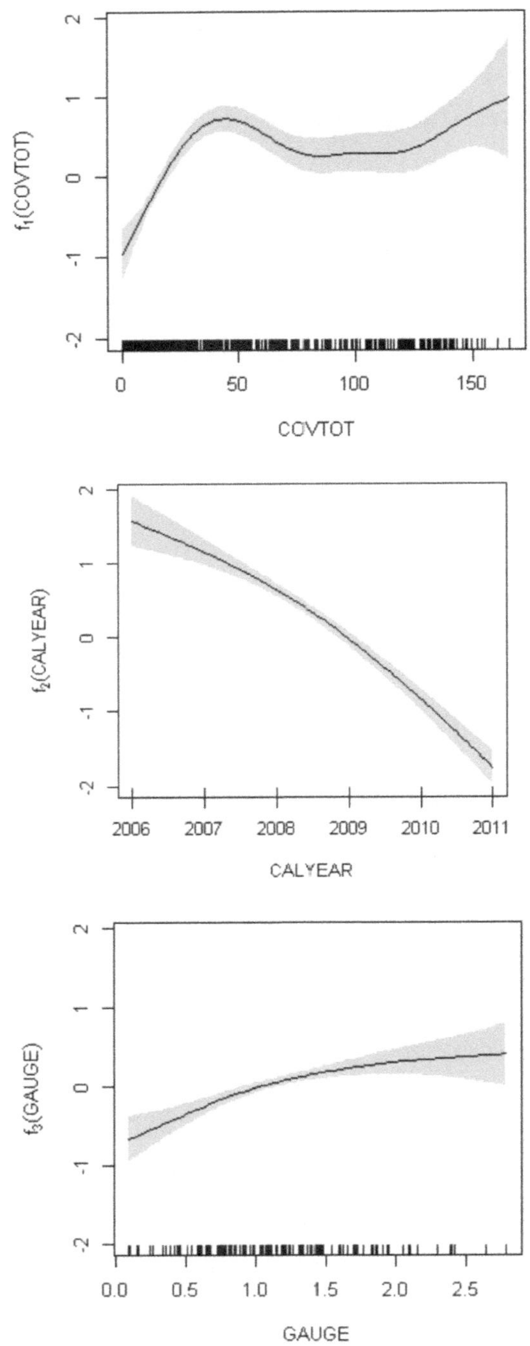

Figure 4. Nonlinear effects of the total degree of coverage (COVTOT), calendar year (CALYEAR), and groundwater level (GAUGE) on the proportion of annual plants (PROPANN) (see Eq. 11). Shaded: 95 %-confidence area.

Table 7. Parameter estimates and significances for the model estimating the proportion of annual plants (PROPANN) as shown in Eq. (11). Significance levels: ***: $p < 0.001$, **: $p < 0.01$, *: $p < 0.05$. The variance of the random effect a in Eq. (11) is 0.1594. Trend: qualitative illustration of linear predictor variables' significant influences. ↑, ↓: PROPANN increases (decreases) with increasing values of the respective predictor variable

Variable	Trend	Parameter	Estimate	Std. Error	Significance
		α	0.2551	0.0868	**
CORGP	↓	β_1	−1.7067	0.4431	***
f_1(COVTOT)		Nonparametric smoother	See Fig. 4		***
f_2(CALYEAR)		Nonparametric smoother	See Fig. 4		***
f_3(GAUGE)		Nonparametric smoother	See Fig. 4		***

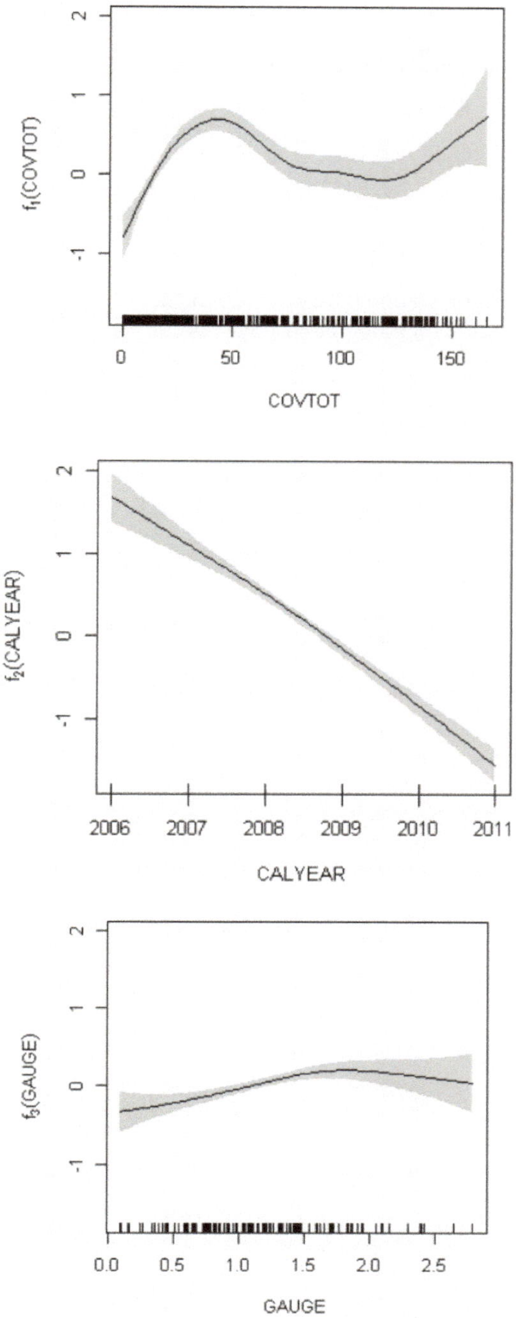

Figure 5. Nonlinear effects of the total degree of coverage (COVTOT), calendar year (CALYEAR), and groundwater level (GAUGE) on the proportion of herbaceous plants (PROPHERB) (see Eq. 12). Shaded: 95 %-confidence area.

Table 8. Parameter estimates and significances for the model estimating the proportion of herbaceous plants (PROPHERB) as shown in Eq. (12). Significance levels: ***: $p < 0.001$, **: $p < 0.01$, *: $p < 0.05$. The variance of the random effect a in Eq. (12) is 0.0855. Trend: qualitative illustration of linear predictor variables' significant influences. ↑, ↓: PROPHERB increases (decreases) with increasing values of the respective predictor variable

Variable	Trend	Parameter	Estimate	Std. Error	Significance
		α	0.6254	0.0685	***
CORGP	↓	β_1	−0.9162	0.3481	**
f_1(COVTOT)		Nonparametric smoother	See Fig. 5		***
f_2(CALYEAR)		Nonparametric smoother	See Fig. 5		***
f_3(GAUGE)		Nonparametric smoother	See Fig. 5		***

Influence of Surface and Substrate Characteristics on Vegetation Structure (H2)

With the proportion of annual plants, PROPANN, as the response variable the following final model resulted (GAMM with logit-link):

$$\ln\left(\frac{E(\text{PROPANN}_{ij})}{1-E(\text{PROPANN}_{ij})}\right) = \alpha + \beta_1 \cdot \text{CORGP}_i + f_1\left(\text{COVTOT}_{ij}\right) + f_2\left(\text{CALYEAR}_j\right) + f_3\left(\text{GAUGE}_{ij}\right) + a_i. \quad (11)$$

With $\beta_1 < 0$, Table 7 shows a negative correlation between CORGP and the proportion of annual plants. The nonlinear effect of COVTOT, the respective vegetation plot's total degree of coverage in percent (Fig. 4) reveals an optimum at roughly 50 % while the increase at very high degrees of coverage is not significant (broad confidence interval). The proportion of annual plants decreases with the calendar year (Fig. 4), while the probability of finding annual plants on a vegetation plot increases with lower groundwater levels (higher GAUGE) roughly in the shape of a saturation curve (Fig. 4).

Table 9. Parameter estimates and significances for the model estimating the proportion of grasses (PROPGRASS) as shown in Eq. (13). Significance levels: ***: $p < 0.001$, **: $p < 0.01$, *: $p < 0.05$. The variance of the random effect a in Eq. (13) is 0.1489

Variable	Parameter	Estimate	Std. Error	Significance
	α	−1.9460	0.0783	***
CORGP	β_1	0.7366	0.3960	$p = 0.0633$
f_1(COVTOT)	Nonparametric smoother	See Fig. 6		***
f_2(CALYEAR)	Nonparametric smoother	See Fig. 6		***

Table 10. Parameter estimates and significances for the model estimating the proportion of woody plants (PROPWOOD) as shown in Eq. (14). Significance levels: ***: $p < 0.001$, **: $p < 0.01$, *: $p < 0.05$. The variance of the random effect a in Eq. (14) is 3.1987. Trend: qualitative illustration of linear predictor variables' significant influences. ↑, ↓: PROPWOOD increases (decreases) with increasing values of the respective predictor variable

Variable	Trend	Parameter	Estimate	Std. Error	Significance
		α	−1118.7522	98.9546	***
COVTOT	↓	β_1	−0.0119	0.0018	***
CALYEAR	↑	β_2	0.5542	0.0493	***

A very similar final model resulted for PROPHERB, the proportion of herbaceous plants (GAMM with logit link).

$$\ln\left(\frac{E(\text{PROPHERB}_{ij})}{1-E(\text{PROPHERB}_{ij})}\right) = \alpha + \beta_1 \cdot \text{CORGP}_i + f_1\left(\text{COVTOT}_{ij}\right) + f_2\left(\text{CALYEAR}_j\right) + f_3\left(\text{GAUGE}_{ij}\right) + a_i \quad (12)$$

The effects observed for PROPHERB are virtually the same as obtained for annual plants (Fig. 5, Table 8), an optimum pattern for COVTOT, a decreasing curve for the calendar year and a saturation curve for the groundwater level, and a negative correlation with CORGP.

The scrutiny of the proportion of grasses (PROGRASS) yielded an even simpler model:

$$\ln\left(\frac{E(\text{PROPGRASS}_{ij})}{1-E(\text{PROPGRASS}_{ij})}\right) = \alpha + \beta_1 \cdot \text{CORGP}_i + f_1(\text{COVTOT}_{ij}) + f_2(\text{CALYEAR}_j) + a_i. \quad (13)$$

Compared to the herbaceous plants, the grasses show an opposing trend. Even though the effect of CORG is just not significant (p = 0.0633, Table 9), the grasses seem to be connected to places with higher organic carbon content in the soil. As Fig. 6 shows, the coverage of grasses decreases strongly with the total degree of coverage (COVTOT), while it increases with the calendar year. This indicates that grasslike plants preferentially establish where the total cover is not too high.

For the proportion of woody plants (PROPWOOD) a simple GLMM resulted:

$$\ln\left(\frac{E(\text{PROPWOOD}_{ij})}{1-E(\text{PROPWOOD}_{ij})}\right) = \alpha + \beta_1 \cdot \text{COVTOT}_{ij} + \beta_2 \cdot \text{CALYEAR}_j + a_i. \quad (14)$$

Only two explanatory variables show a significant relationship in this context. Trivially, the proportion of woody plants increases with time (CALYEAR). Less expectedly, the probability of finding woody plants decreases as the total coverage on a vegetation square increases (Table 10).

The final model for the proportion of Fabaceae (PROPFAB) resulted in the following structure:

$$\ln\left(\frac{E(\text{PROPFAB}_{ij})}{1-E(\text{PROPFAB}_{ij})}\right) = \alpha + \beta_1 \cdot \text{GAUGE}_{ij} + f_1(\text{COVTOT}_{ij}) + f_2(\text{CALYEAR}_j) + f_3(\text{GRAVCONT}_i) + a_i, \quad (15)$$

with GRAVCONT representing the soil's gravel content in percent as measured in 2005.

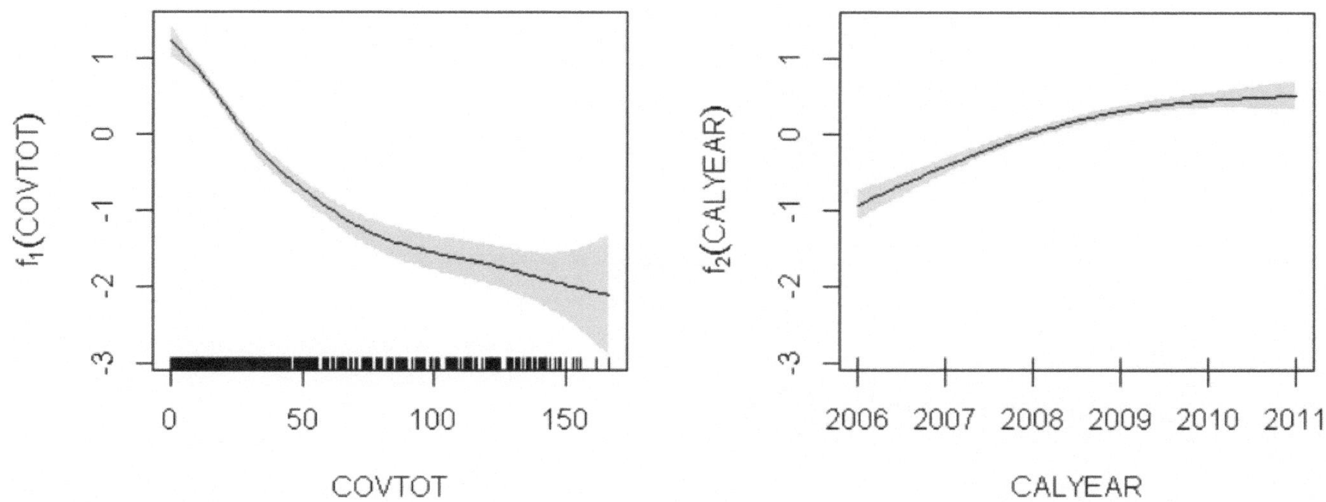

Figure 6. Figure 6 Nonlinear effects of the total degree of coverage (COVTOT), and calendar year (CALYEAR), on the proportion of grasses (PROPGRASS) (see Eq. 13). Shaded: 95 %-confidence area.

As Table 11 shows, there is a positive correlation between lower groundwater levels and the occurrence of Fabaceae (β1 > 0). Up to gravel contents of 10 % and up to total degrees of coverage of about 20 %, their proportion is positively correlated with both variables (Fig. 7). Remarkably, the occurrence of Fabaceae over time showed a maximum in the years 2008 and 2009 (Fig. 7).

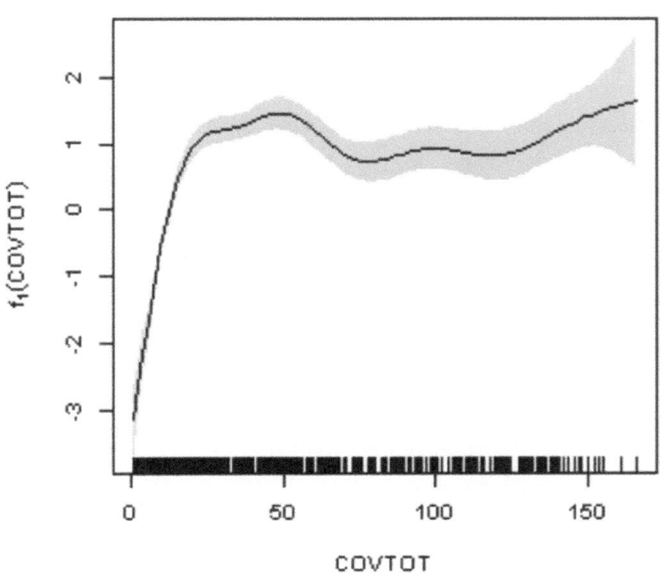

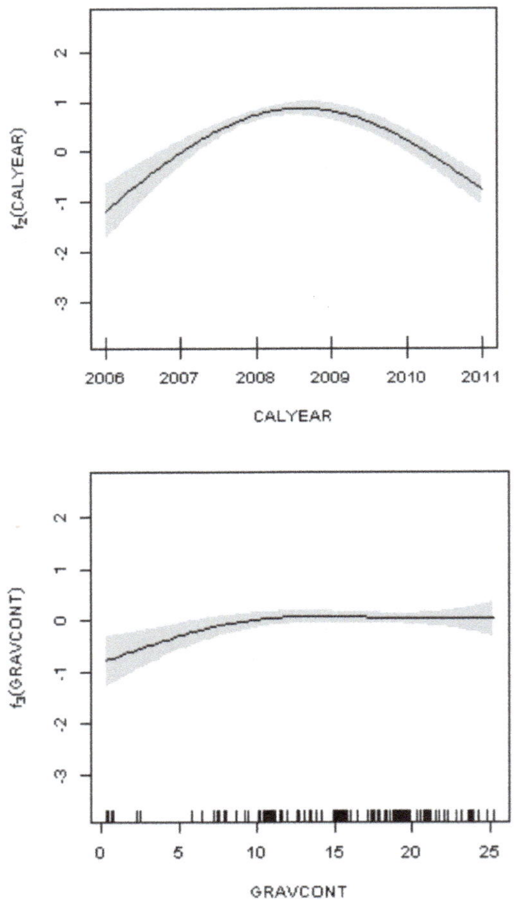

Figure 7. Nonlinear effects of the total degree of coverage (COVTOT), calendar year (CALYEAR), and gravel content (GRAVCONT), on the proportion of Fabaceae (PROPFAB) (see Eq. 15). Shaded: 95 %-confidence area.

Table 11. Parameter estimates and significances for the model estimating the proportion of Fabaceae (PROPFAB) as shown in Eq. (15). Significance levels: ***: $p < 0.001$, **: $p < 0.01$, *: $p < 0.05$. The variance of the random effect a in Eq. (15) is 0.2288. Trend: qualitative illustration of linear predictor variables' significant influences. ↑, ↓: PROPFAB increases (decreases) with increasing values of the respective predictor variable

Variable	Trend	Parameter	Estimate	Std. Error	Significance
		α	−2.6636	0.1723	***
GAUGE	↑	β_1	0.8693	0.1301	$p = 0.0633$
f_1(COVTOT)		Nonparametric smoother	See Fig. 7		***
f_2(CALYEAR)		Nonparametric smoother	See Fig. 7		***
f_3(GRAVCONT)		Nonparametric smoother	See Fig. 7		***

VEGHEIGHT, the mean vegetation height, as obtained from laser scans was described with the following model:

$$E\left(\text{VEGHEIGHT}_{ij}\right) = \alpha + \beta_1 \cdot \text{COVTOT}_{ij} + \beta_2 \cdot \text{GAUGE}_{ij} + f_1\left(\text{CALYEAR}_j\right) + f_2\left(\text{DISTA}_i\right) + a_i. \quad (16)$$

Figure 8 (left) shows that vegetation height increases with time, expectedly in an exponential pattern as it is typical for initial growth processes. The same figure (Fig. 8, right) reveals a clear trend of increasing vegetation height with the distance from grid-row A. Depth to groundwater table and coverage degrees are positively correlated with vegetation height ($\beta_1 > 0$, $\beta_2 > 0$, Table 12). No significant relations of other variables with the laser scanned vegetation density, VEGDENS, could be identified.

Table 12. Parameter estimates and significances for the model estimating the laser-scanned vegetation height (VEGHEIGHT) as shown in Eq. (16). Significance levels: ***: $p < 0.001$, **: $p < 0.01$, *: $p < 0.05$. The variance of the random effect a in Eq. (16) is 0.0074. Trend: qualitative illustration of linear predictor variables' significant influences. ↑, ↓: VEGHEIGHT increases (decreases) with increasing values of the respective predictor variable

Variable	Trend	Parameter	Estimate	Std. Error	Significance
		α	0.3851	0.0367	***
COVTOT	↑	β_1	0.0007	0.0003	*
GAUGE	↑	β_2	0.0879	0.0301	**
f_1(CALYEAR)		Nonparametric smoother	See Fig. 8		***
f_2(DISTA)		Nonparametric smoother	See Fig. 8		***

Table 13. Parameter estimates and significances for the model estimating surface height change (DHEIGHT) as shown in Eq. (17). Significance levels: ***: $p < 0.001$, **: $p < 0.01$, *: $p < 0.05$. The variance of the random effect a in Eq. (17) is 0.0006. Trend: qualitative illustration of linear predictor variables' significant influences. ↑, ↓: DHEIGHT increases (decreases) with increasing values of the respective predictor variable. Thus, ↓ indicates a tendency towards erosion while ↑ indicates trends that counteract erosion

Variable	Trend	Parameter	Estimate	Std. Error	Significance
		α	−0.0427	0.0253	
MSAND	↓	β_1	−0.0024	0.0004	***
GAUGE	↓	β_2	−0.0392	0.0032	***
RILLR	↓	β_3	−0.0215	0.0072	**
DISTA	↑	β_4	0.0004	0.0000	***
TIME	↑	β_5	0.0189	0.0038	***
MSAND*TIME	↑	β_6	0.0007	0.0000	***
RILLR*TIME	↑	β_7	0.1075	0.0016	***
DISTA*TIME	↓	β_8	−0.0002	0.0000	***
f_1(VEGHD, TIME)		Nonparametric smoother	See Fig. 9		***

Influence of Vegetation Structure on Surface Structure (H3)

In order to scrutinize H3 we took the model formulation from Eq. (8) which relates the change of the laser-scanned DHEIGHT, with soil and surface variables, but added different combinations of the laser-scanned VEGHEIGHT, VEGDENS, and TIME, all referring to the beginning of the period during which DHEIGHT occurred. We found that including the product of VEGHEIGHT * VEGDENS = VEGDH in interaction with time yielded the best model:

$$E\left(\mathrm{DHEIGHT}_{ij}\right) = \alpha + \beta_1 \cdot \mathrm{MSAND}_i + \beta_2 \cdot \mathrm{GAUGE}_{ij} + \beta_3 \cdot \mathrm{RILLR}_{ij} + \beta_4 \cdot \mathrm{DISTA}_i + \beta_5 \cdot \mathrm{TIME}_j + \beta_6 \cdot \mathrm{MSAND}_i \cdot \mathrm{TIME}_j + \beta_7 \cdot \mathrm{RILLR}_{ij} \cdot \mathrm{TIME}_j + \beta_8 \cdot \mathrm{DISTA}_i \cdot \mathrm{TIME}_j + f_1\left(\mathrm{VEGDH}_{ij}, \mathrm{TIME}_j\right) + a_i. \quad (17)$$

Compared to the model without vegetation data as explanatory variables (Eq. 10, Table 6), the parameter estimates for β_1, ..., β_8 are very similar (Table 13), thus, there is no contradiction between this and the other model. However, the nonlinear vegetation effect pronouncedly shows a vegetation impact that counteracts erosion. The higher and denser the vegetation (i.e. the greater VEGDH), the stronger this effect, which also changes with time (Fig. 9). In the first observation period (starting September 2008) the correlation is somewhat weak and almost linear but in the following two periods, the curve becomes more and more bent upwards, the VEGDH value where the bend starts moving down to smaller VEGDH values from the second to the third observation period. This indicates that even with the same VEGDH values the stabilizing effect of vegetation increases with time.

DISCUSSION

Surface structure in the study area is mainly influenced by geomorphological characteristics. The slope degree of the area is not the same throughout the area, but shows a maximum in the lower part (rows L to P in Fig. 1). This is reflected (or at least indicated) by the pronounced and partly nonlinear effects of the downhill distance from grid-row A (DISTA) on the frequency of rills (RILLR), the local relief energy (RELEN), and the ground surface height change (DHEIGHT). The pronounced minima for the downhill distance effect (DISTA) for the rill frequency and the local relief energy (Figs. 2, 3) seem to indicate a special local situation: they are located just downhill an area where, due to the construction works of the catchment, many rills take a perpendicular course to the main slope direction (cf. Schaaf et al., 2013).

Gullies channeling surface runoff results in a further incision both in depth and uphill – a positive feedback, which is strongly supported by our model for the ground surface height change (DHEIGHT, Table 6), especially by the significant effects of the rill frequency (RILLR) and the interaction effect of the downhill position (DISTA) with time.

Although the portion of organic carbon (CORGP) is strongly correlated with rill intensity, it seems most likely that this parameter is a proxy for the general substrate properties. The same could be true for parameter MSAND. According to Gerwin et al. (2010), CORGP strongly correlates with clay content. In general, soil texture in the western part of the catchment is dominated by loamy sands, whereas in the eastern part pure sands dominate. These differences are reflected in the spatial extent and form of the erosion gullies being narrower and deeper in the western part, but shallower and wider in the eastern part (Schaaf et al., 2012). Thus, in general, our results indicate positive feedback processes governing rill formation whereby, however, different rill types evolve on substrates with different properties.

Surface runoff dominated catchment hydrology at least in the early years, resulting in gully formation and transport of large sediment amounts that were deposited in the lower part of the hill slope and in the pond (Schaaf et al., 2012). While doing so surface runoff and erosion transported mainly finer textured material, resulting in even coarser texture at the uphill surface substrates. This probably explains the decreasing influence of the substrate proxy variables CORGP and MSAND with time.

Besides such transport effects, the weakening effects of initial surface and soil properties with time reflect the increasing influence of vegetation coverage (see H3).

In summary, our results strongly support H1. Geomorphology (slope degree) and substrate properties like grain size distribution and resulting differences in groundwater levels affect the system strongly and in turn result in fundamental changes in geomorphology all over the area, but with main effects where the slope degree is highest. The developing vegetation cover, however, reduces the influence of these factors in the course of time, leading to an increasing conservation of the structures that developed at the very beginning.

Some of our results concerning the influence of surface and substrate characteristics on vegetation structure (H2) match basic assumptions of successional theory such as the decrease of annuals as well as the increase of grasses and especially woody species in the course of time (Cowles, 1899), and the general increase of vegetation height over time (Prach et al., 1997). Despite a scientific history of well over a century (Cutler, 2010) even this is noteworthy. Little effort has been made to integrate and synthesize the large number of case studies in succession to obtain a coherent generalized understanding. This is certainly due to the fact that rarely comprehensive spatial and temporal data on abiotic and biotic aspects of succession are collected. There are studies in the large body of literature which superbly meet a high temporal resolution but with only a very small spatial coverage (e.g. Rebele, 1992). Projects exclusively using chronosequences (probably the most) lack the spatial coverage for a quantitative analysis on vegetation change, both, in the sense of spatial patterns over time and in the sense of temporal pattern in space (Baasch, 2010). Still vast uncertainty exists about process interactions within the plant community (e.g. dispersal, competition) to processes linking the community to its environment (e.g. soil property dynamics) during succession. In view of the above, a particular contribution of H2 and our study is the detailed integration of biotic and abiotic aspects of succession with both temporal and spatial data.

Annual species – most of them being pioneers – were preferentially and expectedly found at places with less organic carbon, which means with low water and nutrient storage capacities (as mentioned before organic carbon might act as a proxy for general substrate properties), and lower groundwater levels. It was found also in a low-pH system that particularly resource-limited places were almost exclusively colonized by pioneers (Baasch, 2010).

In line with basic assumptions, the annual species' portion decreases with time (Fig. 4). However, during the first years of the succession two annual species (Conyza canadensis, Trifolium arvense) were dominant (Zaplata et al., 2011b) and hence their portion increases with increasing vegetation cover (COVTOT) but does not change with COVTOT values beyond 50 % (Fig. 4). The existence of a succession phase governed by Trifolium arvense (Fabaceae) in 2008– 2009 means, that many (adjacent) vegetation plots had similar covers of this plant species (Zaplata et al., 2013). The "optimum" shortly before COVTOT = 50% (Fig. 4) might re- flect the "preferential cover range", the range of cover which is typical for a given plant species, in this case of T. arvense. Apparently this phase expresses itself in an optimum calendar year effect for the occurrence of Fabaceae (Fig. 7). In previous studies we show a similar temporal trend for both cover (Schaaf et al., 2013) and regularity of the cover (Zaplata et al., 2013) for the two dominant species.

However, the proportion of annual plants' (to which T. arvense belongs) steady decrease means that the perennials on the whole had disproportionately high growth rates, even during the dominance phases of the two annual species. Besides, the previously mentioned bryophytes' cover increase in 2010/2011 incrementally alters (reduces) cover shares of vascular plant groups.

What has been found for annuals is in many respects also true for herbaceous plants and for Fabaceae (Fig. 7), simply because most of the herbaceous plants at the beginning of the successional series are annuals, and a dominating species of this group (Trifolium arvense) belongs to the Fabaceae family.

For the grasses we found two opposing trends. Our model separates a total cover effect and a temporal effect. The latter indicates a strong tendency to increase the grasses' share on the total community (Fig. 6) in contrast to the herbs. Grass species like Agrostis capillaris and Calamagrostis epigejos increased strongly (Zaplata et al., 2011b), and we expect a continued rise in the importance of the grasses. The former reveals that, again in contrast to the herbaceous plants, grasses dominate plots with low total coverage (Fig. 6). Hence, grasses are important contributors on more sparsely covered plots. In such a way they may rule subareas, but not the study system as a whole. This fact supports the assignment of the studied time span to the (first) successional stage of herbaceous vegetation (Zaplata et al., 2013). Only two explanatory variables, total degree of coverage and calendar year, showed a significant relationship in this context (Fig. 6, Table 9, Eq. 13).

The same applies for woody plants, with an increasing

share on the total community and – unexpectedly – a decreasing probability of finding them on vegetation squares with higher total cover (Table 10). We anticipate a reverse situation in the future. At the latest, when there are several crown layers formed by differently sized trees, their joint covers will prevail over the total covers of pure herb and grass plots. Based on considerations about resource limitation-driven competition modes by Hara (1993) and Weiner (1990), we hypothesize that our finding indicates that during the observed time span plant growth in the catchment was mostly limited by the soil bound resources, water and nutrients. Under such conditions the tree competition strategy to invest in vertical size does not pay off so much as it does when the vectorially distributed resource light is limiting. As long as this is not the case, trees may often face disadvantages in competition with non-woody plants.

We also found that overall vegetation height tends to be greater on more densely vegetated plots (Table 12). We may interpret that (i) simply as a consequence of particularly poor growth places, (ii) the high proportion of per se low-growing annuals there, or (iii) as an effect of beginning light competition which comes, however, hardly from trees yet.

Our finding that vegetation height tends to be greater when the distance to the groundwater table is higher probably re- flects a spurious correlation resulting from the coincidental temporal distribution of dry years co-occurring with ongoing succession. On the other hand there are, however, indications that at least some of this effect might be attributable to the intrinsic characteristics of the system. For example, the first dominant plant species, the aforementioned Conyza canadensis, reached the greatest heights in 2006 (known from orienting measurements on the catchment and an adjacent site), which was the driest year so far in the time series. In the course of time a decrease in size was registered even for some perennial species like Artemisia vulgaris, Hypericum perforatum L. and Hieracium umbellatum, as shown by the presence of higher shoots from the previous growing season.

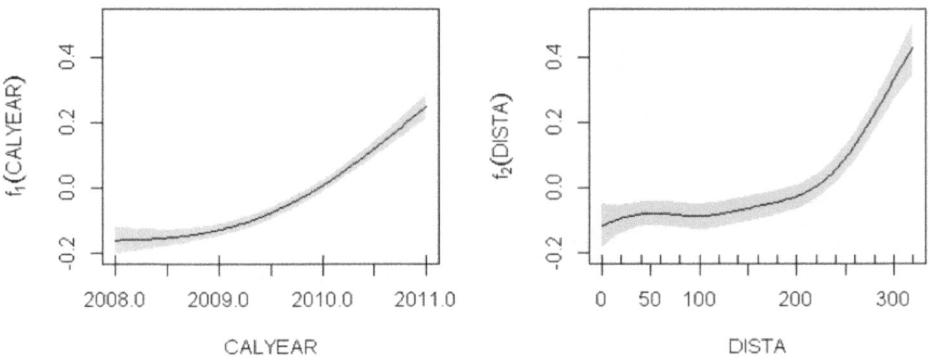

Figure 8. Figure 8 Nonlinear effects of the calendar year (CALYEAR), and distance from grid-row A (DISTA), on the laser-scanned vegetation height (VEGHEIGHT) (see Eq. 16). Shaded: 95 %-confidence area.

In conclusion, the results of the separate analyses of plant functional groups (annuals, herbs, Fabaceae, grasses, woody plants) are in line with successional theory, e.g. the decreasing proportions of annuals and herbs and the increasing importance of grasses and woody species. The latter two were under-represented on denser vegetated plots. This seems paradoxical at first sight, considering that woody species are superior competitors. With such results, this study supports a non-neutral view of ecosystem assembly during succession (cf. Nee and Stone, 2003). Our statistical analyses confirm at least in parts hypothesis 2, that surface and substrate characteristics determine the plant species groups to be found initially. Some few substrate characteristics such as the gravel content and the percentage of organic carbon already correlate in different ways with the occurrence of certain plant groups (cf. Baasch, 2010). Expectedly, woody plants show the weakest connection to substrate properties, the weak but significant connection we found between the occurrence of the N-fixing Fabaceae and the soil gravel content (Fig. 7) is plausible as well, as more gravel content often means more unfavourable conditions in terms of water and nutrient supply. On soils with higher gravel content, plants belonging to the Fabaceae family should possess a pronounced competitive advantage: a symbiotic relationship with root nodule bacteria (Rhizobiaceae) allows for the utilization of atmospheric nitrogen improving the supply of this typically limited nutrient. Introducing vegetation properties into the model for the surface height change (DHEIGHT) clearly confirmed our third hypothesis – the increasing density of vegetation reduces the erosion and therefore the degree of change of terrain surface structure. Remarkably VEGHD, the product of the laser-scanned vegetation height with the laser-scanned vegetation density VEGDENS – a proxy for biomass – performed best as the vegetation-related explanatory variable. The effects of all other substrate and geomorphology variables stayed virtually the same as in the DHEIGHT model without any vegetation variables used in the context of our first hypothesis. As Fig. 9 shows, the terrain-stabilizing effect of vegetation increases in the course of time, even for the same values of VEGHD. This may be interpreted as the result of positive (reinforcing) feedbacks between vegetation and substrate (vegetation establishment stabilizes the surface which supports further vegetation establishment) while vegetation growth itself appears to be also driven by a positive feedback as can be seen from the exponentially shaped calendar year effect on vegetation height in Fig. 8. As typical in many kinds of initial systems, not only in ecosystems (Sterman, 2002), positive feedbacks seem to dominate the surface and vegetation dynamics of the investigated artificial catchment Chicken Creek.

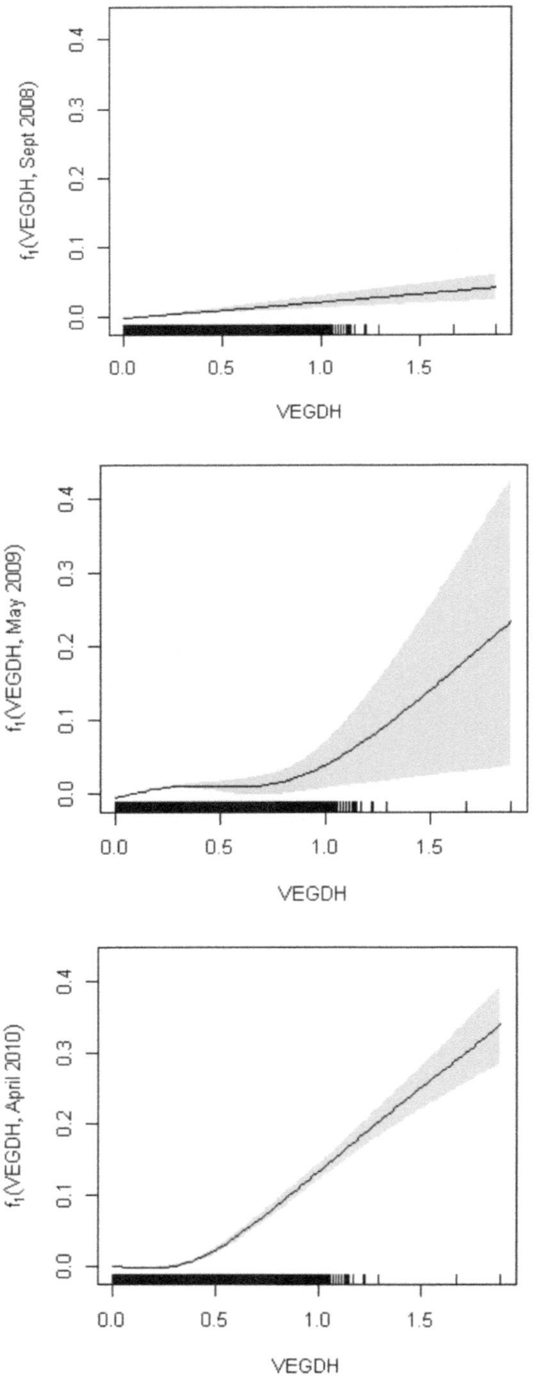

Figure 9. Nonlinear effect of the product of vegetation density and vegetation height (VEGDH) at different times on the surface height change (DHEIGHT) (see Eq. 17). Shaded: 95 %-confidence area.

CONCLUSIONS

The initial geomorphology in terms of position along the slope and the substrate properties, organic carbon and medium sand content have, together with groundwater levels, an effect on the surface and its formation as expressed by local rill density, relief energy, and surface height change. Expectedly, the influence of initial substrate properties diminishes with time (H1). The occurrence of plant functional groups (annual, herbaceous, grass-like, woody, Fabaceae) is connected to initial substrate properties; however this connection does not seem to be particularly close (H2). Signifi- cant relationships could be identified only with organic carbon content and, in the case of Fabaceae, with gravel content. Groundwater level and the local degree of total plant cover more often turned out as important explanatory variables for the appearance of a given plant functional group as well as for local mean vegetation height. For vegetation density no significant variables were found. In general, the time dependencies identified for the occurrence of different plant functional groups are in line with widely accepted views on succession. The product of vegetation height and density, a proxy for vegetation biomass, showed an increasing counteracting effect on surface height decrease (erosion), thus con- firming H3.

Our analysis reveals dynamic processes driving the ecosystem's development. We find positive (reinforcing) feedback mechanisms like the tendency of rills to deepen. However such pure substrate or geomorphology induced effects weaken with time, while vegetation establishes increasingly and differentiatedly introducing new (currently positive) feedbacks. These consist in mutually reinforcing stabilization of surface and vegetation, but also in vegetation growth processes. The analyzed period appears to represent the transition from a geo-hydro-system towards a bio-geohydro-system as defined in Elmer et al. (2013). We strongly believe that continued and broad research in that particular catchment will substantially contribute to an improved understanding of ecosystem functionality – its drivers, mediators, regulators, preservers, and antagonists.

ACKNOWLEDGEMENTS

This study is part of the Transregional Collaborative Research Centre 38 (SFB/TRR 38, "Structures and Processes of the initial ecosystem development phase in an artificial water catchment"), which is financially supported by the Deutsche Forschungsgemeinschaft (DFG, Bonn) and the Brandenburg Ministry of Science, Research and Culture (MWFK, Potsdam). We thank the Vattenfall Europe Mining AG for providing the research site. Thanks are also due to the editors and two anonymous referees for their constructive criticism and helpful comments.

REFERENCES

1. Baasch, A.: Predicting spatio-temporal patterns during succession in a post-mining landscape, Dissertation, Univ. Halle, 63 pp., 2010.
2. Biemelt, D., and Veste, M., Soil water, in: Schaaf, W., Biemelt, D., and Hüttl, R. F. (Eds.): Initial development of the artificial catchment 'Chicken Creek' – monitoring and survey 2005–2008, Ecos. Develop., 2, Cottbus, 57-70, 2010.
3. Bishop, J. G.: Early primary succession on Mount St. Helens: Impact of insect herbivores on colonizing lupines, Ecology, 83, 191–202, 2002.
4. Bradshaw, A. D.: The reconstruction of ecosystems, J. App. Ecol., 20, 1–17, 1983.
5. Brantley, S. L., Goldhaber, M. B., and Ragnarsdottir, K. V.: The Critical Zone: Where Rock meets Life: Crossing Disciplines and Scales to Understand the Critical Zone, Elements, 2007, 307– 314, 2007.
6. Campbell, J. L., Driscoll, C. T., Eagar, C., Likens, G. E.,

Siccama, T. G., Johnson, C. E., Fahey, T. J., Hamburg, S. P., Holmes, R. T., Bailey, A. S., and Buso, D. C.: Long-term trends from ecosystem research at the Hubbard Brook Experimental Forest. Gen. Tech. Rep. NRS-17. Newtown Square, PA: US Dept. Agriculture, Forest Service, Northern Research Station, 2007.

7. Cowles, H. C.: The ecological relations of the vegetation on the sand dunes of Lake Michigan, Botanical Gazette, 27, 95–117, 167–202, 281–308, 361–391, 1899.
8. Cutler, N.: Long-term primary succession: a comparison of nonspatial and spatially explicit inferential techniques, Plant Ecology, 208, 123–136, 2010.
9. Dahlgren, R. A., Ugolini, F. C., and Casey, W. H.: Field weathering rates of Mt. St. Helens tephra. Geochim. Cosmochim. Acta, 63, 587–598, 1999.
10. Dane, J. H. and Topp, C. (eds): Methods of Soil Analysis: Physical Methods. Soil Science Society of America Book Series, Soil Science Society of America, Vol. 5., 866 pp., 2002. del Moral, R. and Wood, D. M.: Early primary succession of the volcano Mount St. Helens, J. Veg. Sci., 4, 223–234, 1993.
11. Ellenberg, H., Mayer, R., and Schauermann, J.: Ökosystemforschung; Ergebnisse des Sollingprojektes 1966–1986, Ulmer, Stuttgart, 1986.
12. Elmer, M., Gerwin, W., Schaaf, W., Zaplata, M. K., Hohberg, K., Nenov, R., Bens, O., and Hüttl, R. F.: Dynamics of initial ecosystem development at the artificial catchment Chicken Creek, Lusatia, Germany, Environ. Earth Sci., 69, 491–505, 2013.
13. Felinks, B.: Primärsukzession von Phytozönosen in der Niederlausitzer Bergbaufolgelandschaft, Dissertation, BTU Cottbus, p. 196 , 2000.
14. Fischer, T., Veste, M., Bens, O., and Hüttl, R. F.: Dew formation on the surface of biological soil crusts in central European sand ecosystems. Biogeosciences, 9, 4621–4628, doi:10.5194/bg-9- 4621-2012, 2012.
15. Fränzle, O., Kappen, L., Blume, H.-P., and Dierssen, K. (Eds.): Ecosystem Organization of a Complex Landscape; Long-term Research in the Bornhöved Lake District, Germany, Ecological Studies, 202, Springer, Berlin, Heidelberg, 2008. Friðriksson, S.: Surtsey, Ecosystems formed. Univ. of Iceland, Reykjavík, 2005.
16. Gerwin, W., Schaaf, W., Biemelt, D., Fischer, A., Winter, S., and Hüttl, R. F.: The artificial catchment "Chicken Creek" (Lusatia, Germany) – A landscape laboratory for interdisciplinary studies of initial ecosystem development, Ecol. Engineer., 35, 1786– 1796, doi:10.1016/j.ecoleng.2009.09.003, 2009.
17. Gerwin, W., Schaaf, W., Biemelt, D., Elmer, M., and Schneider, A.: The artificial catchment „Hühnerwasser“ (Chicken Creek): construction and initial properties, Ecosys. Develop., 1, 55 pp., BTU Cottbus, 2010.
18. Gerwin, W., Schaaf, W., Biemelt, D., Winter, S., Fischer, A., Veste, M., and Hüttl, R. F.: Overview and first results of ecological monitoring at the artificial watershed Chicken Creek (Germany), Phys. Chem. Earth, 36, 61–73, doi:10.1016/j.pce.2010.11.003, 2011.
19. Hara, T.: Mode of Competition and Size-structure Dynamics in Plant Communities, Plant Species Biol., 8, 75–84, 1993.
20. Hüttl, R. F. and Weber, E.: Forest ecosystem development in postmining landscapes. A case study of the Lusatian lignite district. Naturwissenschaften, 88, 322–329, 2001.
21. Jørgensen, S. E., Patten, B. C., and Straškraba, M.: Ecosystem emerging: 4. growth. Ecological Modelling, 126, 249-284, 2000.
22. Müller-Dombois, D., and Fosberg, F. R.: Vegetation of the tropical pacific islands. Vegetation Monographs of the Major World Regions Vol. IX. Springer, New York, 912 pp., 1998.
23. Nee, S. and Stone, G.: The end of the beginning for neutral theory, TRENDS Ecol. Evolut., 18, 433–434, 2003.
24. Pennisi, E.: A groundbreaking observatory to monitor the environment, Science, 328, 418–420, 2010. Pinheiro, J. C., and Bates, D. M.: Mixed-effects Models in S and S-PLUS. Springer, 528 pp., 2000.
25. Prach, K., Pyšek, P., and Šmilauer, P.: Changes in species traits during succession: a search for pattern, Oikos, 79, 201–205, 1997.
26. R Core Team: R: A language and environment for statistical computing. R Foundation for Statistical Computing, Vienna, Austria, ISBN 3-900051-07-0, http://www.R-project.org/, 2012.
27. Rebele, F.: Colonization and early succession on anthropogenic soils, J. Vegetat. Sci., 3, 201–208, 1992.
28. Risse-Buhl, U., Hagedorn, F., Dümig, A., Gessner, M. O., Schaaf, W., Nii-Annang, S., Gerull, L., and Mutz, M.: Dynamics, chemical properties and bioavailability of DOC in an early successional catchment, Biogeosciences, 10, 4751–4765, doi:10.5194/bg-10- 4751-2013, 2013.
29. Rothmaler, W.: Exkursionsflora von Deutschland Gefäßpflanzen: Atlasband. Band 3, 10th Edn., edited by: Jäger, E. J. and Werner, K., Spektrum Akademischer Verlag, Heidelberg, Berlin, 753 pp., 2000.
30. Schaaf, W.: What can element budgets of false-time series tell us about ecosystem development on post-lignite mining sites? Ecol. Eng., 17, 241–252, 2001.
31. Schaaf, W., Bens, O., Fischer, A., Gerke, H. H., Gerwin, W., Grünewald, U., Holländer, H. M., Kögel-Knabner, I., Mutz, M., Schloter, M., Schulin, R., Veste, M., Winter, S., and Hüttl, R. F.: Patterns and processes of initial terrestrial-ecosystem development, J. Plant Nutr. Soil Sci., 174, 229–239, 2011.
32. Schaaf, W., Elmer, M., Fischer, A., Gerwin, W., Nenov, R., Pretzsch, H., Seifert, S., Winter, S., and Zaplata, M. K.: Monitoring the formation of structures and patterns during initial development of an artificial catchment, Environ. Monit. Assess., 185, 5965–5986, 2013.
33. Schneider, A., Gerke, H.H., Maurer, T., Seifert, S., Nenov, R., and Hüttl, R. F.: Evaluation of remotely-sensed DEMs and modification based on plausibility rules and initial sediment budgets of an artificially-created catchment, Earth Surf. Proc. Landforms, 37, 708–725, doi:10.1002/esp.2274, 2012.
34. Sparks, D. L.: Methods of Soil Analysis, Part 3. Chemical

Methods. Soil Science Society of America Book Series, No. 5, American Society of Agronomy-Soil Science Society of America, 1264 pp., 1996.

35. Sterman, J.: Business Dynamics, McGraw-Hill Higher Education, 982 pp., 2002.

36. Walker, L. R. and Willig, M. R.: An introduction to terrestrial disturbances, in: Ecosystems of disturbed ground, edited by: Walker, L. R., Ecosystems of the world, Amsterdam, Elsevier, 16, 1–15, 1999.

37. Weiner, J.: Asymmetric competition in plant populations, Trends Ecol. Evolut., 5, 360–364, 1990.

38. Wood, S. N.: Generalized Additive Models: An Introduction with R. Chapman and Hall/CRC. 2006.

39. Zaplata, M. K., Winter, S., and Fischer, A.: Vegetation dynamics, in: Initial development of the artificial catchment Chicken Creek, monitoring program and survey 2005–2008, edidted by: Schaaf, W., Biemelt, D., and Hüttl, R. F., Ecosys. Develop., 2, 71–96, BTU Cottbus, 2010.

40. Zaplata, M. K., Winter, S., Biemelt, D., and Fischer, A.: Immediate shift towards source dynamics: the pioneer species Conyza canadensis in an initial ecosystem, Flora, 206, 928–934, 2011a.

41. Zaplata, M. K., Fischer, A., and Winter, S.: Vegetation patterning, in: Elmer, M., Schaaf, W., Biemelt, D. Gerwin, W., and Hüttl, R. F.: The artificial catchment "Chicken Creek" – initial ecosystem development 2005–2010, Ecosys. Develop., 3, 67–92, BTU Cottbus, 2011b.

42. Zaplata, M. K., Winter, S., Fischer, A., Kollmann, J., and Ulrich, W.: Species-driven phases and increasing structure in earlysuccessional plant communities, The American Naturalist, 181, E17–E27, 2013.

43. Zikeli, S., Jahn, R., and Kastler, M.: Initial soil development in lignite ash landfills and settling ponds in Saxony-Anhalt, Germany, J. Plant Nutr. Soil Sci., 165, 530–536, 2002.

44. Zuur, A. F., Ieno, E. N., Walker, N. J., Saveliev, A. A., and Smith, G. M.: Mixed Effects Models and Extensions in Ecology with R., Springer, 574 pp., 2009.

CHAPTER 13

REVIEW OF EFFECTIVE VEGETATION MAPPING USING THE UAV (UNMANNED AERIAL VEHICLE) METHOD

Korehisa Kaneko[1], Seiich Nohara[2]

[1]Ecosystem Conservation Society-Japan, Tokyo, Japan

[2]Center for Environmental Biology and Ecosystem Studies, National Institute for Environmental Studies, Ibaraki, Japan

ABSTRACT

We tried more precise mapping of vegetation using UAV (unmanned aerial vehicle), as a new method of creating vegetation maps, and we objected to be clearly the efficient mapping of vegetation using the UAV method by comparing vegetation maps created by analyzing aerial photo- graphs taken by a UAV and an aircraft (manned flight). The aerial photography using UAV was conducted in the Niida River estuary (the secondary river flowing into Minamisoma City in Fuku- shima Prefecture, Japan). The photography period was in August 2013. We analyzed the aerial photographs using ArcGis 9 (Esri Japan Corporation, Tokyo, Japan). The aerial photographs of the main plant communities (Phragmites australis, Typha domingensis, and Miscanthus sacchariflorus) taken by the UAV could clearly discriminate each plant community at the 1/50 scale. Moreover, it could clearly discriminate the shape of a plant at the 1/10 scale. We compared the vegetation maps by analyzing the aerial photos taken by a UAV (2013 shooting) and an aircraft (2011 shooting). As a result, the vegetation map created by the UAV method could clearly discriminate com- munity distributions. We conclude that vegetation surveys using UAV are possible and are capable of a highly precise community division in places where field reconnaissance is difficult. The UAV method is effective and will contribute to the improvement of research methods in the future; this method may reduce research costs associated with a reduction in field survey days and man- power.

Citation: Kaneko, K. and Nohara, S. (2014) Review of Effective Vegetation Mapping Using the UAV (Unmanned Aerial Vehicle) Method. Journal of Geographic Information System, 6, 733-742. doi: 10.4236/jgis.2014.66060.

INTRODUCTION

A vegetation map illustrates the geographical spread of a plant community, which is a vegetation unit. Vegetation maps provide extremely important basic information for environmental planning [1] [2] . In the past, vegetation maps have been created by skilled engineers with knowledge of vegetation. Improvements in efficiency and precision have enabled vegetation to be mapped using aerial photographs and satellite data. Mapping vegetation using satellite images may be more effective than mapping based on aerial photographs [3] . At present, the development of effective mapping and improvements in precision for creating vegetation maps are important technical issues. UAV (unmanned aerial vehicle) methods can be used in various fields and have increased the efficiency of information gathering during disasters, for remote sensing in the forest and agricultural sectors, in the management and monitoring of rivers and embankments, and in the measurement and surveying of structures and topography [4] -[7] . Moreover, UAV system was effective in identifying the present vegetation [8] -[10] [12] , and is possible to generate detailed map of vegetation assemblages at the species level [11] . Experiment obtains the distribution of different vegetation species and detects invasive weeds over two cattle farms of Queensland, Australia.

[13] reported that a resolution of less than 1 cm is required for the precise interpretation of herbaceous plants. This resolution is the limiting factor in the remote sensing capabilities of satellites and aircraft; however, with high-resolution remote sensing using UAV methods, it is possible to discriminate herbaceous species at an individual plant level. No detailed study has yet examined the discrimination of individual plant species and plant communities by UAV.

In this report, we examined the efficient mapping of vegetation using the UAV method by comparing vegetation maps created by analyzing aerial photographs taken by a UAV and an aircraft (manned flight).

STUDY SITE AND METHODS

Study Site

The study site was in the estuarine tidal flats of the Niida River, which is a second-class river located in Minamisoma City, Fukushima Prefecture, Japan (Figure 1).

UAV Aerial Photography Methods

The research was conducted the aerial photography using a UAV and created a vegetation map from the aerial photographs taken by the UAV.

The UAV used a radio-controlled helicopter (MK EASY Quad XL2 1.5 CFK: Edition of HiSystems GmbH Company: Germany, Flachsmeerstr., 2, 26,802, Moormerland). The UAV was equipped with a compact digital camera (Richo GX200) mounted to a small stabilizer (Figure 2 and Figure 3). The schematic diagram of the general process of mapping using the UAV method is shown in Figure 4.

The aerial photography using an aircraft was the altitude of about 2400 m, resolution about 24 cm/pixel, scale 1/20,000 (Photographic camera: DMC), on the other hand, the aerial photography using UAV was the altitude of about 10 m, resolution about 3 mm/pixel, scale 1/546.4 (Photographic camera: Richo NewGR). From above, the scale and resolution ratios of the aerial photograph using UAV were 36.6 and 80.0 times respectively, com- pared with those of aerial photography using an aircraft.

The aerial photography using UAV was conducted over a 1.5 km section upstream from the Niida River estuary (the secondary river flowing into Minamisoma City in Fukushima Prefecture). The photography period was in August 2013, when the vegetation is fully developed. Fixed photography points were set up in a line from the right bank to the left bank at approximately 20 m intervals; using the UAV, we took aerial photographs at 5 m above the ground surface. There were 13 line numbers in the investigated section from upstream to down- stream. Each line included 8 fixed photography points, for a total of 104 points (Figure 5).

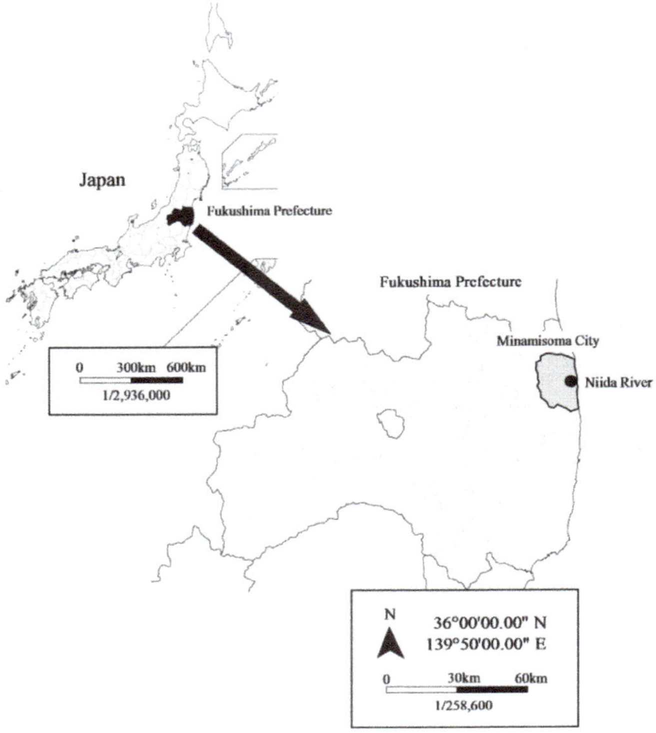

Figure 1. Study site.

Figure 2. UAV (unmanned aerial vehicle) system. *Radio-controlled helicopter used in the study (MK EASY Okto XL2 1.5 CFK Edition of HiSystems GmbH Company).

Figure 3. The landscape where aerial photographs using the UAV (unmanned aerial vehicle) were taken.

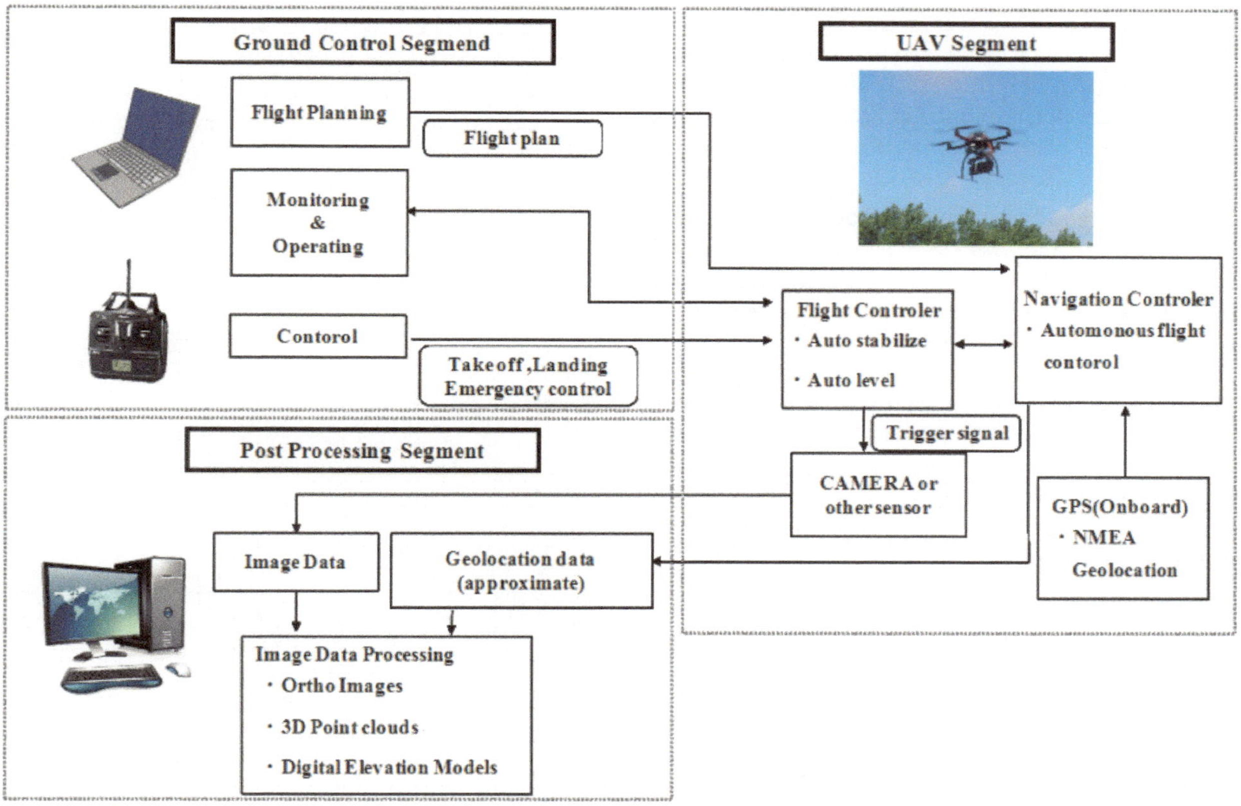

Figure 4. The schematic diagram of mapping using the UAV method.

Positioning of the captured image used GPS for navigation of the equipped UAV. GPS receiver used u-blox, Inc (LEA-6S). Positioning was conducted at 2 Hz (0.5 section intervals), and we were conducted the positioning of the image by matching shooting time (in seconds) that is stored in the log data (latitude, longitude and aircraft direction (camera Richo New GR) and the captured image.

Vegetation Mapping Methods

The vegetation map was geometrically constructed from aerial photographs, and we analyzed the aerial photographs using ArcGis 9 (Esri Japan Corporation, Tokyo, Japan).

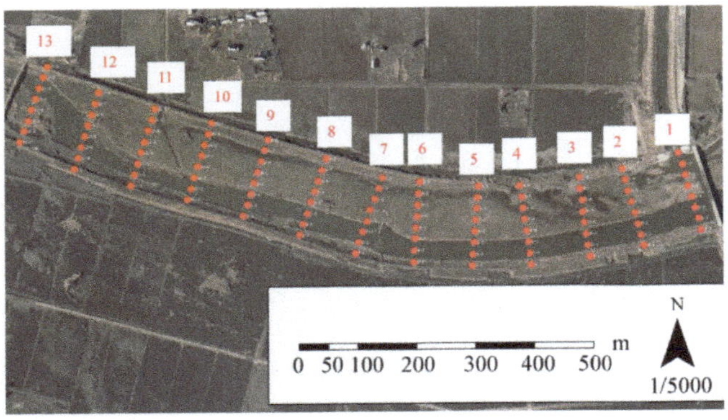

Figure 5. UAV photography lines and points (No. 1-13). *There were 8 fixed photography points on each line.

RESULTS

In the aerial photographs of the main plant communities (Phragmites australis, Typha domingensis, and Miscanthus sacchariflorus) taken by the UAV, a clear discrimination of each plant community was possible at a scale of 1/50. At a scale of 1/10, it was possible to clearly confirm the shape of an individual plant (Figure 6). The borders among the plant communities and mixes of different plant species in the vicinity of the community borders could also be discriminated at this scale. An example of this resolution is the discrimination between the Phragmites australis/Scirpus yagara community and the Phragmites australis/Typha domingensis community (Figure 7).

We compared the vegetation maps that were created using aerial photos taken by the UAV (2013 shooting) and an aircraft (2011 shooting). Although plant communities in these two vegetation maps were slightly different because the shooting period and investigation year were different (Figure 8), a detailed community division was confirmed on the vegetation map created using the UAV method, and the difference in precision was remarkable. The vegetation map created using the UAV method could clearly discriminate community divisions and distributions (Figure 9).

CONSIDERATIONS

Future Developments and Advantages of UAV Methods

UAV provide low operational complexity and costs, compared with the classical manned aerial photogrammetry [6] [14] -[17] , and remote sensing sensors placed on UAV is capacity to combine high spatial resolution and quick turnaround times providing low-cost approaches to meet the critical requirement of spatial, spectral, and temporal resolution [15] .

[18] reported that, as a result of the development of a compact, lightweight, and high-resolution remote sensing system using a radio-controlled helicopter, it was possible to obtain an image with high spatial resolution (1 pixel per 1 cm).

In our survey, plant communities of Phragmites australis and Miscanthus sacchariflorus were clearly dis- criminated (Figure 7). From the aerial photograph taken by the UAV, the shapes and numbers of plants were confirmed in detail, and highly precise discrimination was possible (Figure 6). We could discriminate clearly and precisely the division of many plant communities, resulting in the creation of a highly precise vegetation map (Figure 9). [13] reported that the high-spatial-resolution remote sensing using the UAV method can also be used for monitoring long-term changes. In addition, this technology can be applied to high moors, where ground-based vegetation surveys by exploration are difficult, and the reproducibility of the shooting location can be maintained.

Figure 6. (a) Phragmites australis community; (b) Typha domingensis community; (c) Miscanthus sacchariflorus community photographed by UAV methods (the photo on the right is a close-up of the photo on the left).

UAV reduce the risk of access in harsh environment, still keeping high accuracy potential [19] . UAV appeared to provide a good complement to the current remote sensing platforms because of their promise of low- cost and high resolution [11] .

Therefore, we conclude that vegetation surveys using UAV are possible and are capable of a highly precise community division in places where field reconnaissance is difficult. The UAV method is effective and will contribute to the improvement of research methods in the future and may reduce research costs by reducing field survey days and manpower. Additionally, we conclude that UAV methods are effective for the quantitative evaluation of vegetation community structure because the canopy height can be accurately measured. Moreover, UAV methods are effective in the discrimination of indicator plant communities.

Figure 7. The vegetation division obtained by UAV methods (Top photograph: Phragmites australis community, Scirpus yagara community. Bottom photo- graph: Phragmites australis community, Typha domingensis community).

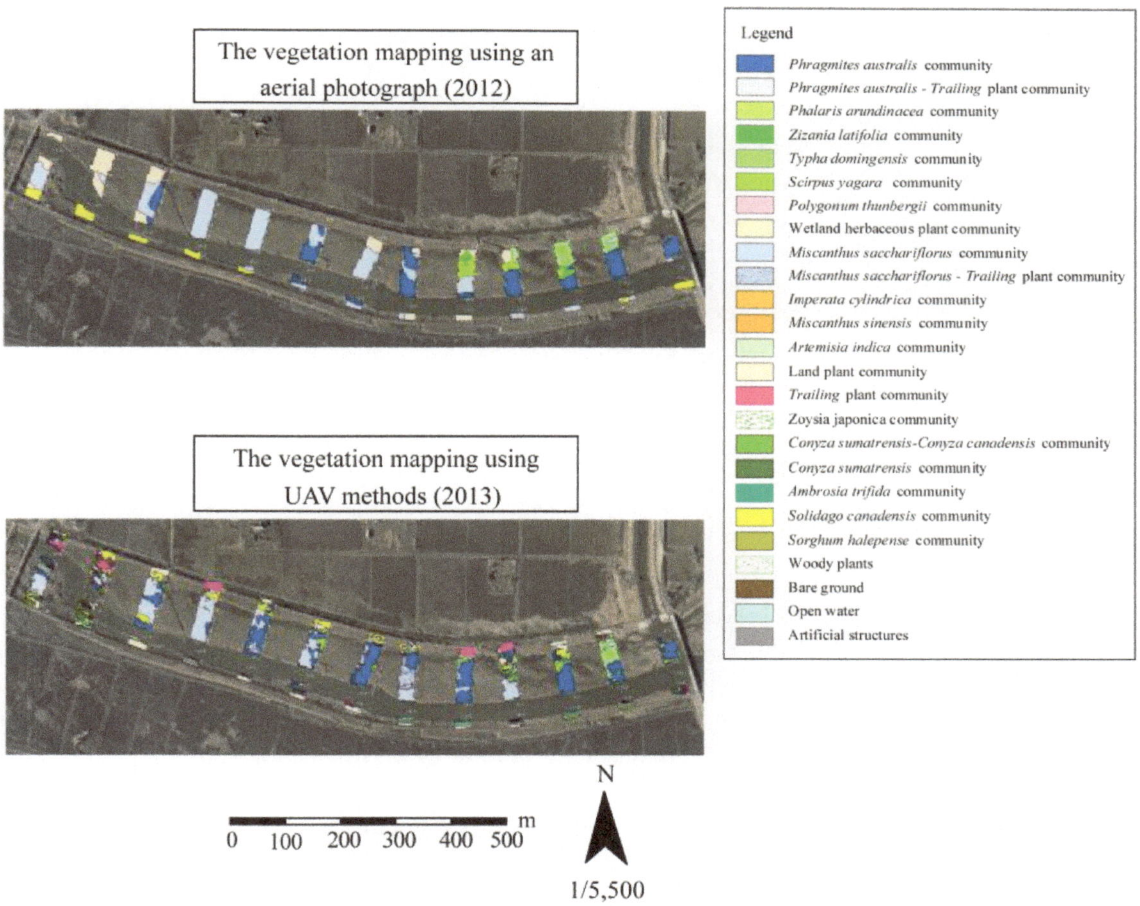

Figure 8. The vegetation map of the Niida River (vegetation mapping using an aerial photograph (above); vegetation map- ping using UAV methods (below).

These methods can also be used for quantitative evaluations of community volume because taxonomists (in insects, amphibians, reptiles, mammals, and birds) can use UAV images to discriminate the morphologies of plants and community habitats.

CONCLUSIONS

We objected to be clear about the efficient mapping of vegetation using the UAV method. The aerial photography using UAV was conducted in the Niida River estuary (the secondary river flowing into Minamisoma City in Fukushima Prefecture, Japan). The aerial photographs of the main plant communities (Phragmites australis, Typha domingensis, and Miscanthus sacchariflorus) taken by the UAV could clearly discriminate each plant community at the 1/50 scale. Moreover, it could clearly discriminate the shape of a plant at the 1/10 scale. The vegetation map created by the UAV method could clearly discriminate community distributions.

We conclude that vegetation surveys using UAV are possible and are capable of a highly precise community division in places where field reconnaissance is difficult.

ACKNOWLEDGEMENTS

This research was supported by JSPS KAKENHI, Grant Number 24310025 (Studies on assessment of ecosystem functions of wetlands using environmental radioisotopes).

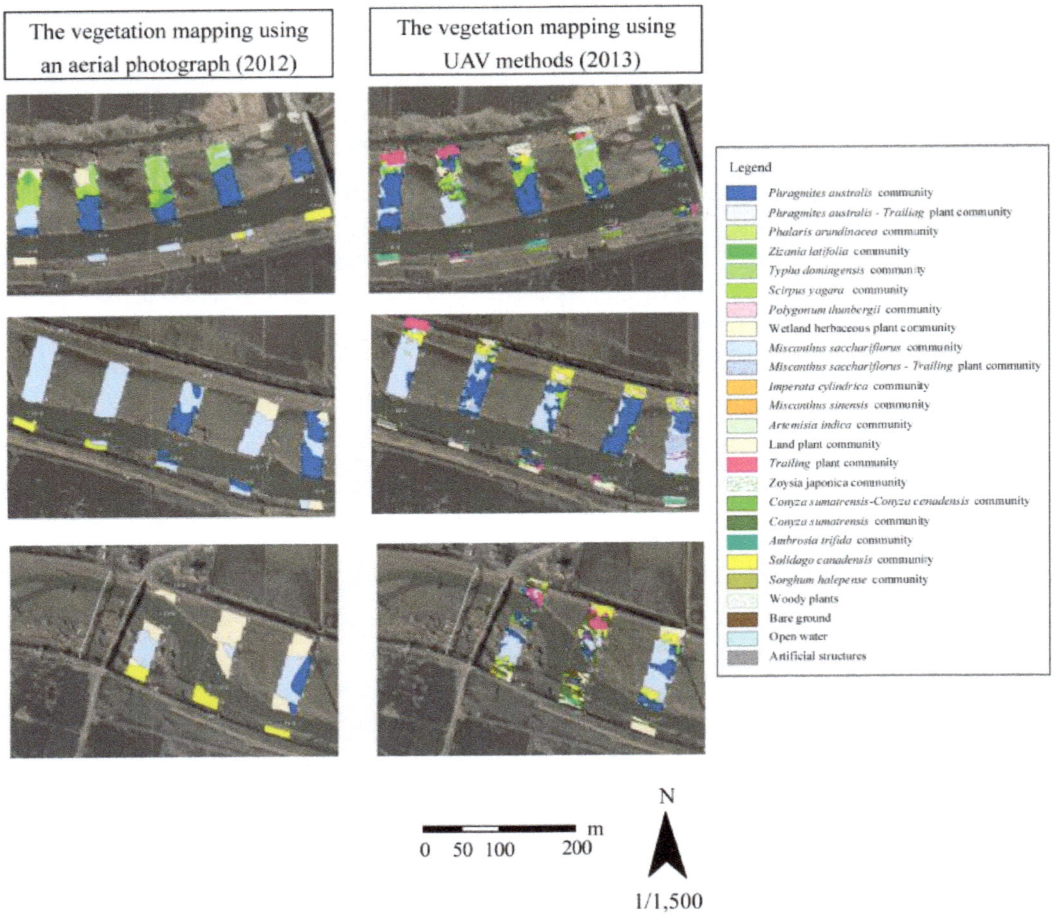

Figure 9. Vegetation maps using UAV methods and aerial photography (Aerial photograph, left; UAV methods, right).

Korehisa Kaneko, Seiich Nohara Moreover, we wish to express our sincere gratitude to Mr. Hiroyuki Fukawa of Ecosystem Conservation Society Japan, who cooperated with the study.

REFERENCES

1. Environment Agency (2000) Technology Assessment of Natural Environment. Ministry of Finance Printing Bureau, Tokyo, 29-30.
2. Organization for Landscape and Urban Green Infrastructure (2000) Ecological Network in the Urban. Gyo-sei, Tokyo , 141-142.
3. Mathubayashi, K., Nemoto, J., Momose, H., Fujuwara, N. and Hioki, Y. (2002) Development of Vegetation Mapping Technique and Mapping Precision Using High Resolution Satellite Data. The Japanese Society of Revegetation Tech- nology, 28, 127-131. http://dx.doi.org/10.7211/jjsrt.28.127
4. Suzuki, T., Hashizume, T. and Suzuki, S. (2012) The Easy Building Measurement by Utilizing Small Autonomous Flying Robot (UAV). Building Construction Planning, 740, 65-69.
5. Salami, E., Barrado, C. and Pastor, E. (2014) UAV Flight Experiments Applied to the Remote Sensing of Vegetation Area. Remote Sensing, 6, 11051-11081.

http://dx.doi.org/10.3390/rs61111051

6. Dunford, R., Michel, K., Gagnage, M., Piégay, H. and Trémelo, M.L. (2013) Potential and Constraints of Unmanned Aerial Vehicle Technology for the Characterization of Mediterranean Riparian Forest. International Journal of Remote Sensing, 30, 4915-4935. http://dx.doi.org/10.1080/01431160903023025

7. Nex, F. and Remondino, F. (2013) UAV for 3D Mapping Applications: A Review. Applied Geomatics, 6, 1-15. http://dx.doi.org/10.1007/s12518-013-0120-x

8. Arnold, T., Biasio, M.D., Fritz, A. and Leitner, R. (2013) UAV-Based Measurement of Vegetation Indices for Envi- ronmental Monitoring. 2013 Seventh International Conference on Sensing Technology, Wellington, 3-5 December 2013, 708-711.

9. Wallace, L. (2013) Assessing the Stability of Canopy Maps Produced from UAV-LiDAR Data. Proceedings of 2013 IEEE International Geoscience and Remote Sensing Symposium (IGARSS), Melbourne, 21-26 July 2013, 3879-3882.http://dx.doi.org/10.1109/IGARSS.2013.6723679

10. Laliberte, A.S. and Rango, A. (2009) Texture and Scale in Object-Based Analysis of Subdecimeter Resolution Un- manned Aerial Vehicle (UAV) Imagery. IEEE Transactions Geoscience and Remote Sensing, 47, 761-770.http://dx.doi.org/10.1109/TGRS.2008.2009355

11. Strecha, C., Fletcher, A., Lechner, A., Erskine, P. and Fua, P. (2012) Developing Species specific Vegetation Maps Using Multi-Spectral Hyperspatial Imagery from Unmanned Aerial Vehicle. ISPRS Annals of the Photogrammetry, Remote Sensing and Spatial Information Sciences, I-3, XXII ISPRS Congress, 25 August-1 September 2012, Mel- bourne.

12. Hung, C., Bryson, M. and Sukkarieh, S. (2012) Multi-Class Predictive Template for Tree Crown Detection. ISPRS Journal of Photogrammetry and Remote Sensing, 68, 170-183.http://dx.doi.org/10.1016/j.isprsjprs.2012.01.009

13. Oguma, H., Usami, M., Shimazaki, H. and Ishihama, F. (2010) A High-Resolution Remote Sensing by Radio Control Helicopter and Apply to Species Discrimination of Individual Level of Wetland Herbaceous Plant. 57th Annual Con- ference of Ecological Society of Japan, Tokyo, 15-20 March 2010, Abstract.

14. Eisenbeiss, H. (2004) A Mini Unmanned Aerial Vehicle (UAV): System Overview and Image Acquisition, Interna- tional Workshop on Processing and Visualization Using High Resolution Imagery, Pitsanulok, 18-20 November 2004. International Archives of Photogrammetry, Remote Sensing and Spatial Information Sciences, XXXVI-5/W1.

15. Berni, J., Zarco-Tejada, P., Suarez, L. and Fereres, E. (2009) Thermal and Narrowband Multispectral Remote Sensing for Vegetation Monitoring from an Unmanned Aerial Vehicle. IEEE Transactions on Geoscience and Remote Sensing, 47, 722-738.http://dx.doi.org/10.1109/TGRS.2008.2010457

16. Berni, J., Zarco-Tejada, P., Suarez, L. and Fereres, E. (2014) Remote Sensing of Vegetation from UAV Platforms Us- ing Lightweight Multispectral and Thermal Imaging Sensors. The International Archives of the Photogrammetry, Re- mote Sensing and Spatial Information Science, 38, 6.

17. Jensen, A.M., Chen, Y., McKee, M., Hardy, T. and Barfuss, S.L. (2009) AggieAir—A Low-Cost Autonomous Multi- spectral Remote Sensing Platform: New Developments and Applications. Proceedings of 2009 IEEE International Geoscience and Remote Sensing Symposium, Cape Town, 12-17 July 2009, IV-995-IV-998.

18. Ishihama, F., Watabe, Y. and Oguma, H. (2012) Validation of a High-Resolution, Remotely Operated Aerial Remote- Sensing System for the Identification of Herbaceous Plant Species. Applied Vegetation Science, 15, 383-389.http://dx.doi.org/10.1111/j.1654-109X.2012.01184.x

19. Remondino, F., Barazzetti, L., Nex, F., Scaioni, M. and Sarazzi, D. (2011) UAV Photogrammetry for Mapping and 3D Modelling—Current Status and Future Perspectives. International Archives of the Photogrammetry, Remote Sensing and Spatial Information Sciences, XXXVIII-I/C22 UAV-g 2011, Conference on Unmanned Aerial Vehicle in Geomat- ics, Zurich, 1-7.

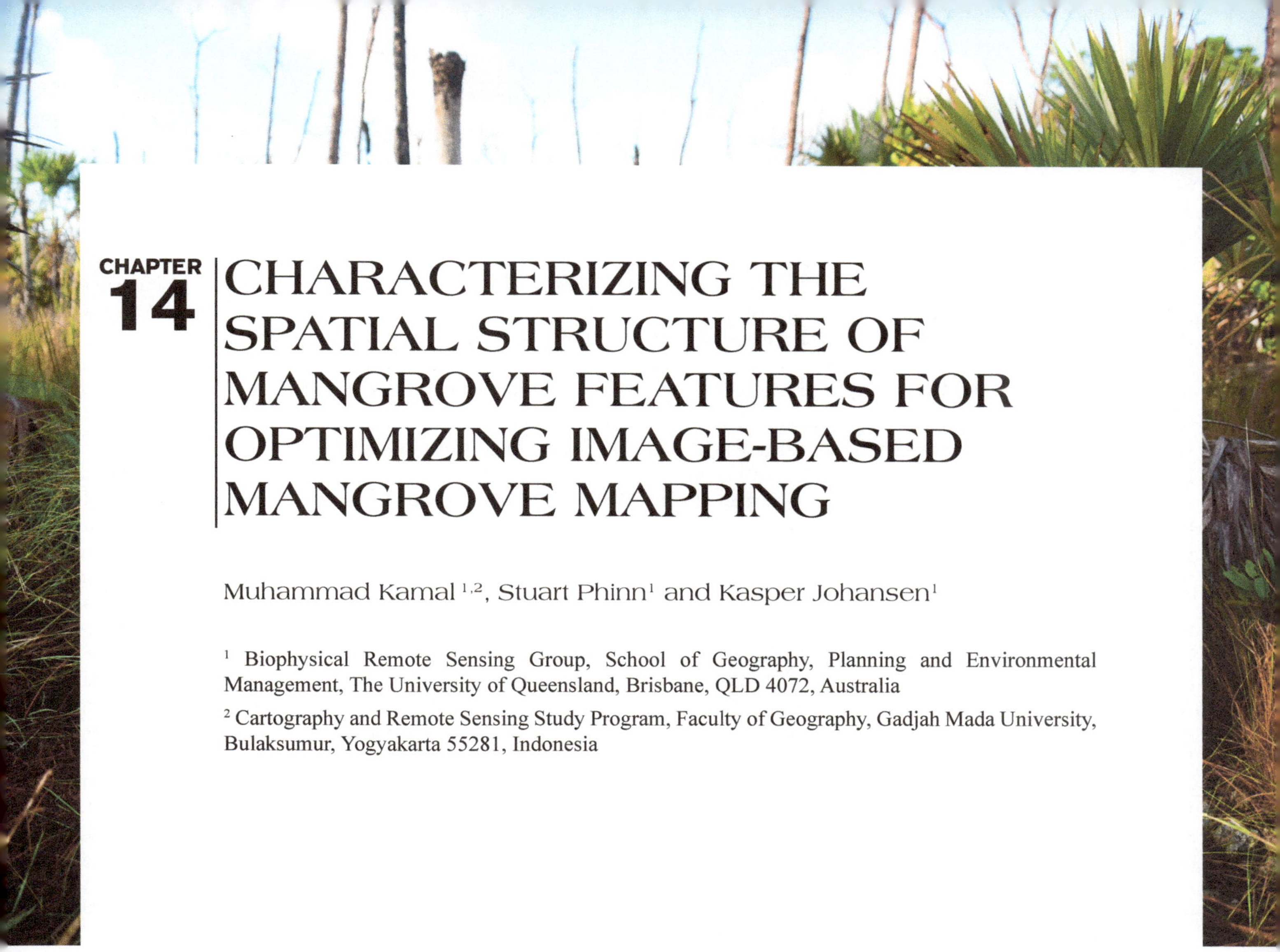

CHAPTER 14

CHARACTERIZING THE SPATIAL STRUCTURE OF MANGROVE FEATURES FOR OPTIMIZING IMAGE-BASED MANGROVE MAPPING

Muhammad Kamal [1,2], Stuart Phinn [1] and Kasper Johansen [1]

[1] Biophysical Remote Sensing Group, School of Geography, Planning and Environmental Management, The University of Queensland, Brisbane, QLD 4072, Australia

[2] Cartography and Remote Sensing Study Program, Faculty of Geography, Gadjah Mada University, Bulaksumur, Yogyakarta 55281, Indonesia

ABSTRACT

Understanding the relationship between the size of mangrove structural features and the optimum image pixel size is essential to support effective mapping activities in mangrove environments. This study developed a method to estimate the optimum image pixel size for accurately mapping mangrove features (canopy types and features (gaps, tree crown), community, and cover types) and tested the applicability of the results. Semi-variograms were used to characterize the spatial structure of mangrove vegetation by estimating the size of dominant image features in WorldView-2 imagery resampled over a range of pixel sizes at several mangrove areas in Moreton Bay, Australia. The results show that semi-variograms detected the variations in the structural properties of mangroves in the study area and its forms were controlled by the image pixel size, the spectral-band used, and the spatial characteristics of the scene object, e.g., tree or gap. This information was synthesized to derive the optimum image pixel size for mapping mangrove structural and compositional features at specific spatial scales. Interpretation of semi-variograms combined with field data and visual image interpretation confirms that certain vegetation structural features are detectable at specific scales and can be optimally detected using a specific image pixel size. The analysis results provide a basis for multi-scale mangrove mapping using high spatial resolution image datasets.

INTRODUCTION

Remote sensing has been used extensively to map and monitor mangrove environments over the past two decades. It offers some key advantages for mangrove studies, including indirect access to mangrove habitats that are usually hard to access [1,2], extrapolation of observation results at specific sample sites over large areas [3], and delivery of data at specific

Citation: Kamal, M.; Phinn, S.; Johansen, K. Characterizing the Spatial Structure of Mangrove Features for Optimizing Image-Based Mangrove Mapping. Remote Sens. 2014, 6, 984-1006. doi:10.3390/rs6020984.

spatial and temporal scales [4]. Recent developments in remote sensing and image processing allow us to explore various types of image datasets, as well as types of mapping techniques, to map mangroves [3,5]. However, there is a need to match the scale of the analysis to the scale of the phenomenon under investigation, as environmental inferences are scale-dependent [6]. Mapping mangroves at specific spatial scales will help scientists to focus their research on the ecological questions that are appropriate to each level of ecological detail [7], and managers to focus on the conservation activities at ecologically relevant spatial and temporal scales [8].

From an ecological perspective, mangrove ecosystems, like other vegetated ecosystems, can be placed within a hierarchical structure [9]. The central concept of this theory focuses on the differences in structure and process rate between hierarchical levels. Based on these differences, ecosystems are viewed as being stratified into discrete levels of interacting subsystems, with attributes and processes occurring at specific spatial and temporal scales [7,10–12]. Remote sensing is a tool able to deliver information on mangrove characteristics at specific spatial and temporal scales. However, the spatial configuration of mangroves, as measured in an airborne or satellite image, is dictated by spatial structures of mangroves in the field, interacting with imaging sensor characteristics. In order to use satellite or airborne image data to extract information at specific scales or features for mangroves, it is essential to understand the control of mangrove spatial structures on their measurement in an image. This raises the question, "what mangrove features are dominant and able to be mapped at specific levels of image resolution, as controlled by pixel size?" Studies linking the spatial structure of mangroves and the image spatial-resolution are limited.

The selection of scale or an appropriate spatial resolution is an important factor that contributes to the successful application of remote sensing. It depends on several factors, including the information to be extracted from the ground scene, the analysis method to be used to extract the information, the spatial structure of features within the image scene [13], the type of environment being investigated and other relevant constraints (*i.e.*, cost and time) [14]. The scale effect is regarded as one of the most important problems in remote sensing studies [15–17]. The "scale" represents the window of perception [16], the ability of observation, and reflects the limitation of knowledge through which a phenomenon may be viewed or perceived [15]. Changing the scale of data collection and analysis impacts the measurements and conclusions able to be drawn for an environment and image data set combination. Consequently, selecting an appropriate or optimal spatial resolution requires information on the spatial characteristics of features within the scene under investigation.

Scientists have developed several methods to select the optimal spatial scale for remote sensing applications, which are tied to selection of pixel sizes and spectral bands, primarily for use in the per-pixel classification. The most widely adopted method is to examine the spatial autocorrelation of the image scene through the analyses of semi-variograms. The semi-variogram is a tool to link models of the ground scene to spatial variation in images [18] and is able to detect the most dominant scale(s) of variation in images [19]. In remote sensing studies, it enables the optimal pixel size for feature mapping in different environments from image data to be specified, for example, in forests [20–23], tropical savannahs [24,25], grasslands [26,27], and wetlands [28]. Semi-variograms can be used to identify the domains of scale where certain features reside [21,24,29], which in turn will provide support in selecting the most appropriate image spatial resolution for a specific mapping purpose.

The current status of remotely sensed data (*i.e.*, various types and resolutions) enables multi-scale information to be derived for mapping and monitoring mangroves across spatial-ecological hierarchies. However, as an increasing number of remotely-sensed data sets with different pixel sizes become available, selecting the most appropriate spatial resolution becomes more difficult. In order to select an appropriate spatial resolution for a specific application, the spatial characteristics of the scene should be examined [30]. The main objectives of this study were to develop methods to estimate the optimum pixel size for mapping mangrove composition and structural properties, and test the applicability of the methods. The examination of spatial characteristics of mangroves was conducted used experimental semi-variograms derived from WorldView-2 images, in Moreton Bay, Australia, to determine the spatial scales of the mangrove features. This information may provide guidance for selecting the most appropriate image spatial resolution for mapping certain mangrove features, and answering the question of what mangrove feature can be mapped from a given spatial resolution, and *vice versa*.

DATA AND METHODS

Study Site

This research was carried out in mangrove areas at the mouth of the Brisbane River, Northern Moreton Bay, Southeast Queensland, Australia (between 153°3′41″–153°11′20″E and 27°19′41″–27°25′31″S), including Whyte and Fisherman Islands, and the Boondall wetlands (Figure 1). This lowland area is classified as sub-tropical, experiencing warm climate (average daily maximum: 28.9 °C, minimum: 20.0°C) with moderate to high rainfall (mean annual rainfall: 1,267.7 mm). The highest rainfall events are associated with monsoonal depression during the summer season (December to February) [31]. Moreton Bay is one of Australia's premier wetlands and a Ramsar Convention listed wetland, with extensive stands of mangroves [32]. *Avicennia marina* is the dominant mangrove species and comprises approximately 75% of the entire mangrove community within this region [33]. However, other species, such as *Rhizophora stylosa*, *Ceriops australis*, *Bruguiera gymnorhiza*,*Excoecaria agallocha*, and *Aegiceras corniculatum*, are occasionally present [34].

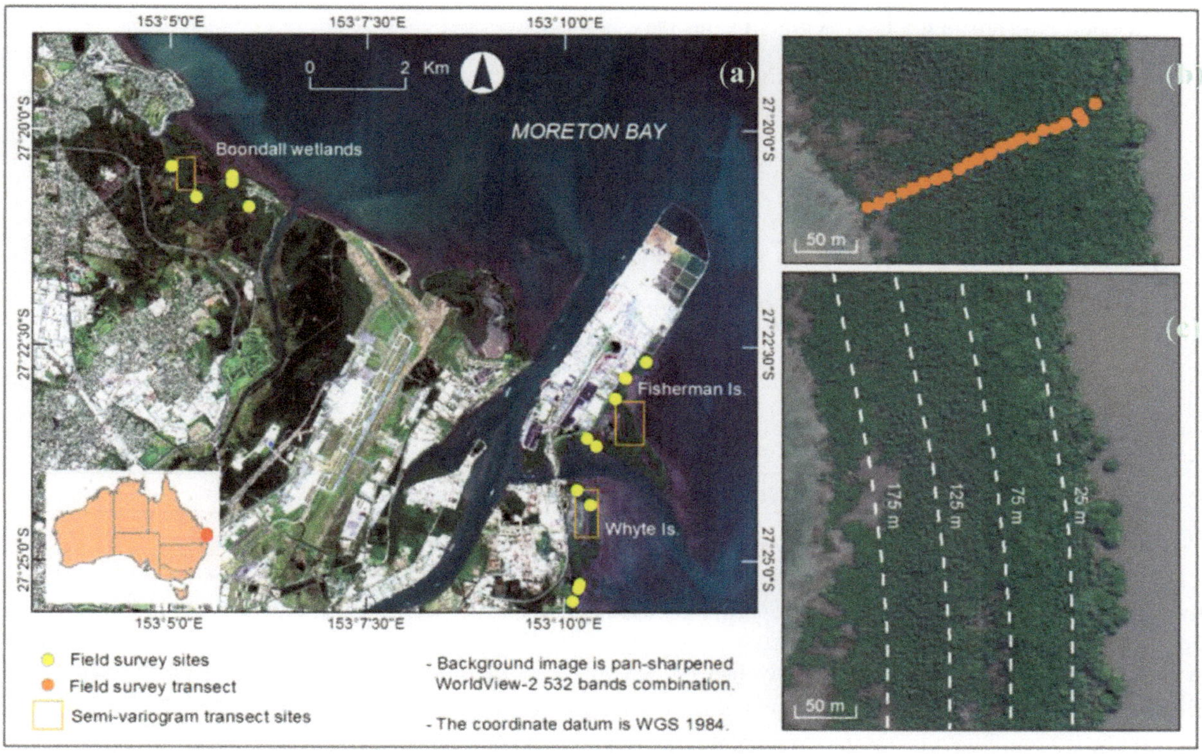

Figure 1. (**a**) Study sites: Mangroves at Moreton Bay area, Brisbane, Australia; (**b**) An example of field survey transect, orange circles represent location of 10 m × 10 m quadrats, and (**c**) Transects of image pixels running parallel to the coastline used to derive the semi-variograms shown for Whyte Island.

There are four different mangrove zonations from the saltmarsh area to the coastline (Figure 1c); starting with open scrub formation (S3), followed by low-closed forest with single-stem (I4a) and mixed-stem (I4b) tree, and finally closed forest (M4), according to the Specht *et al.* [35] classification. This structural zonation is consistent for Whyte Island and Fisherman Island mangroves with some variations in formation width (Figure 1). The zonation is less noticeable for Boondall wetlands, although it has similar vegetation formation gradation from saltmarsh to the river water. The rationale behind selecting these sites was that: (1) they are protected and well-preserved with minimum disturbances, making them ideal for developing and testing mapping approaches for this research; (2) there are distinct mangrove zonation and structural differences from the seaward to the landward locations; and (3) there are multi-resolution image data sets available for these sites.

Image and Field Datasets

The image data used as a basis for mangrove spatial structure examination was a WorldView-2 image of the Brisbane River mouth captured on 14 April 2011 (Figure 1 and Table 1). The image was obtained in an ortho-rectified format, corrected at Level 3X (LV3X); with root-mean-square error (RMSE) 2D of 0.00 [36]. Radiometric correction was applied to the image to convert the pixel values from digital number to top-of-atmosphere spectral radiance ($W/cm^2sr.nm$) using the values provided by Updike and Comp [37]. The atmospheric correction converting top-of-atmosphere spectral radiance to at surface reflectance was performed using the Fast Line-of-sight Atmospheric Analysis of Hypercubes (FLAASH) Atmospheric Correction Model in Environment for Visualizing Images (ENVI) 4.8, with the atmospheric visibility parameter estimated from the *moderate-resolution imaging spectroradiometer* (MODIS) aerosol product [38]. A very high-spatial resolution aerial photograph (Table 1) captured on 14 January 2011 (www.nearmap.com) was used to measure the dimension (*i.e.*, spatial size) of mangrove features investigated (foliage clumping, canopy gaps, tree crown, vegetation formation or community, and vegetation cover type) and as a reference to analyze the image spatial structure.

Table 1. Characteristics of image data used in this study

Image Type	WorldView-2	Aerial Photograph
Acquisition date	14 April 2011	14 January 2011
Acquisition time	00:10:47.59 UTC (10:10:47.59 AEST)	
Product type and level	Ortho, LV3X	Ortho-rectified
Geometric attributes	UTM 56 J in meters	GCS WGS 1984

Radiometric attributes	16 bits per-pixel Coastal (400–450 nm) Blue (450–510 nm) Green (510–580 nm) Yellow (585–625 nm)	
Spectral attributes	Red (630–690 nm) Red edge (705–745 nm) NIR1 (770–895 nm) NIR2 (860–1040 nm) PAN (450–800 nm)	True color image (red, green, blue)
Pixel size	Multispectral 2 m, panchromatic 0.5 m	7.5 cm

Note: AEST—Australian Eastern Standard Time.

The fieldwork was conducted during April 2012, to measure selected mangrove vegetation structure and composition variables in the study area. Fifteen representative 200 to 300 m long field transects were established perpendicular to the shoreline (Figure 1a,b). On each transect, plots of 10 m × 10 m quadrats were sampled along the transect lines as a frame to record mangrove zonation pattern and measure mangrove biophysical properties in the field, including transect and plot positions, vegetation structural information (canopy height and vegetation formation type), dominant species, and field photos.

Positions of each sampling plot (at the start and the end of the plot) were measured using Garmin eTrex Legend H hand-held GPS, using an average reading time for each point between 400 and 600 s (with the positional accuracy of 4–6 m) to maximize the GPS signal inside the mangrove canopy. Additional control points identifiable from the field and image were used to ensure the precise overlay of the transects to the image. Canopy height was measured every 5 m along the transect using a TruPulse 360 laser rangefinder. The mangrove structural formation on each plot was determined using Australian vegetation structural formations table [35] and the dominant species information were identified using Australian mangrove identification wheel [34]. The structural characteristics measured along the transect in each of three mangrove field sample sites are presented in Table 2.

Table 2. Mangrove vegetation structural characteristics of Moreton Bay mangroves, sampled at three sites, derived from field data sampled across the main vegetation zonation boundaries

Distance from Coastline	Whyte Island	Fisherman Island	Boondall Wetlands
175 m (S3)	Open scrub, 1–3 m height, multi stem, gaps > canopy cover, *Sarcocornia quinqueflora*, water or soil understory, low density canopy cover, dominant species *Avicennia marina*.	Open scrub, 2.5–4 m height, single or multi stem, gaps < canopy cover, *Sarcocornia quinqueflora*, water or soil understory, low density canopy cover, dominant species *Avicennia marina*.	Open scrub, 1.5–5 m height, single or multi stem, gaps > canopy cover, *Sarcocornia quinqueflora*, water or soil understory, medium density canopy cover, dominant species *Avicennia marina*.
125 m (I4a)	Low-closed forest1, 4–7 m height, single or multi stem, gaps < canopy cover, *Avicennia marina* seedling or *Aegiceras corniculatum*understory, medium density canopy cover, dominant species *Avicennia marina*.	Low-closed forest1, 4–9 m height, single stem, gaps < canopy cover, *Avicennia marina* seedling or *Aegiceras corniculatum* understory, high density canopy cover, dominant species*Avicennia marina*.	Low-closed forest1, 5–7.5 m height, single stem, gaps < canopy cover, *Avicennia marina* seedling or*Aegiceras corniculatum* understory, high density canopy cover, dominant species *Avicennia marina*.
75 m (I4b)	Low-closed forest2, 6–8 m height, single stem, gaps < canopy cover, clear understory, very high density canopy cover, dominant species*Avicennia marina* with some individual*Rhizophora stylosa*.	Low-closed forest2, 8–10 m height, single stem, gaps < canopy cover, clear understory, high density canopy cover, dominant species*Avicennia marina* with some individual*Rhizophora stylosa* and patches of *Ceriop tagal*.	Low-closed forest2, 7–9 m height, single stem, gaps < canopy cover, clear or *Avicennia marina*seedling understory, high density canopy cover, dominant species *Avicennia marina* with some individual *Rhizophora stylosa*.

25 m (M4)	Closed forest, 10–12 m height, single or multi stem, gaps < canopy cover, clear understory, high density canopy cover, dominant species*Avicennia marina* trees.	Closed forest, 8–11 m height, single or multi stem, gaps < canopy cover, clear understory, high density canopy cover, dominant species*Avicennia marina* trees.	Closed forest, 8–10.5 m height, single or multi stem, gaps < canopy cover, clear understory, high density canopy cover, dominant species*Avicennia marina* trees with some patches of*Ceriop tagal.*

Methods

Figure 2 summarizes the methods used to measure the spatial structure of mangrove features in the study area. The work flow is divided into three steps. Step 1 is necessary to prepare the data and create a series of images with different pixel sizes as a basis for multi-scale level examination of mangrove spatial structure. Step 2 deals with measuring and analyzing the spatial structure of mangrove vegetation from the pre-processed images using semi-variograms. Step 3 applies and evaluates the results from step 2 into image datasets for mapping mangrove features using object-based image analysis (OBIA).

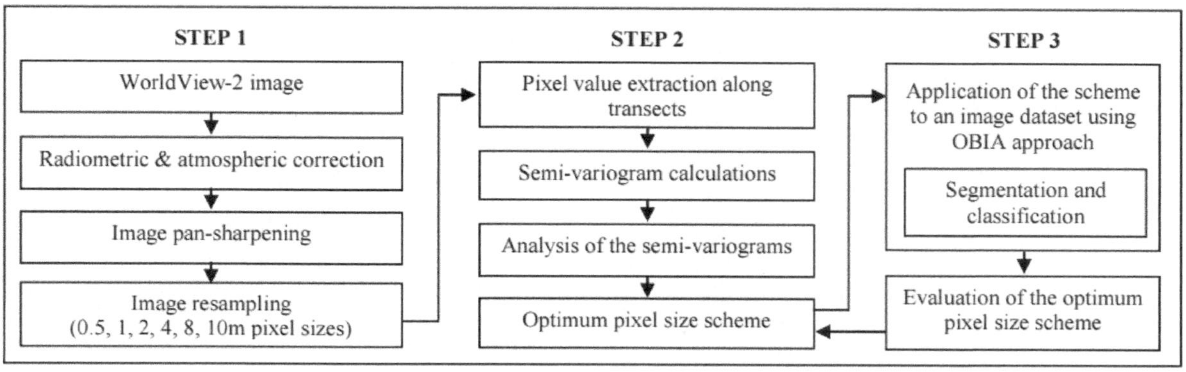

Figure 2. Overview of the methods.

Image Dataset Preparation

Pan-sharpening algorithms were applied to the atmospherically corrected images to obtain a higher spatial resolution image for input to the spatial structure analysis. Six different pan-sharpening algorithms (*principal component, multiplicative, Brovey, wavelet, Gram-Schmidt, and color normalized*) from commercial image processing software were investigated to determine their quality in terms of preserving pixel values of the atmospherically corrected multi-spectral data. Several image quality metrics were applied, including root mean square error, standard deviation, relative shift of the mean, and coefficient of correlation [39,40]. The overall evaluation of the image quality metrics shows that the Gram-Schmidt pan-sharpening algorithm produced the pixel values closest to the original multi-spectral data, and therefore this method was used to produce a sharpened image for spatial structure analysis.

To enable analysis of the spatial structure of mangroves at multiple specific-scales and to examine the effect of different pixel sizes on the derived information, the Gram-Schmidt pan-sharpened images (0.5 m pixels) were resampled to a pixel size of 1 m, and the original multi-spectral image (2 m pixels) were resampled to 4, 8, and 10 m pixel sizes using an averaging algorithm. According to Bian and Butler [41], the pixel aggregation method is more appropriate for processing remote sensing images because a pixel value is assumed to be the averaged value over the associated area on the ground. This process produced a total of six different image pixel sizes (0.5, 1, 2, 4, 8, and 10 m). The reasons for selecting these pixel sizes were to detect specific details of mangrove features, and an approximation of the currently available image datasets. All of processing above was done using ENVI 4.8 image processing software.

Measurement of Mangrove Spatial Structure through Semi-Variogram

The semi-variogram (γ) is a spatial statistical graph of semi-variance, which is the measured difference in variance value between pairs of regionalized variable samples in relation to their spatial separation with a given relative orientation. It provides a concise and unbiased description of the scale(s) and pattern(s) of spatial variability, in both remotely-sensed data and field data [42,43]. If applied to remotely-sensed data; the semi-variogram is used to examine the relation between the digital number (*DN*) or pixel value of *n* pixel pairs at a distance *h* (the lag distance) apart. The equation for semi-variance $\gamma(h)$ is:

$$\gamma(h) = \frac{1}{2n}\sum\{DN(x) - DN(x+h)\}^2 \qquad (1)$$

where $\gamma(h)$ represents half of the mathematical expectation of the squared differences of pixel pair values at a distance, h,and *DN* refers to spectral reflectance or vegetation index in this study. Hence, for image spectral data, $\gamma(h)$ estimates the variability of DNs, as a function of spatial separation.

In creating semi-variograms, the distance of each lag determines the number of the pixel pairs in the transect sample. By increasing the lag distance, the pixel pairs on a transect are fewer; which in turn decreases the confidence level of the analysis [42]. To alleviate this problem, Webster [44] suggested to use lag distances shorter than a fifth of the transect length for the semi-variogram interpretation. The analysis of semi-variograms requires two assumptions: spatial stationarity, which assumes that the correlation between variables is a function of the lag distance between pixels, not because of the variation in spatial position of the transect [21,27,45]; and ergodicity, which assumes that spatial statistics taken over the area of the image as a whole are unbiased estimates of those parameters [46]. Both of these assumptions are appropriate in digital remotely-sensed images; first, because of the variation in scan angle and terrain effects are minimal and, hence, regarded as stationary in increments, and second, the reflectance surface is considered stochastic [46]. However, according to Curran [42], all image bands and directions should be examined in order to define the minimum range of the semi-variogram and therefore the minimum spatial resolution of the feature element. There are two sampling methods used in creating semi-variograms [29,47]; (1) transect method, where the semi-variogram is calculated along a single, representative row or column of pixels from each selected image, and (2) matrix method, where the semi-variogram is calculated for all the row and column pixels in each image. For canopy structure analysis, the transect sampling method provides more detailed sill variation and periodicity, and also smaller range distance compare to the matrix method [29]. It was as a result of averaging the semi-variograms for all row and column directions in matrix method. Therefore, the first sampling method was adopted for this study in order to depict structural information inherent within each of the mangrove zones.

The approach used for examining spatial structure in this study was similar to the one conducted by Cohen *et al.* [29], and Johansen and Phinn [24]. We used several transects created over the image data sets (original and resampled WorldView-2 images) to generate semi-variograms. All eight bands of the WorldView-2 image (Table 1), and the derived normalized difference vegetation index (NDVI) images were used to generate the semi-variograms. According to image visual inspection and semi-variogram evaluation [48], four bands (green, red-edge, near infrared1 (NIR1), NDVI) were sensitive to mangrove vegetation variations and had the least redundant information among other bands. Therefore only these bands were used for further analysis of the semi-variograms. The semi-variogram transects were located parallel to the coastline to represent the homogenous areas within each mangrove zone (*i.e.*, areas assumed to have similar vegetation structural properties). In this study, 12 representative transects located parallel to the coastline at a distance of 25 m, 75 m, 125 m, and 175 m from the coastline were used to develop semi-variograms within zone structures (Figure 1c). The distance of the transects line varied between 500 and 1,000 m, depending on the length of the mangrove zonation. All transects were evaluated at six different pixel sizes (0.5, 1, 2, 4, 8, and 10 m). We used Earth Resources Data Analysis System (ERDAS IMAGINE) 2013 to extract pixel values along the semi-variogram transects and GS+ geostatistical software for semi-variance calculation.

Interpretation of Semi-Variograms

In order to interpret a semi-variogram, it is necessary to understand the term and characteristics associated to the semi-variogram (Figure 3). The range (a) of semi-variograms is a distance at which samples become independent, and it is controlled by the size of dominant objects in an image. The height of the sill (*s*), where the semi-variogram levels off, is considered proportional to the density of objects and scene-scale level of variance. The form of semi-variograms is controlled by the pattern and distribution of objects in an image [18,19,45,49]. These descriptors of semi-variograms are usually used together to interpret the appropriate spatial resolution for mangrove elements.

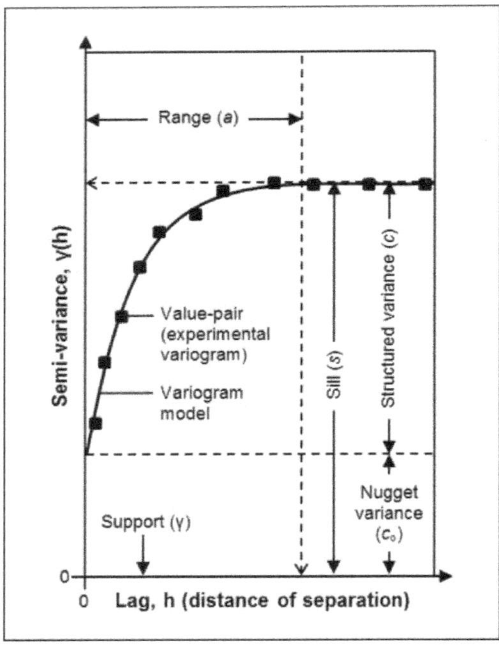

Term	Symbol	Definition
Semi-variance	$\gamma(h)$	Half the averaged variances between pixels at distance h apart.
Support	γ	Area, shape and orientation of space represented by each sample datum.
Lag	h	Distance (and direction in two or more dimensions) between sampling pairs.
Sill	*s*	Maximum value of $\gamma(h)$ model.
Range	*a*	Point on x axis where $\gamma(h)$ model reaches maximum. Places closer than the range are related statistically, places further apart are not.
Nugget variance	c_o	Point where $\gamma(h)$ model intercepts the y axis. Represents a component of the variation that is spatially uncorrelated.
Structured variance	*c*	Sill minus nugget variance.
Gradient	*m*	The slope for $\gamma(h)$ model with a linear component.

Figure 3. An example of variogram with descriptors, and the definition of terms [24,43].

The advantage of using semi-variograms in remote sensing is its ability to relate its descriptors to the spatial characteristics of the scene [50]. The range and sill were extracted from all semi-variograms of transects located parallel and perpendicular to the coastline using the green, red-edge, NIR1, and NDVI bands. Six different image pixel sizes were evaluated to detect the scale at which certain features within each mangrove zone occurred. The range provides a measure of the size of the elements or features in the mangrove environment in the image and has been suggested as a useful indicator for selecting the optimal spatial resolution for discriminating features embedded in the semi-variogram [19,42]. Visual inspection of the field data and visual interpretation of the image were conducted in order to relate the semi-variogram range values to the mangrove vegetation structure dimensions (foliage clumping, canopy gap and tree crown size, vegetation formation, or community and vegetation cover type). This information provides guidance for developing a scheme in selecting the optimum image spatial resolution to extract specific mangrove feature, and serve as a basis for an inversion mapping approach. A total of over 1,500 semi-variograms were analyzed in this study. The findings from the semi-variogram analysis revealed the spatial characteristics of mangrove vegetation features and the optimum pixel sizes to map these features.

Application of the Analysis Results

To test the applicability and validate the results, we used object-based image analysis (OBIA) applied to the original and resampled WorldView-2 images with the segmentation and classification being driven by parameters obtained from the semi-variogram analysis. OBIA offers some fundamental advantages in the context of this study, including: (1) image objects can be created at multiple, yet specific, hierarchical spatial scales (e.g., tree community consists of several single tree canopies) [51,52], (2) numerous attributes can be obtained from image objects, such as object's statistics, geometry and context, (3) better mimics human perception of real-world objects [53], and (4) image objects reduce the salt-and-pepper effect in pixel-based classifications [54].

We applied image segmentation and classification to the original and resampled WorldView-2 images to evaluate the applicability of the semi-variogram results. The segmentation and classification routines were carried out using eCognition Developer software v. 8.7.0. A series of scale parameters from 5 to 100 were tested for the mapping of mangrove features with a weight of 5 applied to the green, red, red-edge, and NIR1 bands to enhance the influence of bands that are sensitive to vegetation reflectance in the segmentation. We used a trial-and-error approach and visual inspection of the segmentation results with the help of a very-high-spatial resolution aerial photograph (7.5 cm pixel size) to determine the mangrove features that can be depicted from different scale parameters and image pixel sizes.

A supervised rule-based classification was applied to implement the optimum pixel size scheme for mangrove feature mapping based on the WorldView-2 images. The rule sets were developed using the objects' spectral information, geometry of the objects, and contextual hierarchy of the classes. The first hierarchy layer was used to separate mangrove and non-mangrove features, the second layer was for dividing the mangrove area into trees and gaps (and shadows), and the third layer, was used for classification of objects within the mangrove tree class. These hierarchical layers were applied to all image pixel sizes being examined. The result of the mapping was used to evaluate the image selection scheme and inversion mapping approach.

RESULTS AND DISCUSSION

Relation of Semi-Variograms to Mangrove Vegetation Structure Properties

Visual characteristics of the images demonstrate the gradual loss of mangrove spatial structure detail by the increasing pixel sizes (Figure 4). The descriptive statistics for each image shows the divergence between original (0.5 and 2 m) and resampled image (1, 4, 8, and 10 m). While the mean values of the bands in the resampled image exhibit some random variations, the coefficient of variance (%CV) decreased with decreasing spatial resolution relative to the original images. The coastal and yellow bands for the first resampling group (from 0.5 to 1 m) and all bands for the second resampling group (from 2 to 10 m) also follow this pattern. Figure 4 shows that an image with lower spatial resolution has lower data variability and therefore contains less information compared to a higher spatial resolution image.

The original and resampled WorldView-2 images.

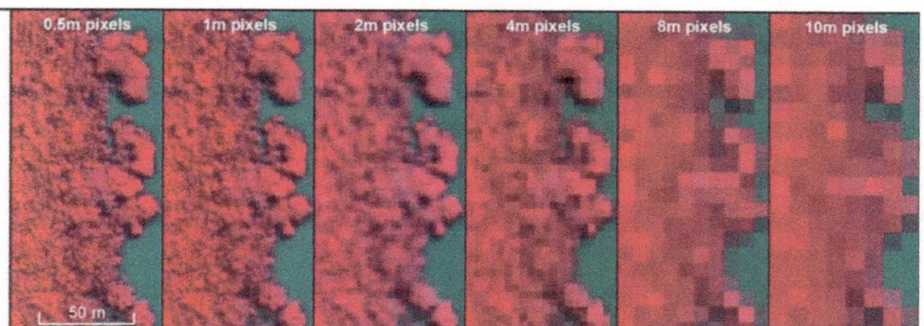

Image statistics		Mean	%CV	Mean	%CV	Mean	%CV	Mean	%CV	Mean	%CV	Mean	%CV
WV-2 image bands	Coastal	0.0605	0.64	0.0605	0.63	0.0628	0.53	0.0628	0.52	0.0625	0.51	0.0628	0.50
	Blue	0.0684	0.68	0.0683	0.68	0.0710	0.58	0.0710	0.56	0.0707	0.54	0.0710	0.53
	Green	0.0876	0.59	0.0875	0.59	0.0908	0.51	0.0908	0.50	0.0904	0.48	0.0909	0.47
	Yellow	0.0879	0.67	0.0879	0.66	0.0915	0.56	0.0914	0.55	0.0911	0.53	0.0915	0.52
	Red	0.0778	0.79	0.0778	0.79	0.0814	0.66	0.0814	0.65	0.0811	0.63	0.0815	0.62
	Red-edge	0.1429	0.50	0.1428	0.50	0.1479	0.54	0.1478	0.53	0.1472	0.52	0.1482	0.51
	NIR1	0.2109	0.60	0.2108	0.60	0.2172	0.67	0.2171	0.66	0.2162	0.65	0.2179	0.63
	NIR2	0.1909	0.63	0.1908	0.63	0.1967	0.70	0.1966	0.69	0.1958	0.67	0.1973	0.66
	NDVI	0.3617	1.11	0.3611	1.10	0.3029	1.34	0.3043	1.33	0.3055	1.32	0.3104	1.29

Figure 4. Subsets of mangrove on Whyte Island at pixel sizes of 0.5, 1, 2, 4, 8, and 10 m displayed with a band combination of R:7, G:5, B:3,

and their associated descriptive statistics.Note: The mean image spectral reflectance and NDVI values were represented at a scale of 0–1, %CV: coefficient of variance

The spatial structural information of mangrove elements (or features) detectable from the original and resampled WorldView-2 images was related to the semi-variogram descriptors (range, sill, and form). Figure 5 (preliminary published in [48]) shows the representative semi-variogram plot of the near infrared1 (NIR1) image band at six different pixel sizes, and the average mangrove feature sizes measured from field-work and the aerial photograph. It reveals similar results to those presented in other studies examining the effects of changes in image pixel size on image information content [19,24,29,55,56]. Specifically, there is a gradual loss of information detail with the increasing pixel size (Figure 4 and 5). The forms of semi-variogram changes with the varying pixel sizes describe the effect of data regularization on the spatial heterogeneity component. The decreasing height of the sill characterizes the loss of spatial variability when the spatial resolution of the image decreases, which is also evident in the image statistics in Figure 4.

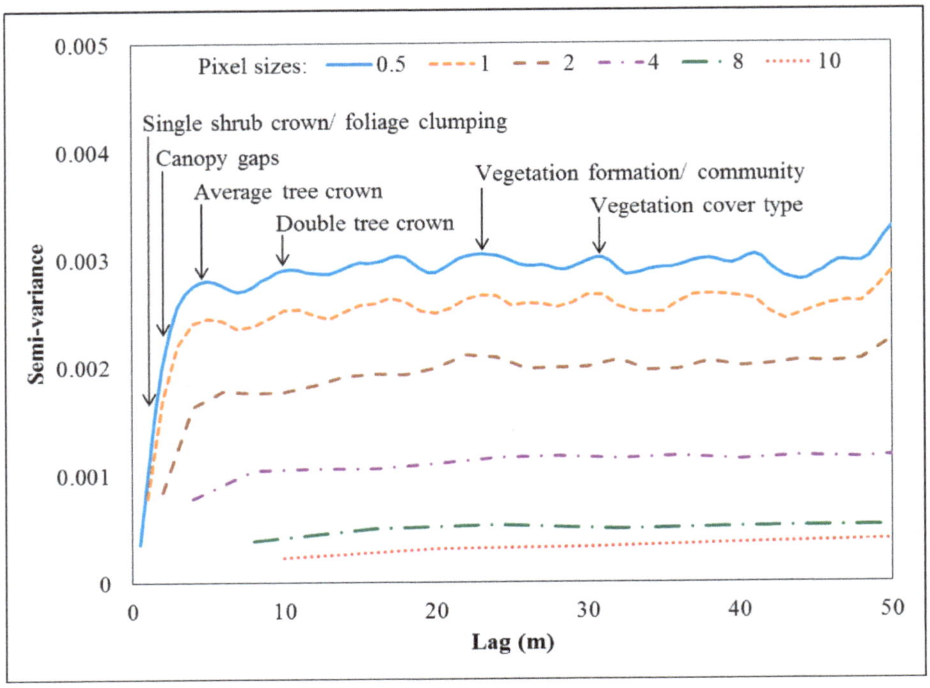

Figure 5. Subset of NIR1 semi-variogram showing the mangrove features responsible for the semi-variogram range and form up to a 50 m lag distance.

At a pixel size of 0.5 m, the form of semi-variograms appears to be controlled by inter-canopy features including individual shrubs and tree crown, foliage clumping and smaller inter-canopy gaps. At a pixel size of 1 m, the individual shrub and tree crown are still distinguishable, but the detailed foliage clumping and inter-canopy gaps become more difficult to identify. Single shrub crowns and foliage clumping identified from the image and field data correspond to the semi-variogram lag distance of 1.5 m, and the inter-canopy gaps at 2 m. Pixel sizes 0.5 and 1 m had similar semi-variogram forms, meaning the variation of information of the pattern and distribution of mangrove features within these sizes are comparable, but they have different sill height which is attributed to the different level of pixel value variance in the image scene (see image statistics in Figure 4). Although these two pixel sizes offer similar capability, detailed information on structural properties and sharper visual image appearance increases the likelihood of correctly identifying mangrove inter-canopy features.

The range of the semi-variograms for pixel sizes 2 m and smaller are all at 5 m, which is approximately equal to the average diameter of single mangrove tree crowns. This indicates that single mangrove tree crowns may be distinguished at a maximum pixel size of 2 m. At a pixel size of 4 m the average single tree crown is no longer apparent, although some individual larger canopies of mangrove tree and gaps that are more than 10 m in diameter can still be identified. Based on field observations, it is likely that the range of about 8 to 10 m corresponds to groups of two tree crowns, and these can be identified at pixel sizes of 4 m. The semi-variogram range of pixel sizes of 8 and 10 m are approximately 20 and 30 m, respectively, which correspond to the average size of mangrove vegetation formation or community at 23 m and larger mangrove cover types at 30 m. At a pixel size of 4 m and larger, the canopy-related information is gradually lost. They are more appropriate for mapping larger mangrove communities, mangrove patches, and separating mangroves from non-mangrove cover types.

Overall, there are noticeable patterns of semi-variogram peaks and troughs with different sill heights across varying pixel sizes. According to Figure 5, the variation of structural information within mangrove stands can only be preserved by pixel sizes of 0.5 to 2 m, where the semi-variogram contains periodic peaks and troughs along the graph. On the other hand, from pixel sizes of 4 to 10 m, these variations are lost and result in relatively flat graphs with minimum or no information on within mangrove structural properties. This finding indicates that pixel sizes of 2 m and smaller are appropriate for mapping small size features and internal variation of mangrove canopy (such as single shrub crown, foliage clumping, canopy gaps, and average tree crown), whereas pixel sizes of 4 m and larger are more appropriate to map larger mangrove features (such as groups of tree crowns, vegetation formations or communities, and cover type).

Relation of Semi-Variograms to Mangrove Zone Features

Variation in spatial structural information between mangrove zones, *i.e.*, areas with consistent structure and composition parallel to the coastline, were examined using transects located parallel to the coastline at varying distances from the three Moreton Bay sites. The semi-variograms were derived from transects of 1000 m in length along the mangrove zonation which were located at distances of 25, 75, 125, and 175 m from the edge of outer mangrove stands. These distances were selected to be as representative as possible of mangrove zonation to depict the within mangrove zonation structural properties. Among the three transect locations; two of them produced similar semi-variograms. Therefore, only two of the transect locations are discussed here, which are Fisherman Island and Boondall wetlands (Figure 6).

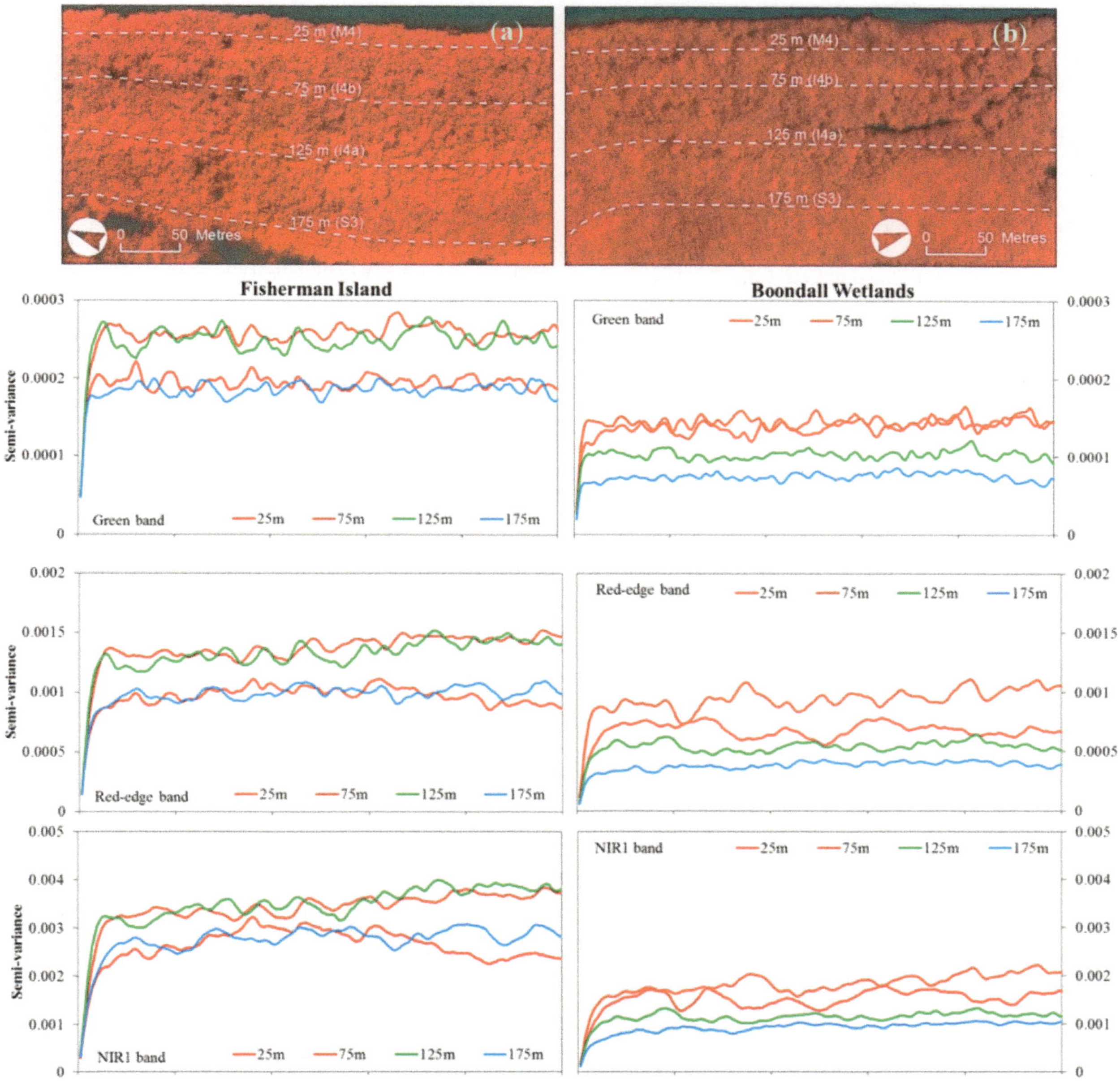

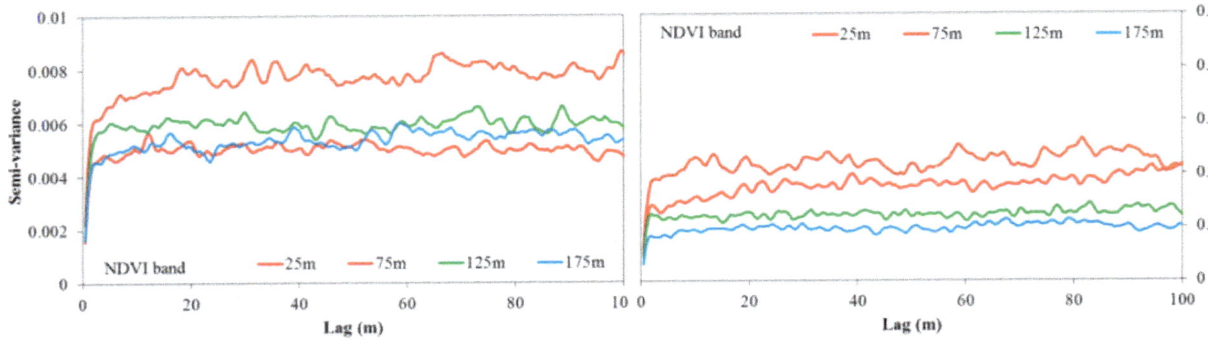

Figure 6. Subset of pan-sharpened image of (**a**) Fisherman Island and (**b**) Boondall wetlands showing the transects along the mangrove zonations, and semi-variograms created from transects along the mangrove zonation at distances of 25, 75, 125, and 175 m from the coastline using the 0.5 m pan-sharpened WorldView-2 image.

The transect at 125 m from the coastline on Fisherman Island had high sill values and was the second highest for all of the bands. It corresponds to a low-closed forest mangrove formation where there are mixed stems of *Avicennia marina* trees (I4a) with high canopy cover and some canopy gaps. For the Boondall wetlands, this transect distance was at the third highest level of the semivariogram sill height. The mixed-stem low-closed forest (I4a) formation in Boondall has a lower canopy cover and more canopy gaps, making it appear to have low contrast in the image. Finally, transects located at a distance of 175 m from the coastline had the lowest semi-variogram sill values and this pattern is noticeable for all bands and all locations. These transects cover the open scrub *Avicennia marina* zone with low density of canopy cover and uniform canopy layers. Some gaps with *Sarcocornia quinqueflora*, water or soil are also frequently found in this zone, which may contribute to decreasing the reflectance intensity of the pixels.

Semi-variograms located along the mangrove zones revealed different spatial structural characteristics of mangrove features within each zonation. As shown in Figure 7, the average range of semi-variograms of all evaluated bands for transects located along the mangrove zonation at the Moreton Bay sites exhibit different patterns. The average range values of open scrub formation (S3) were higher for all of the image bands (2.7 to 8.1 m) compared to the other formations, where low-closed forest1 (I4a) had the lowest average range value (2.5 to 5.2 m), and low-closed forest2 (I4b) together with closed forest (M4) had a similar range value (3.1 to 6.6 and 2.6 to 6.6 m, respectively). The high average range values on the open scrub formation (S3) were attributed to the large variation in the size of *Avicennia marina* scrub patches interleaved with large gaps of ground or water that are frequently found in this zone (see image on Figure 7). Conversely, the low average range values of the low-closed forest1 (I4a) were caused by the foliage clumping and the narrow canopy gaps (1–2 m) between individual tree crowns (approximately 5 m in diameter) that dominate this zone.

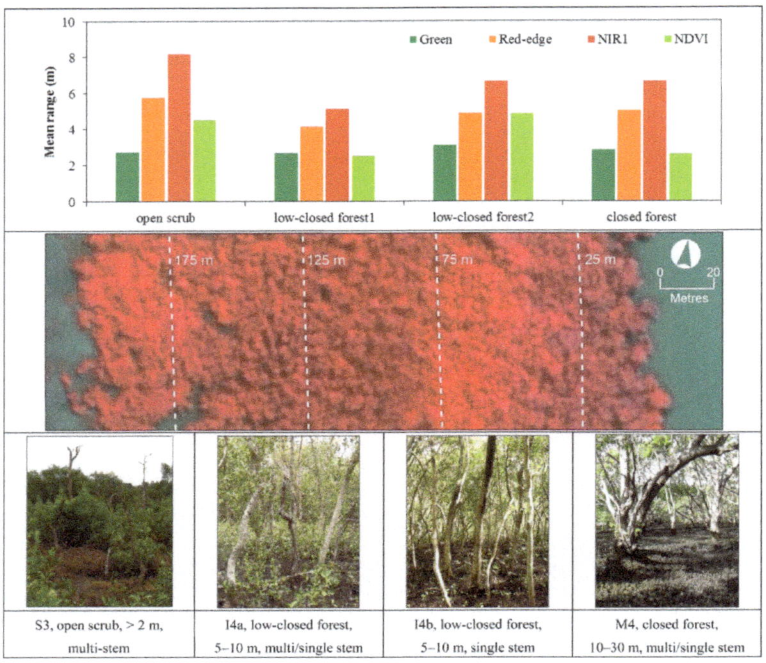

Figure 7. Mean semi-variogram range values sampled at 0.5 m pixel size for the green, red-edge, NIR1, and NDVI bands, based on four transects located at 175 m (S3), 125 m (I4a), 75 m (I4b), and 25 m (M4) from the coastline of Moreton Bay mangroves.

Optimum Pixel Size for Mangrove Mapping

Several transects were established along the mangrove zonation at the Moreton Bay sites to create semi-variograms to depict the variation of spatial size of the structural features of mangroves. The information on the characteristics of spatial structure of mangrove features obtained from semi-variogram interpretation and analysis provides the basis for establishing the relationship between mangrove feature sizes and optimum spatial resolution (*i.e.*, image pixel size) to map these features. According to techniques used in the previous semi-variogram analyses, seven mangrove vegetation structures were apparent in the study area and detectable from the WorldView-2 image data. These included single shrub crown, small foliage clumping within canopy or intra-canopy, smaller canopy gaps, average single tree crown, double tree crown or larger gaps, vegetation structural formation/community, and vegetation cover type. By integrating semi-variogram interpretation results at specific pixel sizes along mangrove zonations, with field data and image interpretation, it was apparent that each of these mangrove features resided within certain range distances and can only be detected at specific image pixel sizes.

Figure 8 illustrates the relationships between the range of mangrove features, the optimum pixel size, and the most appropriate spectral bands able to identify and map these features. The mangrove structural features were plotted on the y axis from the largest scale at the bottom to the smallest scale at the top of the axis. Seven plot boxes inside the graph depict where the mangrove features on the y axis reside along the feature range distance (bottom x axis); and the downward bar charts from the x axis at the top illustrate the influence of the associated pixel size on the plot boxes of mangrove features. The most appropriate spectral bands were placed as dot point indicators beside the associated mangrove feature. For example, the average single tree crown has semi-variogram range values between 4 and 8 m. This feature can be identified using image pixel sizes of 0.5, 1, and 2 m, but is unable to be identified at a pixel size of 4 m or larger. Therefore, in this case, an image with a pixel size of 2 m is the optimum option to map the average mangrove tree crown as it is the largest pixel size able to identify individual tree crowns, and green, red-edge, and NDVI bands will be the most suitable bands for discriminating this feature. For mapping routine purposes, using an image datasets with a smaller pixel size will increase the cost, both for resources and processing, and may produce the result that is similar to a product derived from the optimum pixel size.

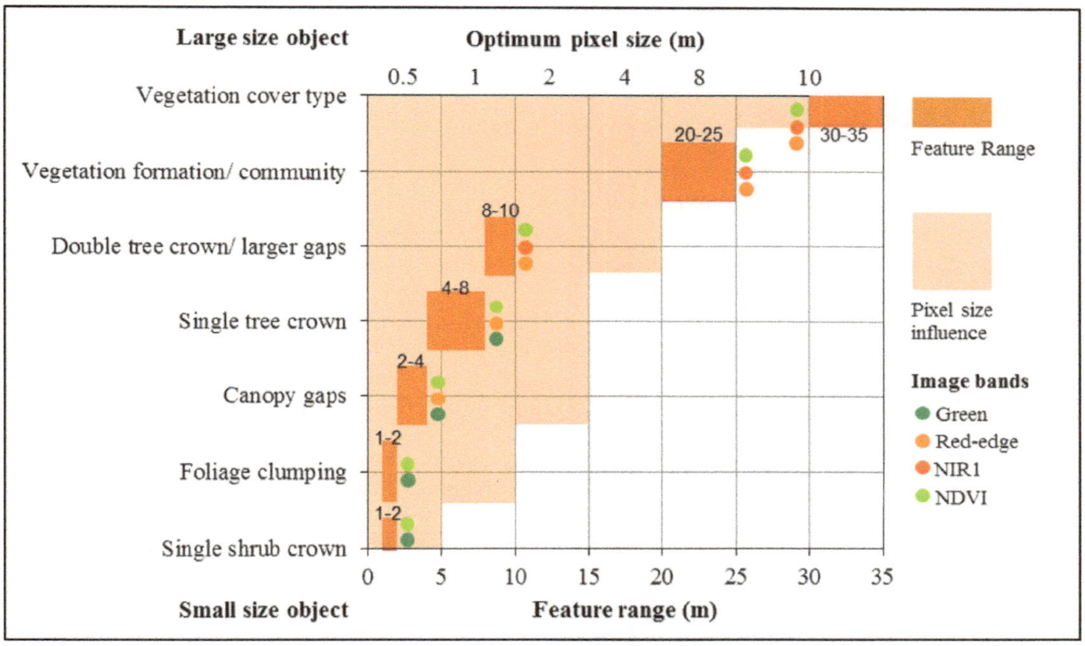

Figure 8. Relationship between mangrove features, feature ranges, optimum pixel sizes, and the most sensitive image bands to map mangrove features.

Application of the Optimum Pixel Size Scheme

The pattern of mangrove feature information obtained from different scale parameters applied to different image pixel sizes was interpreted from the image segmentation results (Figure 9). In accordance to the results of the semi-variogram analysis; the lower value of the scale parameter or the smaller pixel size, the more mangrove information could be extracted from the image. According to the interpretation results, single shrub crowns could only be recognised at a pixel size of 0.5 m with the segmentation scale parameter ≤30, and at a pixel size of 1 m with the scale parameter ≤10. At a pixel size of 2 m, the smallest obtainable mangrove objects were canopy gaps and single tree crowns with a segmentation scale parameter ≤10 and 20–30, respectively. At a scale parameter ≥40 the segment size was too large and failed to identify single tree crowns. Most of the pixel sizes ≥2 m could only differentiate objects larger than the average size of single tree crowns, including double/multiple tree crowns, larger gaps, vegetation formation and community, and vegetation cover types; with the exception of scale parameters ≤10 at a pixel size 4 m, which still allowed discrimination of single tree crowns. In general, there was an obvious relationship between mangrove information detail and the pixel and segmentation scale parameter size, where smaller pixel sizes or segmentation scale parameters will allow more detailed mapping of mangrove features.

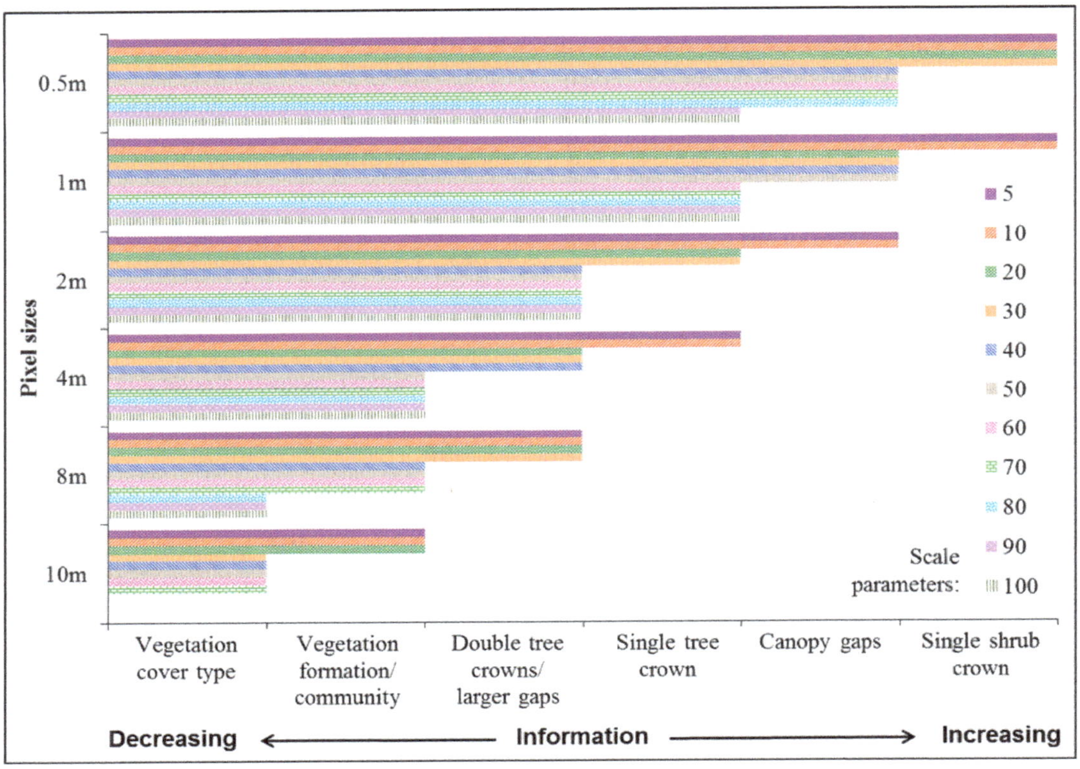

Figure 9. Graph of mangrove features detectable at a number of scale parameters derived from different WorldView-2 image pixel sizes.

We also identified the size of dominant objects able to be delineated in the segmentation process from six different pixel sizes and related the result with the measured feature dimension in the field. The dominant segment sizes and the corresponding pixel sizes were found to correspond well to the result of the optimum pixel sizes from the semi-variogram analysis (Columns c and f in Table 3). The result indicated that the theoretical finding of the optimum pixel sizes from the SV is empirically proven on the image dataset through image segmentation.

Table 3. Optimum pixel sizes interpreted from semi-variogram and image segmentation results

Mangrove Features	Average Features Size (m) in the Field	Optimum Image Pixel Size (m) from SV	OBIA Segmentation		
			Dominant Object Size (m^2)	Estimated Feature Dimension (m)	Optimum Image Pixel Size (m) from OBIA
(a)	(b)	(c)	(d)	(e)	(f)
Single shrub crown	1–2	**0.5**	1.5	1.2	**0.5**
Canopy gaps	2	**1**	4	2	**1**
Single tree crown	4	**2**	16	4	**2**
Double tree crowns/ larger gaps	8	**4**	48	6.9	**4**
Vegetation formation/ community	20–25	**8**	128	11.3	**8**
Vegetation cover type	30–40	**10**	700	26.4	**10**

The classification results showed that the use of each image pixel size enabled the discrimination of the smallest mangrove object according to the previously developed optimum pixel size scheme (Figure 8). Figure 10 shows some selected examples of the classification results applied to the image at pixel sizes of 0.5, 2, and 8 m to depict information of shrub crown, single tree crown, and mangrove formation, respectively. We used a very-high-spatial resolution aerial photograph as a reference to assess and evaluate the quality of mangrove feature mapping. This was used as there was no existing map containing mangrove vegetation features for the study area, and the aerial photograph interpretation was accepted to be correct (due to the features evident at very high resolution) without any form of accuracy assessment [57].

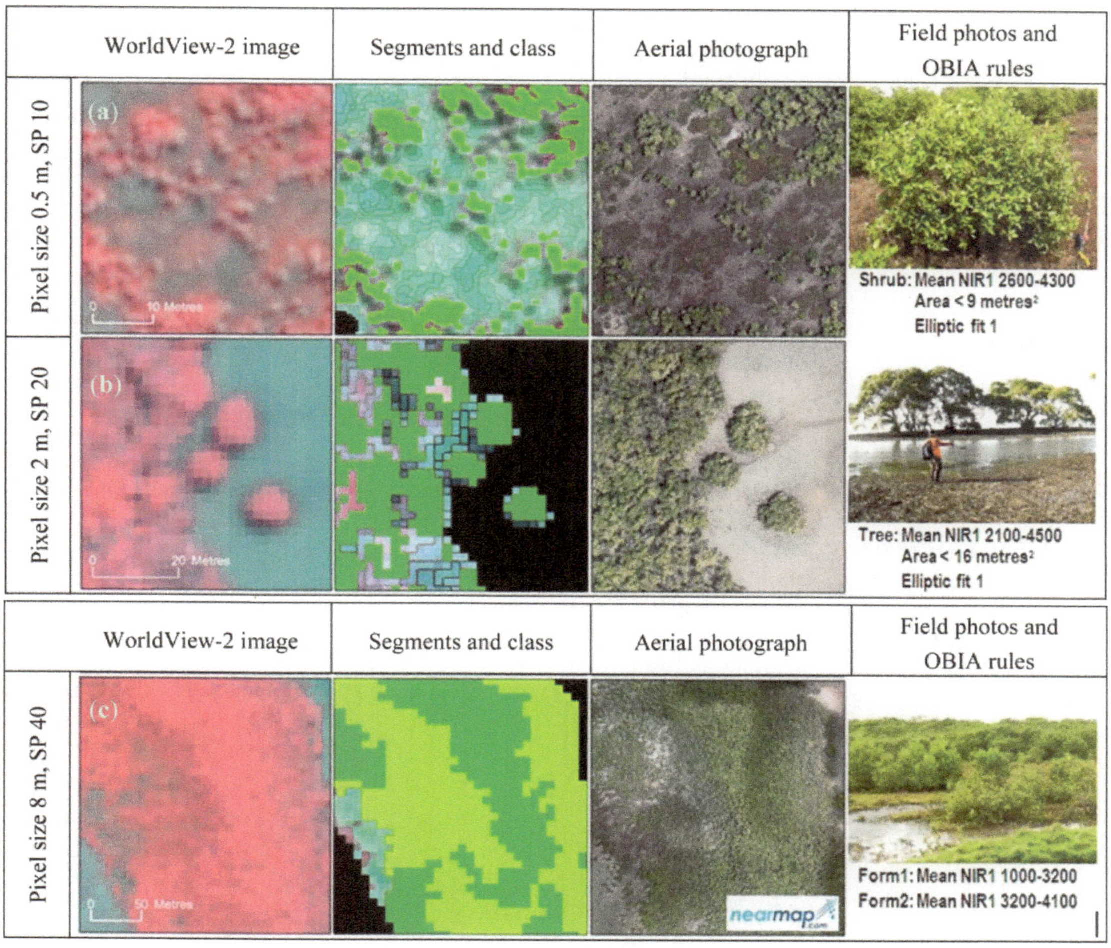

Figure 10. Example of mangrove features segmented and classified (green polygons) from images with different image pixel sizes and scale parameters (SP).

At a pixel size of 0.5 m individual shrub crowns were discriminated from other features, but the classification was unable to identify each single shrub crown in groups of shrub (Figure 10a) due to the short distance between the crowns and the shape similarity. Single stands of tree crowns were well identified at pixel size of 2 m, although the segments were blocky due to the bigger pixel size (Figure 10b). Individual trees in the mangrove forest were difficult to distinguish because of their closed canopy and similarity to the neighbouring objects (*i.e.*, tree crowns). Four mangrove formations found in the study area were clearly separated at a pixel size of 8 m (Figure 10c), with some isolated segments found within the formation. According to the qualitative segmentation assessment, the optimum pixel size scheme worked well when applied to the image datasets used in this study. As the segmentation and classification example was purely used as a proof of concept, further improvements to the segmentation and classification process are still possible to improve the mapping accuracy and repeatability across other areas.

CONCLUSIONS AND FUTURE RESEARCH

This study showed that scale-specific, ecologically relevant information on mangroves can be detected using experimental semi-variogram analysis. This approach indicated the size of dominant mangrove features able to be identified from a specific pixel size and the optimum spectral bands to use to detect and map these features. The results show that there was a gradual loss in mangrove vegetation structural information, as indicated by mangrove features measured, with increasing pixel size. When applied to the original and resampled WorldView-2 images, we found that a pixel size $\leq$2 m was suitable for mapping canopy and inter-canopy related features within mangrove objects (such as shrub crown, canopy gaps, single tree crown). A pixel size of $\geq$4 m was more appropriate for mapping mangrove vegetation formation, communities and larger mangrove features. The green and red-edge bands were optimum for discriminating smaller sized mangrove features (<8 m), such as single shrub crown or foliage clumping, canopy gaps, and single tree crown. The near infrared1 band was more suitable for identifying feature $\geq$8 m (e.g., double tree crown or larger gaps), and the NDVI image was suitable for mapping all targeted features.

The findings of this study provide a basis for an inversion approach to mangrove feature mapping using high-spatial resolution image datasets. The mapping application results demonstrated that the optimum pixel size scheme from the semi-variogram analysis was effectively applied through image segmentation and classification using object-based image analysis. The optimum pixel analysis result from the image segmentation interpretation was highly correlated with the semi-variogram results, and the classification result identified the smallest features that could be discriminated at the corresponding pixel size.

The results of this work were limited to the study site in Moreton Bay, Australia, with sub-tropical mangroves. For future research, the application of the experimental semi-variogram method to other mangrove environments is necessary to assess the consistency of the method. In terms of the mapping application, further development of object-based classification rule sets is essential for improving the detection of multi-scale mangrove features and determining the transferability and general applicability of the findings.

ACKNOWLEDGMENTS

This study is a component of Kamal's PhD project on multi-scale image-based mangrove mapping undertaken at The University of Queensland. The authors would like to thank Digital Globe for providing the WorldView-2 images; Biophysical Remote Sensing Group (BRG) at The University of Queensland (UQ) for providing research facilities and publication funding; School of Geography, Planning and Environmental Management (GPEM) UQ for research funding and fieldwork support; Ralph Dowling and Arnon Accad from Queensland Herbarium for mangrove mapping and fieldwork discussion; and Novi Adi (BRG UQ) for assistance with fieldwork. Authors would also like to thank the anonymous reviewers whose comments were detailed and constructive, improving the overall quality of the paper.

AUTHOR CONTRIBUTIONS

Muhammad Kamal performed the image processing and analysis and prepared the manuscript, Stuart Phinn and Kasper Johansen contributed to the design of the work, discussion and edited drafts.

REFERENCES

1. Davis, B.A.; Jensen, J.R. Remote sensing of mangrove biophysical characteristics. *Geocarto Int* **1998**, *13*, 55–64.
2. Ramsey, E.W., III; Jensen, J.R. Remote sensing of mangrove wetlands: Relating canopy spectra to site-specific data.*Photogramm. Eng. Remote Sens* **1996**, *62*, 939–948.
3. Kuenzer, C.; Bluemel, A.; Gebhardt, S.; Quoc, T.V.; Dech, S. Remote sensing of mangrove ecosystems: A review.*Remote Sens* **2011**, *3*, 878–928.
4. Malthus, T.J.; Mumby, P.J. Remote sensing of the coastal zone: An overview and priorities for future research. *Int. J. Remote Sens* **2003**, *24*, 2805–2815.
5. Heumann, B.W. Satellite remote sensing of mangrove forests: Recent advances and future opportunities. *Prog. Phys. Geogr* **2011**, *35*, 87–108.
6. Wiens, J.A. Spatial scaling in ecology. *Funct. Ecol* **1989**, *3*, 385–397.
7. Delcourt, H.R.; Delcourt, P.A.; Webb, T., III. Dynamic plant ecology: The spectrum of vegetational change in space and time. *Quat. Sci. Rev* **1983**, *1*, 153–175.
8. Schaeffer-Novelli, Y.; Cintrón-Molero, G.; Cunha-Lignon, M.; Coelho, C., Jr. A conceptual hierarchical framework for marine coastal management and conservation: A "Janus-like" approach. *J. Coast. Res* **2005**, 191–197.
9. Feller, I.C.; Lovelock, C.E.; Berger, U.; McKee, K.L.; Joye, S.B.; Ball, M.C. Biocomplexity in mangrove ecosystems.*Annu. Rev. Mar. Sci* **2010**, *2*, 395–417.
10. Farnsworth, E.J. Issues of spatial, taxonomic and temporal scale in delineating links between mangrove diversity and ecosystem function. *Glob. Ecol. Biogeogr. Lett* **1998**, *7*, 15–25.
11. Lee, D.; Grant, W. A hierarchical approach to fisheries planning and modeling in the Columbia River Basin. *Environ. Manag* **1995**, *19*, 17–25.
12. Müller, F. Hierarchical approaches to ecosystem theory. *Ecol. Model* **1992**, *63*, 215–242.
13. Woodcock, C.E.; Strahler, A.H. The factor of scale in remote sensing. *Remote Sens. Environ* **1987**, *21*, 311–332.
14. Phinn, S.R.; Menges, C.; Hill, G.J.E.; Stanford, M. Optimizing remotely sensed solutions for monitoring, modeling, and managing coastal environments. *Remote Sens. Environ* **2000**, *73*, 117–132.
15. Goodchild, M.F.; Quattrochi, D.A. Scale, Multiscaling, Remote Sensing, GIS. In *Scale in Remote Sensing and GIS*; Quattrochi, D.A., Goodchild, M.F., Eds.; CRS Press: Boca Raton, FL, USA, 1997; pp. 1–12.
16. Marceau, D.J.; Hay, G.J. Remote sensing contributions to the scale issue. *Can. J. Remote Sens* **1999**, *25*, 357–366.
17. Raffy, M. Change of scale theory: A capital challenge for space observation of earth. *Int. J. Remote Sens* **1994**, *15*, 2353–2357.
18. Woodcock, C.E.; Strahler, A.H.; Jupp, D.L.B. The use of variograms in remote sensing: I. Scene models and simulated images. *Remote Sens. Environ* **1988**, *25*, 323–348.
19. Woodcock, C.E.; Strahler, A.H.; Jupp, D.L.B. The use of variograms in remote sensing: II. Real digital images. *Remote Sens. Environ* **1988**, *25*, 349–379.
20. Colombo, S.; Chica-Olmo, M.; Abarca, F.; Eva, H. Variographic analysis of tropical forest cover from multi-scale remotely sensed imagery. *ISPRS J. Photogramm. Remote Sens* **2004**, *58*, 330–341.
21. Treitz, P.; Howarth, P. High spatial resolution remote sensing data for forest ecosystem classification: An examination of spatial scale. *Remote Sens. Environ* **2000**, *72*, 268–289.
22. Hyppanen, H. Spatial autocorrelation and optimal spatial resolution of optical remote sensing data in boreal forest environment. *Int. J. Remote Sens* **1996**, *17*, 3441–3452.
23. Wolter, P.T.; Townsend, P.A.; Sturtevant, B.R. Estimation of forest structural parameters using 5 and 10 meter SPOT-5 satellite data. *Remote Sens. Environ* **2009**, *113*, 2019–2036.
24. Johansen, K.; Phinn, S. Linking riparian vegetation spatial structure in Australian tropical savannas to ecosystem health indicators: Semi-variogram analysis of high spatial resolution satellite imagery. *Can. J. Remote Sens* **2006**, *32*, 228–243.

25. Menges, C.H.; Hill, G.J.E.; Ahmad, W. Use of airborne video data for the characterization of tropical savannas in northern Australia: The optimal spatial resolution for remote sensing applications. *Int. J. Remote Sens* **2001**, *22*, 727–740.

26. He, Y.; Guo, X.; Wilmshurst, J.; Si, B.C. Studying mixed grassland ecosystems II: Optimum pixel size. *Can. J. Remote Sens* **2006**, *32*, 108–115.

27. Phinn, S.; Franklin, J.; Hope, A.; Stow, D.; Huenneke, L. Biomass distribution mapping using airborne digital video imagery and spatial statistics in a semi-arid environment. *J. Environ. Manag* **1996**, *47*, 139–164.

28. Wen, Z.; Zhang, C.; Zhang, S.; Ding, C.; Liu, C.; Pan, X.; Li, H.; Sun, Y. Effects of normalized difference vegetation index and related wavebands' characteristics on detecting spatial heterogeneity using variogram-based analysis. *Chin. Geogr. Sci* **2012**, *22*, 188–195.

29. Cohen, W.B.; Spies, T.A.; Bradshaw, G.A. Semivariograms of digital imagery for analysis of conifer canopy structure.*Remote Sens. Environ* **1990**, *34*, 167–178.

30. Chen, D. Multi-Resolution Image Analysis and Classification for Improving Urban Land Use/Cover Mapping Using High Resolution Imagery, Ph.D. Thesis. San Diego State University and University of California, Santa Barbara, CA, USA, 2001.

31. Australian Bureau of Meteorology. Available online: http://www.bom.gov.au/ (accessed on 1 February 2013).

32. Environment Australia. *A Directory of Important Wetlands in Australia*, 3rd ed; Environment Australia: Canberra, ACT, Australia, 2001; p. 157.

33. Dowling, R.M.; Stephen, K. *Coastal Wetlands of South-Eastern Queensland: Maroochy Shire to New South Wales Border*; Queensland Herbarium, Environmental Protection Agency: Brisbane, QLD, Australia, 2001; p. 220.

34. Duke, N.C. *Australia's Mangroves: The Authoritative Guide to Australia's Mangrove Plants*; University of Queensland: Brisbane, QLD, Australia, 2006; p. 200.

35. Specht, R.L.; Specht, A.; Whelan, M.B.; Hegarty, E.E. *Conservation Atlas of Plant Communities in Australia*; Southern Cross University Press: Lismore, NSW, Australia, 1995; p. 910.

36. Digital Globe Digital Globe Imagery Support Data (ISD) Documentation. Available online: http://www.digitalglobe.com/sites/default/files/Imagery_Support_Data_Documentation(1).pdf (accessed on 18 December 2013).

37. Updike, T.; Comp, C. *Radiometric Use of WorldView-2 Imagery*; DigitalGlobe Inc: Longmont, CO, USA, 2010.

38. LAADS Level 1 and Atmosphere Archive and Distribution System. Website of NASA-Goddard Space Flight Center. Available online: http://ladsweb.nascom.nasa.gov/data/search.html (accessed on 5 March 2012).

39. Jayanth, J.; Kumar, T.A.; Koliwad, S. Comparative Analysis of Image Fusion Techniques in Remote Sensing. In*Advanced Machine Learning Technologies and Applications*; Hassanien, A.E., Salem, A.M., Ramadan, R., Kim, T., Eds.; Springer: Berlin/Heidelberg, Germany, 2012; Volume 322, pp. 111–117.

40. Vijayaraj, V.; O'Hara, C.G.; Younan, N.H. Quality Analysis of Pansharpened Images. Proceedings of The IEEE 2004 International Geoscience and Remote Sensing Symposium, Anchorage, AK, USA, 20–24 September 2004.

41. Bian, L.; Butler, R. Comparing effects of aggregation methods on statistical and spatial properties of simulated spatial data. *Photogramm. Eng. Remote Sens* **1999**, *65*, 73–84.

42. Curran, P.J. The semivariogram in remote sensing: An introduction. *Remote Sens. Environ* **1988**, *24*, 493–507.

43. Curran, P.J.; Atkinson, P.M. Geostatistics and remote sensing. *Prog. Phys. Geogr* **1998**, *22*, 61–78.

44. Webster, R. Quantitative spatial analysis of soil in the field. *Adv. Soil Sci* **1985**, *3*, 1–70.

45. Bailey, T.C.; Gatrell, A.C. *Interactive Spatial Data Analysis*; Longman Scientific & Technical: Harlow, Essex, UK, 1995; p. 413.

46. Jupp, D.L.B.; Strahler, A.H.; Woodcock, C.E. Autocorrelation and regularization in digital images. I. Basic theory. *IEEE Trans. Geosci. Remote Sens* **1988**, *26*, 463–473.

47. Feng, Y.; Li, Z.; Tokola, T. Estimation of stand mean crown diameter from high-spatial-resolution imagery based on a geostatistical method. *Int. J. Remote Sens* **2010**, *31*, 363–378.

48. Kamal, M.; Phinn, S.; Johansen, K. Assessment of Mangrove Spatial Structure Using High-Spatial Resolution Image Data. Proceedings of IEEE International Geoscience and Remote Sensing Symposium (IGARSS), Melbourne, Australia, 21–26 July 2013; pp. 2609–2612.

49. Jupp, D.L.B.; Strahler, A.H.; Woodcock, C.E. Autocorrelation and regularization in digital images. II. Simple image models. *IEEE Trans. Geosci. Remote Sens* **1989**, *27*, 247–258.

50. Atkinson, P.M.; Curran, P.J. Choosing an appropriate spatial resolution for remote sensing investigations. *Photogramm. Eng. Remote Sens* **1997**, *63*, 1345–1351.

51. De Jong, S.M.; van der Meer, F.D. *Remote Sensing Image Analysis: Including the Spatial Domain*; Kluwer Academic Publishers: Dordrecht, The Netherlands, 2005; p. 359.

52. Hay, G.J.; Blaschke, T.; Marceau, D.J.; Bouchard, A. A comparison of three image-object methods for the multiscale analysis of landscape structure. *ISPRS J. Photogramm. Remote Sens* **2003**, *57*, 327–345.

53. Morgan, J.L.; Gergel, S.E.; Coops, N.C. Aerial photography: A rapidly evolving tool for ecological management.*BioScience* **2010**, *60*, 47–59.

54. Blaschke, T.; Lang, S.; Lorup, E.; Strobl, J.; Zeil, P. Object-Oriented Image Processing in an Integrated GIS/Remote Sensing Environment and Perspectives for Environmental Applications. In Environmental Information for Planning, Politics and the Public; Cremers, A., Greve, K., Eds.; Metropolis Verlag:

Marburg, Germany, 2000; 2, pp. 555–570.

55. Chen, W.; Henebry, G.M. Change of spatial information under rescaling: A case study using multi-resolution image series. *ISPRS J. Photogramm. Remote Sens* **2009**, *64*, 592–597.

56. Garrigues, S.; Allard, D.; Baret, F.; Weiss, M. Quantifying spatial heterogeneity at the landscape scale using variogram models. *Remote Sens. Environ* **2006**, *103*, 81–96.

57. Congalton, R.G. A review of assessing the accuracy of classifications of remotely sensed data. *Remote Sens. Environ* **1991**, *37*, 35–46.

SECTION V

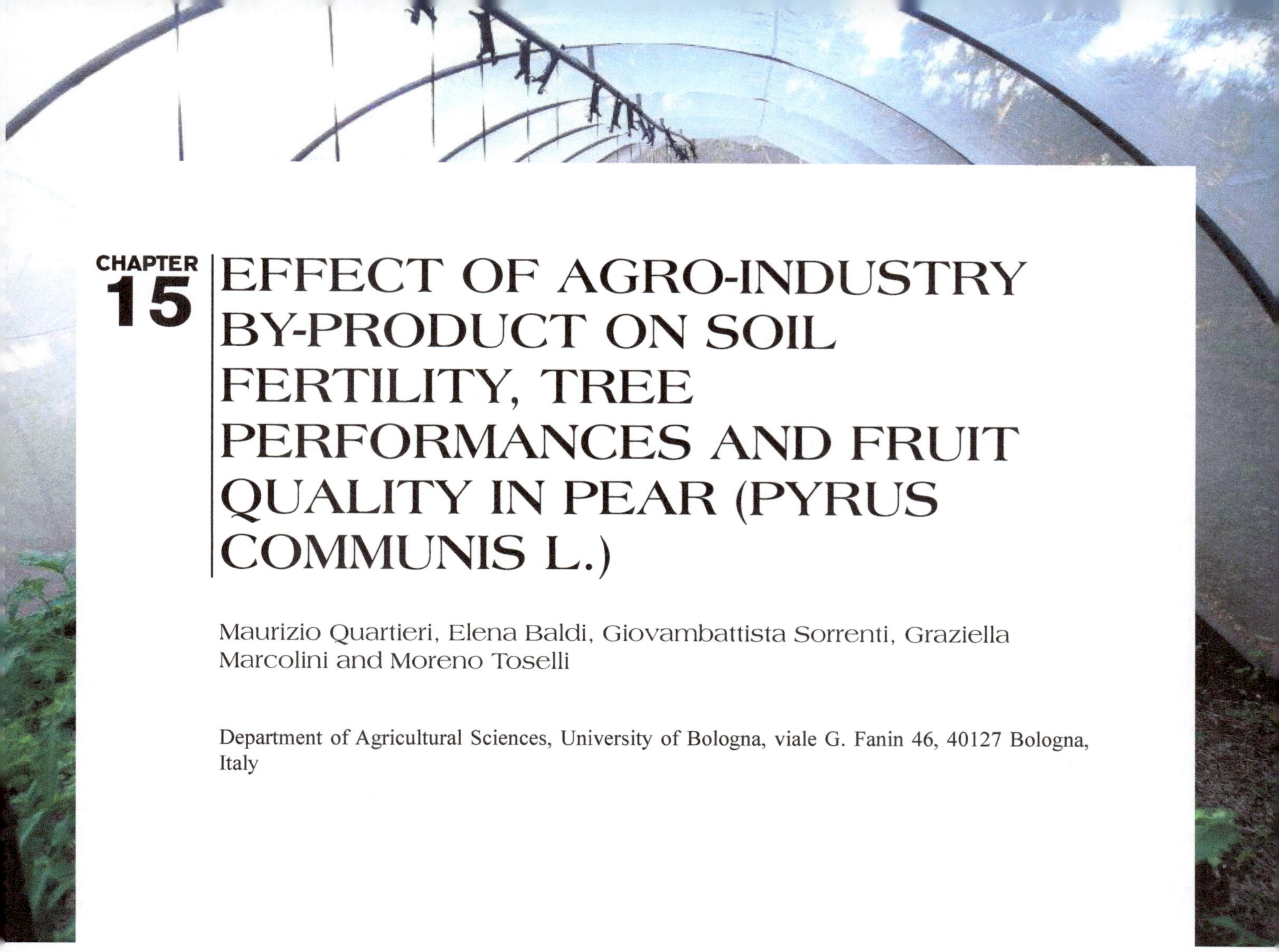

CHAPTER 15 EFFECT OF AGRO-INDUSTRY BY-PRODUCT ON SOIL FERTILITY, TREE PERFORMANCES AND FRUIT QUALITY IN PEAR (PYRUS COMMUNIS L.)

Maurizio Quartieri, Elena Baldi, Giovambattista Sorrenti, Graziella Marcolini and Moreno Toselli

Department of Agricultural Sciences, University of Bologna, viale G. Fanin 46, 40127 Bologna, Italy

ABSTRACT

Organic materials from agro-industry processes can be used in agriculture as a way to recycle materials that still maintain a high fertilizing value. The aim of the experiment was to evaluate the value of soil applied apple juice by-product as fertilizer for pear trees. A 3-year experiment was carried out in a mature pear orchard (cv Abbé Fétel grafted onto quince MC) in the Po valley (Italy), where the following treatments were compared: 1) unfertilized control; 2) mineral N fertilization (60 kg N ha^{-1} year^{-1} split in two spring applications); 3) apple juice by-product (1.3 t ha^{-1} year^{-1}, equal to 60 kg N ha^{-1}), fully supplied at petal drop; 4) apple juice by-product, at twice the rate of the previous treatment. Apple juice by-product soil decomposition accounted for 12% in the first 6 months. At the end of the 24-month-assay, the decomposition accounted for 24% on total dry weight that makes 28% of initial C and 36% of initial N. Soil nitrate-N concentration was increased by the mineral N fertilizer, while the application of apple juice by-product increased microbial carbon. Tree growth, yield and fruit quality were not affected by treatments, while mineral N fertilization raised leaf and fruit N concentration. In conclusion, in our conditions the use of apple juice by-product did not show negative effects on tree performances and fruit quality, with some advantages related to the recycling of organic wastes in agriculture.

INTRODUCTION

Soil organic matter (OM) plays an important role in long-term soil fertility preservation, due to the improvement of soil physical, chemical and biological properties [1]. However, most of the agricultural soils of the Eastern part of the Po Valley of Italy is showing a low concentration of OM[2], related to the decreased availability of traditional organic fertilizers, the increasing of farm specialization toward fruit tree cultivation [3,4] and the increasing of soil mineralization rates due to soil

Citation: Maurizio Quartieri, Elena Baldi, Giovambattista Sorrenti, Graziella Marcolini and Moreno Toselli. Effect of agro-industry by-product on soil fertility, tree performances and fruit quality in pear (Pyrus communis L.). AIMS Agriculture and Food, 2016, 1(1): 20-32. doi: 10.3934/agrfood.2016.1.20

tillage and warmer climate [5]. The high environmental impact of conventional farming practices has increased interest in strategies that help soil OM preservation such as crop rotations [6], conservative soil tillage systems [7], green manure crops [8] and organic waste application [9]. The production of organic wastes (e.g. municipal activities, farming and agro-industrial processes, animal manures, composted residues, etc.) is increasing worldwide; their disposal processes imply considerable social costs, may be responsible for detrimental environmental impacts and represent a loss of valuable biomass and a sourse of nutrients for crops. The recycling of such biomass in agriculture (e.g. as soil amendments or fertilizers) could represent an interesting way to reduce disposal costs, recycle OM and supply mineral nutrients to the soil [10,11]. The use of by-products from orange processing as fertilizer has been shown to improve soil fertility in an orange orchard [12] and in durum wheat [13]. The mixture of exhausted olive-cake and poultry manure produced an organic fertilizer that improved soil fertility and yield in potato [14]. By-products from the apple processing industry are usually used for the production of pectin [15]; however, they can be potentially treated and used as organic fertilizers.

The aims of this experiment were to evaluate the effect of soil-applied by-products from the apple juice industry in a commercial pear orchard on 1) the dynamics of release of carbon (C) and nitrogen (N) under field conditions; 2) soil fertility; 3) tree growth, yield and nutritional status and 4) fruit quality.

MATERIALS AND METHODS

Experimental Site and Plant Material

The trial was carried out over three years (2008-2011) in a mature high-density pear (*Pyrus communis* L.) orchard of cv Abbé Fétel grafted onto quince (*Cydonia oblonga* Mill.) MC, located on a Haplic Calcisol soil (FAO classification) (Table 1) at the experimental station of Bologna University (Italy, 44°33-N; 11°25 - E).

Table 1. Physical and chemical soil properties

Soil parameters	Soil depth (cm)	
	0–40	40–80
Sand (%)	8.9	7.4
Silt (%)	51.7	53.7
Clay (%)	39.4	38.9
pH	7.9	8.0
Ca carbonate (%)	2.0	3.0
Active lime (%)	1.0	2.0
Organic matter (%)	1.9	1.5
Total N (‰)	1.4	1.1
C/N ratio	7.9	7.9
C.E.C. (meq 100 g^{-1})	24.9	26.0
Exchangeable K (meq 100 g^{-1})	0.5	0.4
Exchangeable Ca (meq 100 g^{-1})	30.4	27.4
Exchangeable Mg (meq 100 g^{-1})	3.3	2.7

Trees were spaced 3.8 m between rows and 0.9 m along rows (2924 trees ha^{-1}) and trained as spindle bush. Irrigation water was daily supplied during vegetative season by drip irrigation, using 3.8 L h^{-1} emitters, to return moisture lost through evapotranspiration. Tree rows were sprayed with herbicide while the alleys were maintained with a grass cover, which was regularly mown, 4-5 times per year.

The following soil-applied treatments were compared in a randomized complete block design with six replicates: 1) unfertilized control; 2) mineral nitrogen (N) fertilization (60 kg N ha^{-1} $year^{-1}$) split in two equal applications (50:50) at petal drop and 40 days after the first application; 3) by-product of apple juice production (AJBP) supplied at a rate of 1.3 t ha^{-1} $year^{-1}$ (equal to 60 kg total N), at petal drop; 4) AJBP supplied at a rate of 2.6 t ha^{-1} $year^{-1}$ (equal to 120 kg total N) at petal drop. In 2009 and 2010, the rate of N supplied per year was increased to 80 kg ha^{-1} in treatments 2 and 3 and to 160 kg ha^{-1} in treatment 4, to satisfy the increased tree N demand according to a higher fruit set. Regarding phosphorus (P), potassium (K) and magnesium (Mg), each plot, including control, was fertilized in order to reach the same amount per hectare and per year. Mineral fertilizer (urea) and AJBP were localized along the tree row and tilled into the top soil (10 cm).

The AJBP, provided by ILSA SpA (Arzignano, Vicenza, Italy), was made from sludge generated by the apple juice industry, after screening, drying, grinding and homogenizing the sludge; chemical characteristics of AJBP used are summarized in Table 2.

Table 2. Physical and chemical properties of apple juice by-product used in the trial

Parameter	Value
Dry matter (%)	93.5
Total organic C (%)	38.1
Total N (%)	4.5
C/N ratio	8.5
Total S (SO_3) (%)	2.6
Total P (P_2O_5) (%)	1.2
Total K (K_2O) (%)	1.2
Total Ca (CaO) (%)	1.2
Total Mg (MgO) (%)	1.0
Total Na (Na_2O) (%)	0.1
Total Fe (%)	1.5
Total Zn (mg kg^{-1})	304
Total Mn (mg kg^{-1})	171
Total Cu (mg kg^{-1})	163
Total Cr (mg kg^{-1})	89
Total Ni (mg kg^{-1})	42
Total Pb (mg kg^{-1})	33
Total Al (mg kg^{-1})	18

Mineralization Assay of AJBP

The study was conducted from 2009 to 2011, using the litter bag technique [16,17] to assess the decay dynamics of AJBP. Portions of non-woven fabric were cut and sewn in 18 cm × 18 cm bags; 4 g of apple juice by-product, corresponding to the amount (in g m^{-2}) of AJBP applied in treatment 3 trees were placed into each bag. Thirty bags in total were prepared and placed on 16 April 2009 along tree row into the soil, at 10 cm depth.

Collection of five bags each sampling time randomly chosen along the row was performed in 2009 on 16 May (+1 month), 16 July (+3 months) and 16 October (+6 months), in 2010 on 13 April (+12 months) and 14 October (+18 months) and in 2011 on 13 April (+24 months). Decomposing AJBP litter was cleaned, oven dried at 65 °C and milled (0.2 mm mesh) for chemical analysis. Concentration of C and N was evaluated by C/N elementar analyzer (Carlo Erba, Milan, Italy).

Soil Analysis and Microbial C Biomass

To evaluate the effect of the treatments on soil fertility, ammonium-N (NH_4^+-N), nitrate-N (NO_3^--N), pH, soil OM, total N, humic and fulvic acids (HA+FA) were measured throughout the experiment. From 2008 to 2010, soil cores were collected 4 times per year at 0-40 cm depth to monitor NH_4^+-N and NO_3^--N concentration before and 40 days after first fertilization, in mid-July and mid-October (3 and 6 months after first fertilization, respectively). Soil NH_4^+-N and NO_3^--N were determined by extracting 10 g of soil with a 2 mol KCl solution (1:10 w/v) [18] and using an auto analyser (Auto Analyzer AA-3; Bran+Luebbe, Norderstadt, Germany). Soil samples collected in October were also analysed for pH, OM, total N[18] and (only in 2010) HA+FA [19].

Microbial C biomass was determined in soil samples taken from a depth of 5-15 cm on the same dates of mineral-N sampling, using the substrate induced respiration method [20]. This involved the sieving (2 mm mesh) of 50 g of fresh soil, placing the screened soil in a 500 mL glass jar and allowing it to equilibrate at room temperature for at least 24 hours. The soil was then mixed with 200 mg of glucose and incubated at 22 °C for 3 hours. Carbon dioxide evolution was measured by an infrared gas analyser (EGM-4; PP system; Hitchin, UK) and converted into microbial C [20].

Tree Performances and Nutritional Status

The effect of treatments on tree performances was evaluated by measuring, at the end of growing season, trunk cross-sectional area (TCSA) 15-20 cm above the grafting point; tree yield and fruit weight were also determined at commercial harvest. To assess the effect of treatments on tree nutritional status, leaves were sampled in summer from annual shoots, washed, oven dried at 65 °C for 72 h and milled (0.2 mm mesh). At harvest, two slices were taken from each of the 20 fruits of the sample, lyophilized and milled for analyses. Nitrogen was determined by Kjeldahl method [21], while phosphorus was spectrophotometrically quantified at 700 nm after extraction with acid mineralization [22]. Finally, calcium (Ca), K, Mg, iron (Fe), manganese (Mn), zinc (Zn) and copper (Cu) were determined by atomic absorption spectrometry (SpectrAA-200, Varian, Mulgrave, Australia), after acid digestionby a microwave lab station (Ethos TC-Milestone, Bergamo, Italy) [23].

Fruit Quality

At commercial harvest, a representative sample of fruits was collected from each plot and used to determine soluble sugars and organic acids[24] on lyophilized fruit flesh by high performance liquid chromatography (HPLC). Analyses of sugars were performed using a Jasco PU-1580 HPLC (Jasco Inc., Easton, MD, USA) with a refractive index detector Jasco RI-930 (Jasco Inc., Easton, MD, USA), using an Aminex HPX-87-C, 300 × 7.8 mm (Bio-Rad, Hercules, CA, USA) column. The analysis was performed maintaining the column at 85 °C, using ultra pure water as mobile phase, flux at 0.6 mL min^{-1} while the running time was set at 25 min. Soluble sugars were identified and quantified by the comparison of the retention time with those of standard solutions (Sigma-Aldrich Co. LLC., St. Louis, MO, USA) of known concentration. Soluble organic acids were measured by Jasco PU-1580 (Jasco Inc., Easton, MD, USA) HPLC coupled with a UV-visible detector (MD-1530, Jasco Inc., Easton, MD, USA) using a Phenomenex Rezex ROA-Organic Acid H+ 300 mm × 7.8 mm column (Phenomenex, Torrance, CA, USA). The analysis was performed maintaining the column at 25 °C, using sulphuric acid 0.08M as mobile phase, flux at 0.6 mL min^{-1} while the running time was set at 25 min. Soluble organic acids were identified and quantified by the comparison of the retention time with those of standard solutions (Sigma-Aldrich Co. LLC., St. Louis, MO, USA) of known concentration.

Antioxidant activity was determined according to Miller [25] evaluating the relative ability of pear flesh to scavenge the radical cation of 2,2□-azinobis 3-ethylbenzothiazoline-6-sulfonate (ABTS+) in aqueous phase in comparison to a standard rate of the synthetic antioxidant 6-hydroxy-2,5,7,8-tetramethylchroman-2-carboxylic acid (Trolox, Sigma-Aldrich, Co., LLC., St. Louis, MO, USA). 0.5 g of lyophilized fruit flesh were added to 20 mL of MeOH (60%), homogenized by ultraturex, centrifuged 15 min at 3000 rpm then 1.5 mL of supernatant was collected. Successively, 196 µL of MeOH (60%) and 20 µL of sample were added to 2 mL of stabilized (0.700 ± 0.010) ABTS+ solution. Absorbances were obtained spectrophotometrically (CPS-240 UV-Visible spectrophotometer, Shimadzu Corp., Kyoto, Japan) after interval of 2□30″ from sample addition at 734 nm.

Statistical Analysis

Data were submitted to analysis of variance as in a complete randomized block design. When analysis of variance showed statistically significant effects of treatments (*p* £ 0.05), means were separated by the Student Newman Keuls (SNK) test.

RESULTS AND DISCUSSION

Mineralization Assay of AJBP

The loss of mass during the AJBP decomposition was more rapid during the first 6 months, when about 12% of the original mass was lost; between October 2009 and April 2010 the mass remained relatively stable and the decrease was negligible (Figure 1). During summer 2010 a small but significant decrease of litter mass was recorded (−3% at October sampling), while a consistent and significant mass decrease was measured during winter 2010-2011: after October 2010, the mass decreased by 9.3% and at the end of the study (+24 months after placement) was 76% of initial value (Figure 1). The C and N loss dynamics over the 2 years were similar except in the first month (Figure 1), when the release of N was twice that measured for C (17% and 8% of original amounts, respectively). Starting from the second sampling (+3 months), the dynamic of C and N release showed a similar trend and the amount of nutrient found in the litter gradually decreased until April 2011 (+2 years after bags placement); at the end of study the C and N still present in the litter were respectively 72% and 64% of the amount originally set (Figure 1). Regarding N, the loss was statistically significant only at the beginning (first 6 months) and at the end (last 6 months) of the trial (Figure 1).

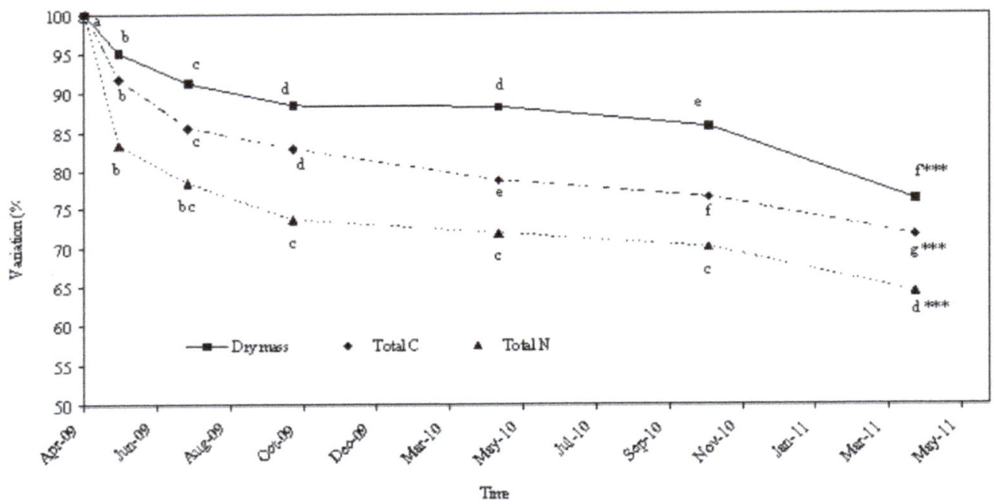

Figure 1. Variation of mass, carbon and nitrogen measured 1, 3, 6, 12, 18 and 24 months after AJBP litter bags placement (April 16, 2009). ***: significant at $p < 0.001$. Within line, values followed by the same letter are not significantly different (SNK test, $p = 0.05$).

Our findings are partially in agreement with results obtained in a similar study on apple leaves [26]; however, the mass, C and N lost by apple leaves after 2 years were $\geq$ 50% of initial values, higher than that observed in AJBP litter bags of our experiment. Moreover, the N dynamic found in AJBP litter bags was different than that observed in apple leaves litter, where N concentration in litter constantly increased during the first year after deposition, while decreased slightly only during the second year [26]. The incorporation of external nitrogen in the decomposing litter was observed by other authors and was associated with microbial immobilization of N from external sources [27,28,29].

Soil Parameters

No treatment effects were observed on soil NH_4^+-N (data not shown). Soil NO_3^--N concentrations showed a similar trend during the three-year-study, so that in Table 3 the 3-year averages for each of the four annual sampling dates are reported. In April, before fertilizer application, nitrate-N concentrations were similar (9 to 11 mg NO_3^--N kg^{-1} DW) in all treatments (Table 3). The use of mineral N fertilizer significantly increased NO_3^--N availability in May, July and October compared with the control and AJBP treatments (Table 3). Soil nitrate-N availability was fairly even in the absence of fertilizer supply throughout the year. Nitrate concentrations in spring (April-May) ranged between 9 and 10 mg NO_3^--N kg^{-1} DW, while values in summer and autumn were somewhat higher (17-18 mg NO_3^--N kg^{-1} DW) (Table 3) due to nutrient release during the decomposition of soil OM.

Table 3. Effect of fertilization on soil NO_3^--N concentration (mg N kg^{-1} DW) at 0-40 cm depth

Treatment	April	May	July	October
Control	10.2[2]	9.61 b	18.2 b	17.2 b
Mineral N	10.9	42.4 a	81.0 a	69.5 a
AJBP 1.3 t	10.4	13.8 b	16.2 b	27.2 b
AJBP 2.6 t	10.2	22.5 b	20.3 b	33.3 b
Significance[1]	*ns*	***	***	**

[1]ns, ** and ***: not significant and significant at $p < 0.01$ and $p < 0.001$, respectively.

[2]Average of 2008-2009-2010 yrs.

The main parameter to quantify the risk of nitrate-N leaching is the concentration of soil nitrate-N at the end of the growing season. In our experiment, fertilization with mineral N resulted in high NO_3^--N levels in the soil (70 mg NO_3^--N kg^{-1} DW in the 0-40 cm layer) at the end of the growing season, when tree is entering in the dormancy stage and root uptake is low, and the risk of nitrate leaching is considerable in our conditions [30]; the experimental site, in fact, is located in the Po Valley, a vulnerable zone for nitrate directive. The use of AJBP had no influence on the post-harvest soil nitrate-N and hence did not increase substantially the risk of nitrate leaching during winter [4,31]. The lower nitrate-N concentration observed in AJBP plots than in Mineral N treatment was the consequence of high stability in the soil of the by-product (Figure 1) used in this study; however, the soil amendment could stimulate a development of soil microbial biomass that absorbed part of the mineral N released by the soil OM mineralization process [4,32]. As a matter of fact, the application of apple juice by-product increased microbial C biomass only in 2010 (Table 4), when the application of AJBP at highest rate significantly increased microbial C, compared to the unfertilized plots (Table 4). In May, microbial C was similar in all treatments (data not shown), but in July and October AJBP application significantly increased microbial C compared to the control and mineral N treatment (Table 4). In 2009 and 2011 the organic fertilization did not modify microbial biomass (data not showed). Our results are partially in agreement with literature, which reported that the application of compost [4] and other organic materials [10] can increase soil microbial C. In our study, however, the highest microbial biomass was found in summer, in contrast with those reported by Baldi and co-authors, who found the highest values in spring and autumn [4], because of the different climatic conditions of the two experimental sites.

Table 4. Effect of fertilization strategies on soil pH, organic matter, total N, HA and FA, and microbial C measured in 2010

Treatment	pH	OM	Total N	HA+FU	Microbial C ($\mu g\ C\ g^{-1}$ DW)	
		(%)	(‰)	(%)	July	October
Control	7.24	1.60 b[2]	1.27	0.46	239 b	170 b
Mineral N	6.96	1.67 ab	1.29	–	301 ab	167 b
AJBP 1.3 t	7.28	1.83 a	1.29	0.52	307 ab	214 a
AJBP 2.6 t	7.03	1.83 a	1.29	0.51	366 a	209 a
Significance[1]	*ns*	*	*ns*	*ns*	*	*

[1]ns and *: not significant and significant at $p < 0.05$, respectively.

[2]Within columns, values followed by the same letter are not significantly different (SNK test, $p = 0.05$).

Soil pH and total N were unaffected by treatments, but after three years soil amended with AJBP showed higher organic matter concentrations than the control (Table 4). Considering the data of OM concentration at the beginning (Table 1) and at the end of the trial, AJBP has maintained or even increased soil OM, while it decreased significantly in the control and mineral N plots (Table 4). Soil humic and fulvic acid (HA+FA, Table 4) concentrations were not affected by AJBP supply, as expected in a short-term experiment, and ranged between 0.46% (control) and 0.52% (AJBP). Our results are in agreement with other reports [4,33] and confirm that the use of organic residues in horticulture is an effectiveness strategy to preserve and eventually ameliorate soil fertility. In this study, soil OM was maintained with relatively small amounts of organic fertilizer (1.6 t ha^{-1} in 2009 and 2010) and soil NO_3^--N release was compatible with the N requirements for pear orchard, which range between 60 and 80 kg N ha^{-1} $year^{-1}$ depending on the graft combination [34,35]. The use of AJBP in pear fertilization can also contribute to soil carbon sequestration, in agreement with literature [4,36,37].

Tree Growth, Yield and Nutritional Status

Tree growth and crop yield were not affected by the different fertilization strategies, as previously published [38]. The absence of significant effects of the N rate on tree growth and yield can be explained by considering the tree age and grafting combination. The use of dwarfing rootstock (i.e. quince MC) and the age of the trees contributed to achieve an optimum balance between tree vigor and production.

Trees fertilized with mineral N showed significantly higher leaf and fruit N concentrations compared with untreated trees, while AJBP treatments showed intermediate values (Tables 5-6). No statistical differences were observed in leaf P, K, Ca and Mg concentrations (Table 5). Regarding fruit mineral composition, trees fertilized with mineral N produced fruits with significantly higher N concentration (0.21%) than unfertilized trees, while K levels (0.50%) were significantly lower (Table 6). Regardless the rate, both fruit and leaves of tree treated with AJBP showed a similar K concentration of control trees. Leaf and fruit micronutrient concentrations were not affected by use of different fertilizer treatments (Tables 5-6).

Table 5. Effect of fertilization strategies on leaf nutrient concentration (means of 3 years)

Treatment	Macronutrients (% DW)					Micronutrients ($mg\ kg^{-1}$ DW)			
	N	P	K	Ca	Mg	Fe	Mn	Cu	Zn
Control	1.50[1] b	0.15	0.74	1.47	0.45	95	112	35	101
Mineral N	1.76 a	0.15	0.61	1.50	0.49	99	112	35	87
AJBP 1.3 t	1.61 ab	0.16	0.64	1.52	0.46	99	122	35	107
AJBP 2.6 t	1.65 ab	0.14	0.63	1.47	0.46	99	114	34	106
Significance[2]	**	*ns*	*ns*	*ns*	*ns*	*ns*	*ns*	*ns*	*ns*

[1]Within columns, values followed by the same letter are not significantly different (SNK test, $p = 0.05$).

[2]ns and **: not significant and significant at $p < 0.01$, respectively.

Table 6. Effect of fertilization strategies on fruit nutrient concentration (means of 3 years)

Treatment	Macronutrients					Micronutrients			
	N (%)	P (‰)	K (%)	Ca (‰)	Mg (‰)	Fe (mg kg^{-1} DW)	Mn	Cu	Zn
Control	0.18[1] b	0.56	0.56 a	0.45	0.40	15	3.7	2.4	9.3
Mineral N	0.21 a	0.53	0.50 b	0.47	0.35	13	3.3	2.1	7.2
AJBP 1.3 t	0.20 ab	0.55	0.56 a	0.49	0.40	16	3.7	2.1	7.8
AJBP 2.6 t	0.20 ab	0.55	0.55 a	0.51	0.39	15	3.7	2.0	10.0
Significance[2]	*	*ns*	*	*ns*	*ns*	*ns*	*ns*	*ns*	*ns*

[1]Within columns, values followed by the same letter are not significantly different (SNK test, $p = 0.05$). Values are the mean of 3 years.

[2]ns and *: not significant and significant at $p < 0.05$, respectively.

Data of leaf mineral composition were generally in line with the reference values considered optimal for cv Abbé Fétel grown in the same area[39], except for N which was below the optimum range (2-2.4%) in all treatments, probably due to the grafting combination, that included the very dwarfing quince MC.

Fruit Quality

Soluble carbohydrates concentrations and organic acids were not increased by mineral or organic N supply (Table 10). Fructose resulted the sugar most abundant in the fruit (32% of DW), followed by glucose and sorbitol (15-20% of DW), and sucrose (less than 5%, Table 7). Regarding organic acids, succinic acid was the most representative (2.3% of DW) followed by malic acid (1.7-2% of DW); citric acid was the least representative, with a concentration ten times lower than the succinic acid (Table 7). The concentration of fruit soluble sugars and organic acids were in agreement with data reported in literature [36,40].

Table 7. Effect of fertilization strategies on fruit soluble carbohydrates and organic acids concentration (means of 2008-2009-2010 yrs.)

Treatment	Carbohydrates content (g 100 g^{-1} DW)				Organic acids (g 100 g^{-1} DW)		
	Fructose	Sorbitol	Glucose	Sucrose	Succinic acid	Malic acid	Citric acid
Control	32.4	20.1	15.7	4.22	2.36	1.91	0.21
Mineral N	32.7	19.9	18.5	3.70	2.36	1.73	0.22
AJBP 1.3 t	31.4	19.0	16.5	3.75	2.22	1.74	0.22
AJBP 2.6 t	32.1	19.8	17.5	3.38	2.32	2.00	0.23
Significance[1]	*ns*	*ns*	*ns*	*ns*	*ns*	*ns*	*ns*

[1]ns: not significant at $p< 0.05$.

The antioxidant activity of fruits, an important aspect for defining fruit quality because of the increasing attention to functional food, was not also modified by treatment in 2008, 2009 and 2010 (Table 8), confirming the low reactivity of fruit crop to soil management in term of fruit composition and functional activity [36,41].

Table 8. Effect of fertilization treatment on fruit antioxidant activity (AA)

Treatment	AA (mM Trolox g^{-1} DW)		
	2008	2009	2010
Control	31.4	51.3	54.3
Mineral N	23.9	41.8	42.2
AJBP 1.3 t	36.7	36.2	38.5
AJBP 2.6 t	32.5	42.9	42.7
Significance[1]	*ns*	*ns*	*ns*

[1]ns: not significant at $p < 0.05$.

CONCLUSIONS

Our results indicate that the use of by-products from apple juice production in pear fertilization can be an interesting strategy for maintaining or improving soil OM content and soil fertility. In addition, the use of AJBP promoted soil microorganisms activity, increased soil N availability for plant uptake during the first months after its addition to the soil without increasing the risk of nitrate leaching. At the rates supplied in this study, no effects were observed on tree performances and fruit quality.

Pear fertilization can be managed with the use of organic fertilizers made from fruit processing residues and this study confirms that by-products from the apple juice industry is an effectiveness strategy to preserve soil organic matter and can replace mineral fertilizers as a source of N, without negative effects on tree performances and fruit quality. The recycling of these organic wastes in agriculture can also reduce disposal costs and environmental pollution.

ACKNOWLEDGEMENTS

Research supported by a grant from Regione Emilia-Romagna and ILSA S.p.A. (Vicenza, Italy). Project 'Riciclo in frutticoltura di residui di processi agro-industriali'.

REFERENCES

1. Sequi P (1989) Le funzioni agronomiche della sostanza organica. In: Patron (Ed.), *Chimica del suolo*, Bologna, Italy, 279-292.
2. Ungaro F, Calzolari C, Tarocco P, et al. (2002) Soil organic matter in the soils of Emilia Romagna plain (Northern Italy): knowledge and management policies. Ottawa OEDC Expert Meeting on Organic Carbon indicators for Agricultural Land, Canada, Volume 85, Supplement Special Issue.
3. Komatsuzaki M, Ohta H (2007) Soil management practices for sustainable agro-ecosystems. *Sustain Sci* 2: 103-120.
4. Baldi E, Toselli M, Marcolini G, et al. (2010) Compost can successfully replace mineral fertilizers in the nutrient management of commercial peach orchard. *Soil Use Manag* 26: 346-353.
5. Montemurro F, Maiorana M, Convertini G, et al. (2007) Alternative sugar beet production using shallow tillage and municipal solid waste fertilizer. *Agron Sustain Dev* 27: 129-137.
6. Watson CA, Atkinson D, Gosling P, et al. (2002) Managing soil fertility in organic farming systems. *Soil Use Manag* 18: 239-247.
7. Ismail I, Blevins RL, Frye WW (1994) Long-term no-tillage effects on soil properties and continuous corn yields. *Soil Sci Soc Am J* 58: 193-198.
8. Giovannini D, Scudellari D, Aldini A, et al (2001) Esperienze di conduzione del terreno in un pescheto biologico. *Frutticoltura* 1: 21-29.
9. Montemurro F, Maiorana M (2008) Organic fertilization as resource for a sustainable agriculture. In: Elsworth L.R. et al. (Eds.). *Fertilizers: properties, applications and effects*. Nova Science Publishers, Inc, 123-146.
10. Diacono M, Montemurro F (2010) Long-term effects of organic amendments on soil fertility. *A review. Agron Sustain Dev* 30: 401-422.
11. Miller DM, Miller WP (2000) Land application of wastes. In: Sumner M. (Ed.), *Handbook of soil science*, CRC Press, Chap. 9.
12. Intrigliolo F, Allegra M, Torrisi B (2005) Application of organic fertilizers in orange orchard in southern Italy. *Geophys Res* Abstract 7. 286.
13. Tuttobene R, Avola G, Gresta F, et al. (2009) Industrial orange waste as organic fertilizer in durum wheat. *Agron Sustain Dev* 29: 557-563.
14. Sellami F, Jarbouia R, Hachichaa S, et al. (2008) Co-composting of oil exhausted olive-cake, poultry manure and industrial residues of agro-food activity for soil amendment. *Bioresource Technol* 99: 1177-1188.
15. Schieber A, Stintzing FC, Carle R (2001) By-products of plant food processing as a source of functional compounds — Recent developments. *Trends Food Sci Tech* 12: 401-413.
16. Magill AH, Aber JD, (1998) Long-term effects of experimental nitrogen additions on foliar litter decay and humus formation in forest ecosystems. *Plant Soil* 203: 301-311.
17. Harmon ME, Nadelhoffer KJ, Blair JM, (1999) Measuring decomposition, nutrient turnover, and stores in plant litter. In: Robertson, G.P., Coleman, D.C., Bledsoe, C.S., Sollins, P. (Eds.), Standard Soil Methods for Long-term Ecological Research. Oxford University Press, New York, 202–240.
18. Violante P (2000) Metodi di analisi chimica del suolo. Franco Angeli, Milano.
19. Ciavatta C, Govi M (1993) Use of insoluble polyvinylpyrrolidone and isoelectric focusing in the study of humic substances in soils and organic wastes. *J Chr An* 643: 261-270.
20. Anderson J, Domsch K (1978) A physiological method for the quantitative measurement of microbial biomass in soils. *Soil Biol Bioch* 10: 215-221.
21. Schuman G, Stanley M, Knudsen D (1973) Automated total nitrogen analysis of soil and plant samples. *Soil Sci Soc Am J* 37: 480.
22. Saunders WM, Williams EG (1955) Observations of the determinations of organic phosphorus in soils. *J Soil Sci* 6: 254-267.
23. Kingston HM (1988) Environ. Protection Agency IAG DWI-393254-01-0 January 1-March 31, Quarterly Report.
24. Sturm K, Koron D, Stampar F, (2003) The composition of fruit of different strawberry varieties depending on maturity stage. *Food Chem* 83: 417-422.
25. Miller NJ, Rice-Evans BH, Davies MJ, et al. (1993) A novel method for measuring antioxidant capacity and its applications to monitoring the antioxidant status in premature neonates. *Clin Sci* 84: 407-412.
26. Tagliavini M, Tonon G, Scandellari F, et al. (2007) Nutrient recycling during the decomposition of apple leaves (*Malus domestica*) and mowed grasses in an orchard. *Agric Ecosyst Environ* 118: 191-200.
27. Colpaert JV, vanTichelen KK, 1996. Decomposition, nitrogen and phosphorus mineralization from beech leaf litter colonized by ectomycorrhizal or litter-decomposing basidiomycetes. *New Phytol* 134: 123-132.
28. Joergensen RG, Scheu S, (1999) Depth gradients of microbial and chemical properties in moder soils under

beech and spruce. *Pedobiologia* 43: 134-144.

29. Tonon G, Ciavatta C, Solimando D, et al. (2007) Fate of 15N derived from soil decomposition of abscised leaves and pruning wood from apple (*Malus domestica*) trees. *Soil Sci Plant Nutr* 53: 78-85.

30. Ventura M, Opstad N, Scandellari F, et al. (2006) Soil nitrogen losses through leaching from tree row strips and grassed alleys in a pear orchard. In: Shröder J.J., Neeteson J.J. (eds.). *N management in agro-ecosystems in relation to the Water Framework Directive*. Proceedings of the 14th N workshop, October 2005, Maastricht, Netherlands, 241-243

31. Van Erp PJ, Evers MA, et al. (1997) De gebruikswaarde en perspectieven voor de afzet van GFTcompost (The value and perspectives of VFG compost sales). Publicatie no. 97051C.R Vereniging van Afvalverwerkers (VVAV), Utrecht, 36. (in Dutch).

32. Leifeld J, Siebert S, Kogel-Knabner I, (2002) Changes in the chemical composition of soil organic matter after application of compost. *Eur J Soil Sci* 53: 299-309.

33. Sanchez PA, Palm CA, Szott LT, et al. (1989) Organic input management in tropical agroecosystems. In: Coleman D.C. et al. (Eds.).*Dynamics of Soil Organic Matter in Tropical Ecosystems*. NifTAL Project. University of Hawaii, Honolulu, HI, 125-152.

34. Tagliavini M, Quartieri M, Millard P (1997) Remobilised nitrogen and root uptake of nitrate for spring leaf growth, flowers and developing fruits of pear (*Pyrus communis* L.*) trees. Plant Soil* 195: 137-142.

35. Tagliavini M, Quartieri M (2008) I fabbisogni nutrizionali del pero per la definizione del piano di fertilizzazione. *Italus Hortus* 15: 131-137.

36. Sorrenti G, Toselli M, Marangoni B (2012) Use of compost to manage Fe nutrition of pear trees grown in calcareous soil. *Sci Hort* 136: 87-94.

37. Triberti L, Nastri A, Giordani G, et al. (2008) Can mineral and organic fertilization help sequestrate carbon dioxide in cropland. *Eur J Agron* 29: 13-20.

38. Quartieri M, Baldi E, Toselli M, et al. (2014) Use of agro-industrial byproducts in pear fertilization: preliminary results. *Acta Hort* 1018: 263-270.

39. Toselli M, Mazzanti F, Marangoni B, et al. (2002) Determination of leaf standards for mineral diagnostics in pear orchards in the Po Valley, Italy. *Acta Hort* 596: 665-669.

40. Bosetto M, Arfaioli P, (1992) Time changes of free sugar and polyol content in seven Italian pear (*Pyrus Communis*) varieties with different ripening periods. *Agric Med* 122: 59-65.

41. Baldi E, Toselli M, Marcolini G, et al. (2012) Effect of organic fertilization, ripening stage and storage on quality of fruits of 'Stark RedGold nectarine. *Acta Hort* 933: 593-600.

CHAPTER 16

IMPROVING RICE-BASED RAINFED PRODUCTION SYSTEMS IN SOUTHEAST ASIA FOR CONTRIBUTING TOWARDS FOOD SECURITY AND RURAL DEVELOPMENT THROUGH SUSTAINABLE CROP PRODUCTION INTENSIFICATION

Abha Mishra[1], Prabhat Kumar[1], Jan Willem Ketelaar[2]

[1] Asian Center of Innovation for Sustainable Agriculture Intensification, Asian Institute of Technology, Klong Luang, Pathumthani 12120, Thailand

[2] FAO Inter-Country IPM/Pesticide Risk Reduction Programme, FAO Regional Office for Asia and the Pacific, 39 Phra Athit Road, Bangkok 10200, Thailand

ABSTRACT

Continuing degradation of the environment and the cumulating food, energy, water and financial crises have led to a situation where many people's access to sufficient, nutritious food is affected as well as their livelihoods, income, and ultimate food and nutrition security. In the wake of these stresses and crises, there is an emerging interest to find efficient, easily accessible and sustainable approaches that can address these crises. One candidate for this is the System of Rice Intensification (SRI) with its "less can produce more" prescription. A regional collaborative project currently underway is being implemented in rainfed areas of the Lower Mekong River Basin (LMB) countries. This involves smallholder rice farmers, researchers, extension personnel, and development professionals, together with staff of relevant government ministries (http://www.sri-lmb.ait.asia/). The project objective is to produce healthier and profitable rice crops under rainfed conditions using SRI methods, evaluated and refined through farmers' participatory action research (FPAR). As part of the action-research, more than 120 sets of field experiments have been carried out at 60 FPAR sites in Cambodia and Thailand, directly involving 3600 farmers. The experiments have ranged from the integration of many SRI principles with farmers' current local practices or improved practices which was termed as "SRI-transition" to full demonstrations and assessments of SRI methodology, i.e., SRI demonstration. The initial calculation of yields has showed an average paddy yield of 5.03 t/ha with SRI-transition, whereas with SRI-demonstration the average yield was 6.41 t/ha. These yields were 60 and 100% higher than the average baseline yield in the region, 3.14 t/ha, for the same farmers and same locales. Productivity gains (dollars gained/dollars spent per ha) were calculated for both rainfed and irrigated production areas. In comparative terms, the economic gains for farmers were found to be higher in rainfed areas when using the new methods. This paper addresses the potential of new strategies to promote food security in rainfed areas in the LMB region by managing household and natural resources more productively.

Citation: Abha Mishra, Prabhat Kumar, Jan Willem Ketelaar. Improving rice-based rainfed production systems in Southeast Asia for contributing towards food security and rural development through sustainable crop production intensification. AIMS Agriculture and Food, 2016, 1(2): 102-123. doi: 10.3934/agrfood.2016.2.102.

INTRODUCTION

Continued degradation of the environment and the various food, energy, water and financial crises which have been caused by multiple drivers, including prevailing economic models for production, consumption and development, have led to a broader realization that these crises, if not addressed successfully, impede people's access to sufficient, nutritious food, affect their livelihoods, income and food prices, and ultimately worsen food and nutrition security. Although the repercussions of these multiple crises are global in nature, the vulnerability of rural communities whose livelihood is directly linked to agriculture and who are mostly food-insecure is a major concern for predominantly local agrarian economies. In particular, this is relevant in the Lower Mekong River Basin (LMB) region where a majority of the countries' poor households live in rural areas and depend on agriculture for their livelihoods.

The LMB region encompasses the Lao PDR, Vietnam, Thailand and Cambodia. Out of 65 million inhabitants of this large basin (604,300 km^2), 60 million people in the LMB reside mostly in rainfed areas. Agriculture along with fishing and forestry employs 90% of the people living in the Lower Basin, mostly on a subsistence level. Over 10 million hectares of land are devoted to rice cultivation in the LMB, out of which 6 million hectares are rainfed, characterized by seasonal flooding and drought, low cation exchange capacity and low organic matter in the soil, and low availability of phosphorus, all resulting in lower productivity. Farmers operate with limited cash flows so they must carefully weigh the risks of adopting any costlier farming techniques, given the level of uncertainty.

There is widespread poverty in the basin, and almost one-third of its population are living on less than one dollar per day. For example, in Northeast Thailand, the largest rice-producing area in the country with the greatest number of farms (2.7 million); income per capita is less than 40% of the national average, and the incidence of poverty is high, affecting more than 37% of the population [1]. Though the situation has relatively improved in recent years, still almost three out of five poor people in the country reside in this region.

The situation is even more depressing and chronic in Cambodia, where rainfed lowland rice is the single most important crop, occupying 69% of the total cultivated area [2] and contributing 75% toward people's per capita calorie supply; almost 36% of this population is food-insecure [3]. Likewise, in Laos the average GNP per capita is just US$280—lowest among all LMB countries. Rice constitutes almost 70% of the calorie and protein intake of Lao households, and chronic malnutrition affects up to 47% of the population. A similar situation prevails in upland areas of Vietnam, where the rainfed rice-based production system predominates and is associated with widespread food insecurity.

These figures show that the rice-based production system is the mainstay for most households' survival, although presently an inadequate one, constraining both livelihoods and food security in the region. Major and common constraints to rice production in the region are poor fertility of the soil, highly seasonal and unreliable patterns of rainfall, frequent droughts, and limited access to inputs and credit [4], plus limited farm management skills. The population of the LMB is expected to increase from 60 million to 90 million by 2050 [5] without any real increase in availability of production resources. This will further increase the demand for food and put greater adverse pressure on land and other natural resources [6]. This means that large increases in the amount of water and other physical inputs would be required to achieve the goal under a "business as usual" scenario. No economic sector consumes as much fresh water as agriculture. Current irrigation water withdrawals already cause stress in many of the world's major river basins [7]. There is stiff competition between industries and agriculture for water, so very little room for expansion of large-scale irrigation is available.

An additional, exacerbated threat to food supply is expected from climate change. The rainy season from May to October is expected to intensify with an increase in rainfall in the wettest months, whereas in the dry season from November to April the basin is expected to receive slightly less rain than now. Thus, both seasonal water shortages and floods are expected in the region. Farmers of the basin have already been experiencing early drought spells for the last 2-3 years, although the currently ongoing El Niño phenomenon complicates attribution.

While food security at a national level is not an issue in the Lower Mekong region if there is sufficient rain during the growing season and if there is a stable national and international market, at the household and individual levels food and nutrition security is still a major concern in Cambodia and Laos and in some parts of Vietnam and Thailand. For example, in Thailand, national per capita availability of rice is the highest in the world; however, farmers of the Northeast region still struggle to escape from temporary and seasonal food insecurity. Thus, agriculture faces increased demands for food, on one hand, and threats to production due to resource constraints and climate change variability, on the other. A step towards addressing the challenges of increasing food demand and reducing poverty is to increase sustainably the agricultural productivity especially of smallholder farms.

Several interdisciplinary and integrated modes of enquiry have been established at the plot level with a "less can be more" prescription. Recently the System of Rice Intensification (SRI) has been gaining momentum at farmers' field level, capturing their attention by enabling them to get higher yields with reduced external inputs, and fuelling their capacity for innovation (e.g., SRI concepts and methods are being extrapolated to other crops, such as wheat, finger millet and sugar cane [8].

It is believed that SRI could be instrumental in developing sustainable solutions to local agricultural problems, especially for smallholders who have fewer economic inputs but who have better control over their resources [9,10]. The SRI management principles—transplanting young seedlings, giving plants more space, avoiding continuous flooding—when implemented together have in many instances resulted in substantial increases in yield while reducing input use [10,11,12,13]. Most of the good results, more specifically *evaluation results*, have been reported from irrigated areas, however; whether SRI idea can be useful in achieving higher yields with greater resource use efficiency in rainfed areas is still to be understood.

To understand this and to learn more about SRI's usefulness for contributing towards achieving food security at household level, a regional collaborative project, funded by the European

Union, is being implemented in rainfed areas of the Lower Mekong River Basin (LMB) countries involving smallholder farmers (including women and landless), researchers, extension personnel, and development professionals, together with staff of relevant government ministries (http://www.sri-lmb.ait.asia/). The project objective is to produce healthier and profitable rice crops under rainfed conditions using SRI methods, evaluated and refined through farmers' participatory action research (FPAR). This paper shares the results of more than 120 sets of field experiments, which were carried out at 60 FPAR sites located in five provinces in Cambodia and Thailand directly involving 3,600 farmers during wet season of 2014 from June to December. It addresses the potential for introducing new strategies for managing household resources more productively for rainfed agriculture so as to contribute towards enhanced food security in the LMB region using farmer-participatory learning methods.

MATERIALS AND METHODS

General Approach

Five food-insecure provinces, three in Cambodia and two in Thailand, were selected in consultation with relevant government ministries, respectively, the Ministry of Agriculture, Forestry and Fisheries (MAFF) in Cambodia and the Ministry of Education (MoE) in Thailand. In the latter selection, provinces were identified by the National Food Insecurity and Vulnerability Mapping System (FIVIMS). Then in each province, three districts were selected, making a total of 15 districts in the region (Figure 1). Out of these 15 districts, 13 districts were entirely rainfed, whereas two districts, both located in Northern Thailand, have partial access to irrigation facilities.

Figure 1. Geographical map showing the action research site, 9 districts and 3 provinces in Cambodia and 6 districts and 2 provinces in Thailand.

Baseline Survey

Prior to commencement of farmer's participatory field experimentation, a baseline survey was done in each identified FPAR district with the objective of visualizing the pre-project scenario and for establishing criteria for selecting village and farmer, landless and women participants. With this information, a representative sample of farms was selected from each identified FPAR district in terms of farm resources, area, and crop type.

The respective farmers were surveyed using a semi-structured questionnaire designed for the purpose. The questionnaire was developed and adapted as per country-specific needs prior to pre-testing at village level. Questions were asked about general household information, farm characteristics, and socio-economic conditions, access to productive capital, agricultural practices, production and income, migration patterns, and familiarity with SRI. Pre-testing was done, and accordingly questions were modified and finalized. The final questions were translated into the respective local languages prior to the survey.

The baseline findings were used to assess and benchmark the existing parameters for comparison, and also for design of training interventions, including FPAR curricula and training module design, and the design of field experimentation.

General biophysical and socio-economic characteristics for sampled villages and households in all five provinces are given in Table 1, which were gathered through participatory rural appraisal and baseline survey.

Table 1. Biophysical and socio-economic characteristics of five provinces of Cambodia and Thailand (Values shows mean ± standard error)

	Cambodia			Thailand	
	Kampot	Kampong Speu	Takeo	Surin	Uttaradit
Soil characteristics	Sandy loam *(Chumkiri district.)*: Loam *(Angkor Chey and Chhouk districts.)*	Sandy loam (Krong Pisey district); Loam *(Samrong Trong and Borset districts.)*	Sandy loam (Tram Kak district); Loam (Prey *Kabas and Bati districts.)*	Sandy loam (Tha Tum and Srikoraphum districts); Clay loam (Chumponburi dictrict)	Clay (Ban khok, Pichai and Tron districts)
Average annual rainfall (mm)	1409±110.21	1297±105.63	1318±102.35	1294±111.51	1243±103.54
Topography	Lowland	Lowland	Lowland	Lowland	Lowland (Tron and Pichai) and Upland (Ban Khok)
Average landholding (ha/household)	0.9±0.008	0.64±0.003	0.84±0.004	3.04±0.002	3.4±0.003
Average age of farmers (years)	54±2.16	55.25±2.78	57.5±2.58	55.25±2.21	57.5±2.57
Average household members	5.6±0.22	6±0.31	5.9±0.21	4.63±0.19	4.53±0.11
Average household labour	4.76±0.17	5.1±0.19	4.42±0.18	2.3±0.09	2.2±0.07
Total household income (US$)	1906.13±115.5	2473.3±112.8	1939.34±145.12	5976.57±239.04 2162.67±64.86*	6178.98±216.16 2627.01±112.96*
% income from rice farming with respect to total household income	19 ±0.66	21 ±0.84	22 ±0.92	48.66±1.70 22.33±0.77*	76.84±3.15 32.66±*1.07
% income from **non-rice** farming	15±0.66**	8±0.18**	6±0.24**	10±0.25	9±0.29
From livestock *(only 30-40% household in Cambodia)*	67±%2.68	51±%1.9	62±2.48%	--	--
% Off farm income with respect to the total annual income	40.1±1.8	38.21±1.68	42.3±1.22	51.34±2.00 56.97±2.16*	15.82±0.52 41.92±1.55*
Average rice yield (t/ha)	2.01±0.07	3.04±0.10	3.00±0.12	2.48±0.09	4.98±0.12
% rice insufficiency***	19±0.79	14±0.30	17±0.45	Sufficient production (but not stored for self-consumption)	Sufficient production (but not stored for self-consumption)
Household loans (US$)	1042±45.84 (34% household)	937±43.16 (45% household)	844±38.82 (60% household)	2294 ±100.93 (91.67% household)	3442±134.23 (95% household)

*Figures are for smallholders in project area

**only 10% household

***rice shortage with reference to rice food demand: 143 kg of milled rice/person in Cambodia and 110 kg rice/person in Thailand

Action Research Group and Farm Selection

Using standard selection criteria [10], in each province three districts were selected, and in each district, four FPAR sites were identified for farmer-led field experimentation. At province level, a representative catchment area—centrally-located for a farmers' participatory action research site, also known as the CFPAR site—was selected for common meetings and for conducting season-long training and field experiments.

The training involved farmers selected from all three districts, district trainers, provincial coordinators, and resource persons from the project's regional coordination unit. From each district, eight farmers, two landless, and one district trainer from the local agriculture ministry were selected. The selection process ensured participation of at least 50% women farmers from each district.

The selection process itself commenced with an inception meeting for the whole community with the objective of discussing the purpose of the action research, creating ownership of the project, and discussing criteria for participation, i.e., representativeness for the catchment, definition of SRI, development of low-cost and profitable location-specific technology, farmers evaluating different options, etc.

Following that, a season-long training around field experiments, i.e., Central Farmers Participatory Action Research (CFPAR), was conducted with the objective to build and strengthen their capacity in experimentation and to develop training curricula on scientific, technical, social and managerial aspects. As a part of training on designing, conducting and evaluating field experiments, SRI and conventional practices were applied at each CFPAR site for comparison and evaluation purposes. The experiments were designed using randomized block design where each experiment had at least three treatments and four replications.

In successive seasons, two CFPAR-trained farmers, functioning as farmers' trainers, initiated FPAR experimentation at each FPAR site. On an average, two or more experiments were set up at each FPAR site using the same field layout practiced at CFPAR site, i.e., randomized block design with four replications. Each experiment involved 25-30 farmers. The size of the field experiments varied from 700 to 1000 m^2.

Prior to the CFPAR at provincial level, a regional training workshop was organized for all provincial coordinators and country coordinators with an objective to familiarize and strengthen the scientific capacity for experimental design, setting, data observation, analysis and recording; and to agree upon the country-specific criteria for selection of districts, villages, the CFPAR site, and farmer and landless participation.

Identification, Testing and Adaptation of SRI Practices

The FPAR commenced with a rehearsal of the problem diagnosis. The common issues and interests expressed by farmers in both countries were to achieve higher yield with reduced costs of production by reducing input use for cost saving and for making rice cultivation more efficient and profitable. Using various group-dynamic tools such as sub-group discussion, visual tools, and brainstorming sessions, a range of options were selected for each of the target areas that revolved around the integration of a few SRI principles with existing conventional practices to be applied on a learning plot for location-specific adaptation, but also to have application of the full set of SRI principles on a demonstration plot which would serve as a "test site" to test and show the full potential of SRI methods at smallholder farmers' field level. For comparison purposes, the practices that were applied were categorized into:

- Conventional practices (CP) - the existing management practices generally followed in the target area as identified through the baseline survey.
- SRI-demonstration (SRI-D) where the full set of SRI practices was applied.
- SRI-transition (SRI-T) where a few principles of SRI were applied in combination with modified or existing conventional practices.

The word "transition" was used because the practices are generally transitioning towards SRI with different degrees of SRI adoption and types. To gain further insight on the practices being applied by farmers, SRI-T was further broken down into: 1) SRI-innovation (SRI-I), where a few principles of SRI were integrated with modified conventional practices, i.e., practices that do not fall in either category of SRI or CP, but that are improved and better than CP, and 2) SRI-locally adapted practices (SRI-LAP) where a few principles of SRI were integrated with CP.

CP is taken as the baseline for purposes of comparison and evaluation. Details of the SRI-D, SRI-T, and CP alternatives are given in Table 2.

Table 2. Crop management practices followed in SRI-demonstration (SRI-D), SRI-transition (SRI-T) and conventional management practices (CM)

Crop management practices	Conventional practices (CM)*	SRI-transition (SRI-T)**	SRI-D
Seedbed	Wet seedbed with high seeding rate (more than 150 kg/ha)	Wet seedbed with less seeding rate (20–30% less than CM)	Dry raised seedbed with less seed rate (5–10 kg/ha)
Seedling age	More than 30-day-old	16–30-day-old	8–15-day-old
Seed rate/if doing direct seeding/broadcasting	More than 200 kg/ha	20–30% less than CM	5–20 kg/ha or less than 5 kg
Transplanting spacing	Random/less than 10×10 cm	10×15 cm–19×19 cm	20×20 cm–30×30 cm
Planting/hill	>5–6 seedlings/hill	4–5 seedlings/hill	1–3 seedlings/hill
Soil condition	Flooded (or no effort in maintaining aerobic soil condition	Relatively aerobic soil condition with respect to CM either through shallow water level or through intermittent drying	Maintaining aerobic soil condition at least for a week during tillering stage
Compost application	Less than 5t/ha	6–9 t/ha	More than 9 t/ha
Weed management	Chemical and manual	With rotary hoe 1–2 times	With rotary hoe more than 2 times
Pest management	Chemical	Apply IPM	Apply IPM with emphasis on plant environment management

*Established by baseline survey; **SRI-T includes both: SRI-I and SRI-LAP

SRI-I: SRI innovation—few principles of SRI + *modified* conventional practices

SRI-LAP: SRI locally adopted practices—few principles of SRI integrated with existing conventional practices.

As a part of FPAR, on an average four training-cum-observation sessions were conducted during the entire crop growth stages, corresponding to the transplanting/sowing, tillering, flowering and harvesting stages of the crop cycle. The first session was 4-5 days, and the other three sessions were 2 days each.

Data were recorded at three stages: tillering, flowering, and harvesting, and were recorded at two levels: 1) at farmer trainer level, which was recorded using farmers' diaries (input use, yield and cost-benefit, number of sessions conducted, number of FPAR participants, number of women participants, number of landless) and was cross-checked during backstopping visits by researchers, and 2) at researcher level, by both national and regional researchers, with an objective to study the effects of applied management practices on yield and income and other socio-economic implications, and also to learn about the SRI adaptation responses from FPAR farmers in their own fields. For monitoring and learning on SRI adaptation/adoption patterns at FPAR farmers' field, participants of both CFPAR and FPAR were surveyed.

Final results of the field experimentation were discussed and deliberated at the provincial workshops organized at the end of the FPAR involving farmers, researchers, local ministries, national universities, and development professionals from international organizations/institutions. Final data were entered in the project database created for each FPAR site. Data were compiled at provincial, national, and regional levels for further analysis and for presentation and deliberations at national and regional workshops.

Data Analysis

The data were analyzed using Kruskal Wallis rank ANOVA first. Computed F values for SRI-D and SRI-T and CP for selected parameters, yield, fertilizer use efficiency, and economic productivity gain, are presented in Table 3. Followed to that Meta-analysis was conducted to test the trends of the following: how SRI-I, SRI-LAP and SRI-D respond to yield in comparison to CP across the region, and also specifically in rainfed environments by excluding the yield data from irrigated areas and data from SRI demonstrations.

For meta-analysis, initially fixed effects model was considered. In the fixed model, if value of Q (heterogeneity in effect size) was statistically significant, then a random effects model was used. The random effects model assumes that the studies were drawn from populations that differ from each other in ways that could have an impact on the treatment effect. It follows that the effect size will vary from one study to the next for two reasons: the first is random error within studies, and the second is true variation in the effect size from one study to the next. Further, comparative analyses were done to evaluate the economic productivity gain (net return/cost of cultivation) with SRI-T practices in rainfed and irrigated areas with respect to baseline.

Table 3. Computed F values, degree of freedom (*df*) from analysis of variance (anova) of rice yield, fertilizer use efficiency and economic productivity in SRI demonstration (SRI-D), SRI-transition (SRI-T) and conventional practices (CP) across the five provinces of Thailand and Cambodia

	SRI-demonstration (SRI-D)	SRI-transition (SRI-T)	Conventional practice (CP) (baseline)
Rice yield (t/ha)	6.41±0.16 a	5.02±0.09 b	3.14±0.06 c
Fertilizer use efficiency (kg)	40.46± 2.82 a	37.68±1.39 a	27.66±1.44 b
Economic productivity	1.67±0.11 a	1.46±0.07 a	0.38±0.03 b
Rice Yield	*F Value = 109.98** (df = 2, 299)*		
Fertilizer use efficiency	*F value = 10.80** (df = 2, 299)*		
Economic productivity gain	*F value = 52.25** (df = 2, 299)*		

Value in row shows mean ± standard error. Mean values sharing same letter in row are not statistically significant.
**shows F value is significant at $p < 0.001$

In order to understand the adaptation response of farmers for SRI practices in FPAR groups, cluster analysis was performed. Clusters were defined according to five parameters: 1) seedling raising method, 2) seedlings/hill, 3) seedling age, 4) spacing, and 5) aerobic soil conditions at least for a week at vegetative stage. For all five variables, the SRI-D definition, presented in Table 2, was used.

Data are presented as means ± standard error (*s.e).* In most cases, the analysis was performed across the country; and in some cases it was performed by production system, comparing rainfed with irrigated systems.

RESULTS

Productivity Gains with SRI and SRI-T Practices Across the Region

The results across the region showed that in the SRI full demonstration plots (SRI-D), the average yield increment was 100% higher than with conventional practices (CP) while in SRI-T plots, it was 60% greater. This resulted in increased fertilizer use efficiency (kg grain produced/kg inorganic fertilizer applied), which was 46 and 36% higher in SRI-D and SRI-T, respectively, compared to conventional practices (CP).

The economic productivity gain (the ratio of net return in dollars/dollars spent per ha) was calculated using the local costs of production and the prices for produce, using the producer prices reported from all FPAR sites (US$0.30 and 0.43 per kg in Cambodia and Thailand, respectively). It was found that the economic productivity gain was 339 and 284% higher with SRI-D and SRI-T, respectively, compared to CP. There was no statistical difference between SRI-D and SRI-T.

Meta-Trends of SRI-I, SRI-LAP and SRI-D Yield with Respect to Conventional Practice

Data on SRI-I, SRI-LAP and SRI-D all indicated significantly higher yield benefits with respect to conventional practices. The overall effect size was positive and significant (Figure 2). The percentage of variation (I^2) across the studies due to heterogeneity (Q) was zero and therefore confirming the consistency of the studies and provides confidence in the trend obtained. Meta-trends were also observed to evaluate the yield gain or reduction in rainfed environments excluding yield data from the irrigated areas of two districts (Tron and Pichai in Thailand) and also excluding the yields from SRI demonstration plots (SRI-D) from both countries. Figures 3 shows the distribution of effect size across the studies for SRI-I and SRI-LAP, respectively under fixed model analysis. The results showed large effect size though with significant heterogeneity (Q) and variation (I^2) and so the data was further analyzed for random effects. Figure 4 shows the summary effect size for both SRI-I and SRI-LAP under random effects. An effect size of 0.2 is considered small, 0.5 moderate, and 0.8 large [14]. The overall effect size was large with insignificant heterogeneity (Q). Therefore, yield advantage under SRI-I and SRI-LAP with respect to CP can be confirmed.

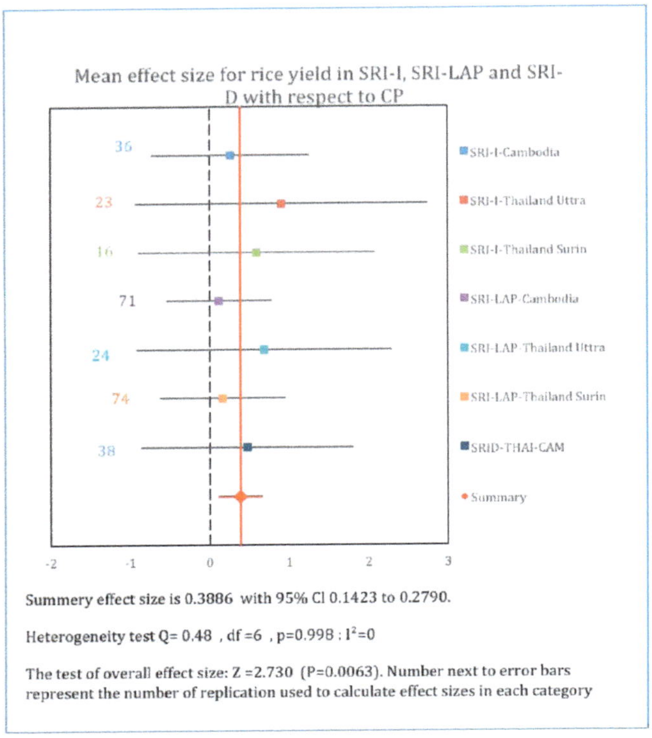

Figure 2. Standardized mean difference for rice yield in SRI-T (SRI-I and SRI-LAP) and SRI-D with respect to CP. X axis shows the effect size. The effect size shows yield benefit in SRI-I, SRI-LAP and SRI D. Overall effect size is positive and significant.

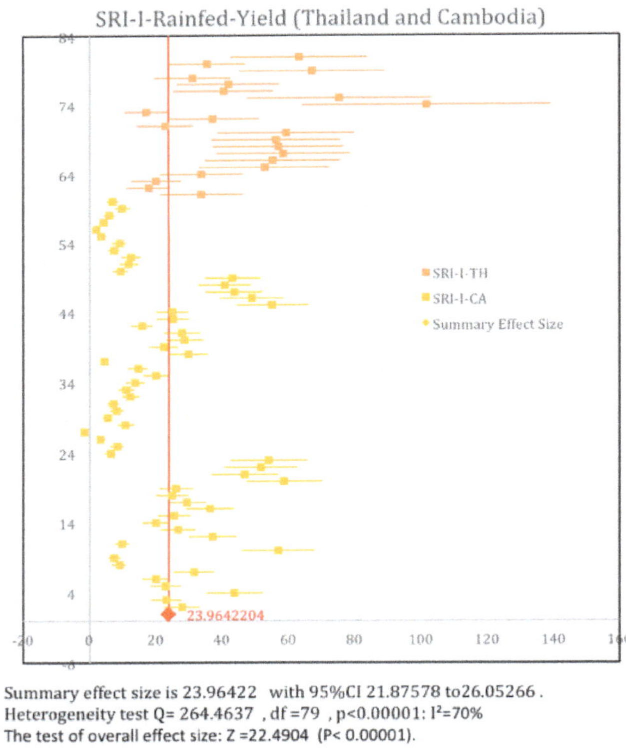

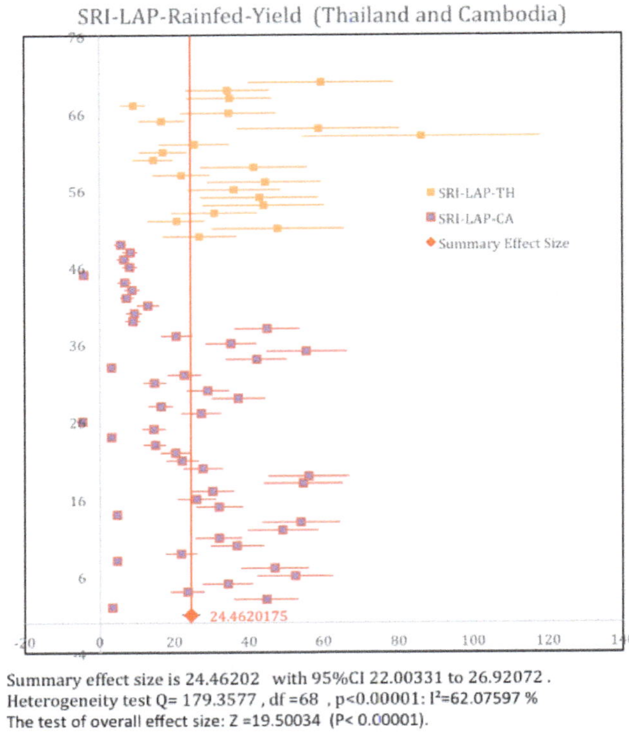

Figure 3. Distribution of effect sizes across the studies measuring the effect of SRI-I and SRI-LAP practices in rainfed environment. Effect sizes below zero indicate decrease in yield and above zero indicate increase in yield with respect to baseline yield (CP).

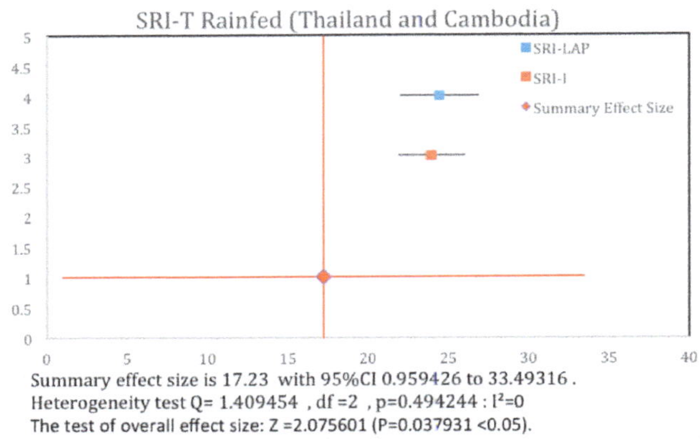

Figure 4. Standardized mean difference for rice yields in SRI-T (SRI-I and SRI-LAP) in rainfed environment. Effects sizes shows yield benefit in SRI-I and SRI-LAP with respect to baseline (CP). Overall effect size is positive and significant.

Both SRI-I and SRI-LAP showed yield benefits with respect to CP. Positive and significant large effect sizes confirmed the benefit of applying SRI-T practices even in rainfed environments.

Economic Implications of SRI-T Practices

The economic productivity gains reflected in Table 3 were reported from FPAR sites. Further analyses of the economics of SRI-T were done for different scenarios utilizing the producer prices and domestic rice prices in both countries, and also the paddy seed price in Thailand since most of the FPAR farmers reported that they were able to sell their grain as seed at a higher price than for consumption. The input costs were also "actualized" based on the actual field conditions but also valued to get in-depth ideas on the factors involved and their cost that would affect micro-economics of rice cultivation.

Four different scenarios were considered as represented in Table 4. In the first scenario, the costs of all inputs were valued, including water cost and family labour, were considered. For rice price, the farmer's/producer's reported price was used, which was 0.30 and 0.43US$/kg in Cambodia and Thailand, respectively. With this scenario, it was found that overall economic productivity gain was more than three times higher in Thailand compared to Cambodia (Scenario 1a). This was due to the higher yield along with higher price of paddy.

Table 4. Computed F value, degree of freedom (*df*) from analysis of variance of economic productivity with SRI-T practices (dollar earn/dollar spent per hectare) in rainfed and irrigated rice production system of Cambodia and Thailand (value in rows show mean ± standard error)

	Economic productivity with SRI-T	Cambodia rainfed	Thailand rainfed	Thailand irrigated
Scenario 1a	With producer price (0.3 and 0.43US$/kg for Cambodia and Thailand respectively) when all inputs were valued	0.769±0.03 c	2.655±0.12 a	2.066±0.08 b
		(0.05±0.01)	(0.4±0.01)	(0.05±0.01)
Scenario 1b	With producer price (0.3 and 0.43US$/kg for Cambodia and Thailand respectively) when labour input for	2.10±0.06 b	2.9±0.13 a	2.06±0.08 b
	transplanting, weeding and harvesting cost were excluded from Cambodia and cost of water are excluded from Cambodia and Surin province of Thailand	(0.66±0.04)	(0.54±0.01)	(0.05±0.01)
Scenario 2a	With domestic rice price (0.3 and 0.38US$/kg for Cambodia and Thailand respectively) when all inputs	0.769±0.03 c	2.17±0.11 a	1.79±0.08 b
	were valued	(0.05±0.01)	(0.40±0.01)	(0.05±0.01)
Scenario 2b	With domestic rice price (0.3 and 0.38US$/kg for Cambodia and Thailand respectively) when labour input	2.08±0.06 b	2.4±0.12 a	1.79±0.08 c
	for transplanting, weeding and harvesting cost were excluded from Cambodia and cost of water excluded from Cambodia and Surin province of Thailand	(0.66±0.04)	(0.54±0.01)	(0.05±0.01)
Average rice yield at FPAR sites		4.3±0.73 c	5.82±1.29) b	6.26±0.82 a
Scenario 1a	*F* = 230.73** (*df* = 2, 184)			
Scenario 1b	*F* = 22.09** (*df* = 2, 184)			
Scenario 2a	*F* = 149.34** (*df* = 2, 184)			
Scenario 2b	*F* = 7.00** (*df* = 2, 184)			

Mean value sharing same letter in row are not significantly different. Value in bracket shows baseline figures.

**Shows F value is significant at $p < 0.001$

Note: For baseline yield economy calculation, rice price US$3/kg is used for both countries as reported in the baseline survey.

Farmers of Thailand reported that SRI practices produced higher quality of grain that enabled them to sell their paddy as seed with a higher premium price. Accordingly, it was found that within Thailand, the economic productivity gain in rainfed areas was 28.5% higher compared to irrigated areas. Above all, in both countries and in both kinds of systems, both irrigated and rainfed, the productivity gain under SRI-T practices was much higher compared to baseline situation (CP).

In scenario 1b, we excluded the cost of labour for transplanting, weeding and harvesting in Cambodia as these operations are mostly done by family labour in the project areas (85-90%) and thus do not involve cash expenditure. In addition, we excluded the cost of water in Cambodian sites because all FPAR sites there are completely rainfed. Similarly, water cost was excluded from Surin province data in Thailand because the production system there is completely rainfed. In this scenario (scenario Ib), we found that productivity gain was highest in rainfed areas of Thailand compared to irrigated areas. Further, the economic productivity gains in rainfed areas of Cambodia and from irrigated areas of Thailand were similar.

In scenario 2a, we used the domestic rice price [15] for calculation. In this situation, when all inputs were valued, the productivity gain under SRI-T practices was again higher in the rainfed areas of Thailand compared to the irrigated ones. When we actualized the cost of inputs as reflected in scenario 2b, we found that the economic productivity gain was higher in rainfed areas of both countries compared to the irrigated areas of Thailand. Indeed, the productivity gains were again highest in Surin, a rainfed province of Thailand. Overall, this analysis showed that economic productivity gains were increased in both irrigated and rainfed system under SRI-T practices and were highest in rainfed areas compared to irrigated ones.

Further, as evident from Table 3, the gains under SRI-D were still more attractive given that the yield was 28% higher from SRI-D compared to SRI-T and 100% more than the baseline (CP). Accordingly, net profit was also higher in the SRI-D plots. The economic productivity gain was similar under SRI-D and SRI-T as shown in Table 3. The SRI-D's yield and profit showed the potential of SRI principles at the test sites when all practices were applied as recommended.

However, we have excluded these results from our discussion of productivity gains under SRI-D given that the field conditions of the SRI-D trials may or may not represent the actual field conditions under which farmers operate. Therefore, for our analysis, we focused on the results obtained from learning sites where farmers were encouraged to adopt and adapt SRI methods to develop their own technological options.

Adaptation Responses from Farmers for SRI Practices

On farmers' responses to learning SRI practices, which was characterized as SRI in transition (SRI-T), it was found that single seedling transplant (1-2 seedlings/hill) and SRI spacing (20×20-30×30 cm) were the most preferred practices in both countries and were applied by almost 80% of FPAR farmers (Figure 5). SRI seedbeds (dry seedbed) and young seedlings (8 to 15-days-old) for transplanting were followed by relatively few farmers (only 15%). Keeping paddy soil aerobic for a week or so during the vegetative growth stage was least followed. However, most of the farmers responded that aerobic soil management was not followed as a part of their crop and soil management because this occurred naturally, given that they are working in a rainfed environment. Almost all FPAR districts experienced early drought last year due to a late monsoon. Farmers who followed aerobic soil conditions for some period were mainly from Tron and Pichhai districts where supplementary irrigation facilities were available. In those two districts, they reported 2-3 less irrigations with SRI compared to conventional practices.

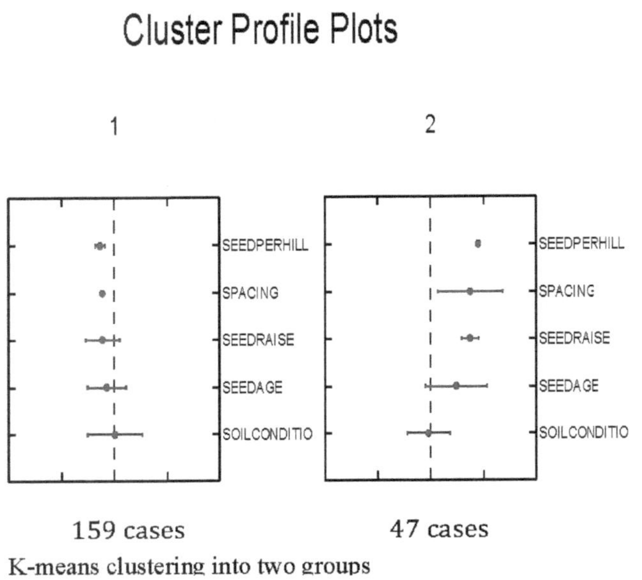

SRI practice	F ratio
Seedling/hill	7307.98
Spacing	412.39
Seedling raising method	398.12
Seedling age	80.23
Aerobic soil condition at least for a week in vegetative growth stage	0.358

Figure 5. Adaptation response on SRI practices by FPAR farmers in 15 districts of five provinces of Cambodia and Thailand.

DISCUSSION

Effectiveness of SRI Management Practices for Achieving Higher Yield and Profit with Higher Resource Use Efficiency at Rainfed Smallholder Farms

Various literatures have identified specific linkages between economic growth, poverty alleviation and political stability which have direct consequences on achieving food security [16,17]. The explanation given in these literatures is that greater stability in food economy contributes to faster economic growth and in turn to political stability. And greater stability contributes in turn to more equity and poverty alleviation by reducing the vulnerability of the poor to sudden shocks in food prices or availability. This occurs when economic growth raises the poor above a meaningful poverty line and when stabilization of the food economy prevents external shocks from threatening their food intake, which is not the case with smallholder farmers of this region. Therefore, for addressing food security concerns, it is important to increase the productivity of the poor farmers themselves so that they can protect themselves from external shocks.

It is evident that rice cultivation is the main farming occupation in both countries (Table 1). At macro level, both countries have achieved rice production sufficiency in terms of production, and Thailand has some surplus production even though this sufficiency and surplus is inadequate to achieve food sufficiency and security at the household level in many rural areas there. Particularly, smallholder rice farmers are more prone to this insecurity living in a rainfed areas.

The baseline figures (Table 1) show the average rice productivity in each province in pre-project scenario and also rice insufficiency and sufficiency status. It also reveals that the contribution from conventional rice farming to household income in Cambodia is not very significant compared to Thailand. In rainfed areas of Cambodia, rice farmer produces only 20% of the total household income from rice. This is mainly because in Cambodia, rice insufficiency is still a concern for the farmers (Table 1) and that is why most of the rice produced is stored for self-consumption.

In Thailand, the contribution of rice production to total household income is 49 and 77% in Surin and Uttradit, respectively. The reason is that a majority of the rice farmers in these parts of Thailand sell their produce after harvest and often depend on the market for their own food needs. It is also evident in both countries that food availability from the farm is linked mainly to rice production because of the limited diversification of farming activities since farmers do not grow many other crops. Although in Cambodia, livestock is another means for increasing household incomes, only 30-40% of households in project areas are engaged in livestock production (Table 1). This means that they need to rely on markets to meet their other food requirements.

Our results from five provinces of Cambodia and Thailand, involving approximately 3600 farmers, clearly demonstrated higher yield with SRI methods compared to the regional baseline yield of 3.14 t/ha. In the same districts and with the same farmers, the average yield with STI-T was 5.03 t/ha and with fuller use of SRI methods (SRI-D) it was 6.41 t/ha. The higher yield obtained under SRI management practices is in line with our previous research findings [9,10,13,18,19,20]. The scientific explanation for higher yield under SRI practices is already well established within the scientific domain and reviewed and researched by many [9,21,22] so there should not be any surprise.

The interesting results in this study was that in addition to higher yield, higher net returns and higher fertilizer use efficiency were also achieved with SRI methods. The economic productivity gain was higher with SRI-T compared to conventional systems in both countries, and with the both systems, rainfed and irrigated. Importantly, the gain was much higher in rainfed systems compared with irrigated ones.

Thus, SRI-guided practices can help smallholder farmers to increase their productivity and higher resource use efficiency with less input use. The productivity gains can ensure more income, almost double, from rice farming even with the existing market scenario. It can be further improved if better incentives are provided to them for the healthy (pesticide free) and higher-quality produce (bold and good filled grain) by ensuring a favorable market price for rice grown with agro-ecological approach, e.g., organic rice production.

SRI Adaptation Response from Farmers in the Context of Existing Bio-Physical and Socio-Economic Environment

To gain more understanding of farmers' adaptation response, discussions were held with them on the agricultural practices and the factors that drive adaptation and adoption decisions for any new technology. The responses were different between the two countries.

In Cambodia, farmers operating in rainfed areas are very much dependent on favourable weather conditions for a good harvest. Often these households become food-insecure immediately if the weather is unfavourable. Although the extent of food insecurity is difficult to quantify and varies considerably from one area to another, and from one household to another, in areas where agricultural production has increased in recent years, a significant proportion of the incremental production has been consumed by the producing households, indicating continued significant inadequacies in meeting household food needs, as evident from baseline findings (Table 1).

Only twenty percent (20%) of household income in rural Cambodia comes from rice farming (Table 1), which is non-significant realizing that the rice is the only crop that they grow. This indicates that households are mainly growing rice to ensure sufficient food production for consumption by utilizing their own resources, mainly family labour and land under rainfed conditions. Working in rainfed and insecure environment makes subsistence farmers further insecure about making gains from farming, and therefore they do not want to expend cash and extra efforts in farming. Further, the farm-gate price of paddy in Cambodia is the lowest compared to other countries of the Lower Mekong River basin, which makes rice cultivation a relatively unattractive option to them compared to off-farm employment opportunities. So, increasingly most of them are being employed as construction labourers or in factories with higher wages rate, especially the young generation and male members of the farming community.

Further, in Cambodia as the prices of agriculture inputs are increasing, the costs of cultivation are generally becoming higher in recent years. Moreover, the opportunity (and risks) associated with rainfed farming have not been adequately addressed by government, and so the current policy by and large tends to favour agriculture in the irrigated areas. While the reach of micro-credit schemes has widened over the years in rainfed areas and farmers are taking loans as evident from our baseline study (Table 1), these loans are primarily used for physical construction and for social functions. This has resulted in increasing migration from rural to urban areas for better livelihood options, and it has changed the structure of agriculture. From the action research it became evident that more than 70% of the farmers in the project area are women, and a majority of them are older than 50 years [23].

Within the context of the existing bio-physical and socio-economic environment, we discussed with farmers SRI adoption in the project area. They are quite interested in increasing yield and reducing their costs of cultivation, especially for seeds and fertilizers, and in making higher profits. So there is a considerable increment in the adaptation rate for SRI-recommended practices that help them to get higher yield without any extra input cost.

Since manual transplanting is a common feature in most regions of Cambodia and given that family labour is fairly available, for which households do not need to pay, it turns out that the SRI-recommended practices of single-seedling transplants and transplanting with wider spacing are "taking root" in rice farming fairly easily. Younger seedling transplanting can enhance the yields with these other new methods even more, but due to early drought spells which is increasingly increasing in recent years, this practice has been difficult for farmers to follow unless there is some supplementary irrigation support available for the early growth stage of the rice crop.

This is also applicable with the SRI water management practice that is least adopted by FPAR farmers. Lack of assured water supply limits opportunity either for maintaining shallow water level with drying period for a week or draining excess water from the rice field, creating aerobic soil condition at least for a week during vegetative growth stage. Part of the higher yield from SRI-D was attributable to transplanting young seedlings as recommended, along with availability of supplementary irrigation water.

Discussion was also held on the matter of water availability and on the costs associated with it. It was found that in areas which are near to a canal, water can be pumped to the fields if this facility is available. The cost of pumping water is about US$48-50/ha. Further discussion was held on labour requirements. Regarding labour requirements with SRI, 40% of the FPAR farmers reported that their labour requirement was less, whereas 33% reported no effect (data not shown here). Three-quarters thus experienced no increase in their labor inputs or even had a reduction.

Results were also discussed with respect to the income for landless labourers that they could earn from rice employment. It was reported that currently in Cambodia the wage for labourers working in rice fields is much less than in Thailand, reflecting part the lower price received for rice in Cambodia. Landless labourers who work in rice fields thus reported that they get only 10% of their total household earnings from rice field labour [23]. Also, 85-90% of the labour inputs for smallholder production is available from family members, so hiring external labourers is not a part of rice cultivation. However, recently the increasing migration of male members to urban areas from rural areas has created a shortage of labour in the latter, and this has put an extra burden on women farmers. With increasing feminization and aging of agriculture society, these issues will become more prominent in the near future. Further research is warranted on this emergent trend in rice farming.

On the other hand, in Thailand, the situation is quite different. First, in rice-farming households, the share of net food-buying by households is higher among those with smaller landholdings and more particularly with rice-farming households. As can be seen from Table 1, a major proportion of household income comes from rice farming as the rice produced is not kept for self-consumption but rather is sold. About 87% of those who were affected by food poverty in 2007 were agricultural households, mainly in the north and northeastern parts of Thailand. If global food prices rise again as they rose in 2007-2008, this will induce a sharp increase in domestic food prices causing a high rate of overall inflation. For the poor farm households, food constitutes a considerable portion of their expenditure, and therefore any rise in food prices makes them food-insecure.

Smallholder farmers are hard-hit by soaring agriculture input prices and by rising production costs. So this kind of inflation is a big issue for them. For example, most chemical fertilizer used in Thailand is imported. Between 2003 and 2010, the cost of urea increased by 2.5 times. In general, agriculture input markets are free, although public policies on chemical fertilizer make fertilizer available at a reduced price. However, smallholder farmers often cannot benefit from this policy because they do not meet the eligibility requirements for subsidized fertilizer.

Another factor is that labour wages have been increasing many-fold, making the costs of production even higher. This is the reason why labour use in agriculture has declined from 392 man-hours per hectare in 1980 to less than 56 man-hours per hectare in 2008 [24]. The level of man-hours/ha has been further decreased after a further increase in labour wages in 2010-2011 following the 2011 flood.

Further, an unstable market for rice and price volatility at the time of harvesting further constrains farmers' marketing opportunities. In addition, many smallholders rent land, and the cost of renting land becomes higher when the market price for rice is high. Indeed, these forces in combination create an unfavorable environment for smallholders and put them on risk.

These are the driving forces for migration of farmers from rural to city areas, and they explain why mostly older people and women who are left behind in the villages to engage in rice cultivation (Table 1). In addition, more than 90% of farmers in both provinces of Thailand take out loans, and these loans are taken mainly for agricultural activity (Table 1). The amount of loans taken by farmers is more in irrigated areas than in rainfed area because the costs of production are higher in irrigated than in rainfed cultivation.

To address all these issues, farmers are seeking alternatives to reduce their cost of production by reducing labour, fuel and fertilizer inputs. That is also why direct seeding, broadcasting, and indigenous fertilizer production and application are all part of common rice cultivation practices in Thailand. Our survey report indicated that almost 90% of farmers in Thailand now follow broadcasting and direct-seeding in their rice cultivation, given labor shortages and wage costs for transplanting. However, for their SRI experimentation, almost 70% farmers agreed to go with transplanting (single seedlings and wider spacing, see Figure 5).

When we discussed the reasons for this, farmers informed us that with SRI practices they can get higher quality grain, and that grain can be sold as seed for almost double the price compared to the existing market price of rice. For expanding these practices in a larger area, they are seeking some substitute for labour. There is a considerable interest among farmers to experiment with SRI ideas used direct-seeding methods with reduced seed rate and to develop organic SRI production for better marketing opportunities. This is an area where further research is required, especially in rainfed areas where low seed rates increase weed infestation during the early vegetative growth stage, which coincides usually with an early drought spell in the region.

Potential of SRI for Addressing Food Insecurity by Actively Involving Smallholder Farmers

The above analysis showed that at ground level the situation for rice production and innovation is rather different in both countries, although the repercussions of these prevailing situations are the same, i.e., out-migration of farmers. Our learning shows that SRI can boost yield and income and resource use efficiency without any additional cost. Further, the productivity gain is higher in rainfed areas compared to irrigated ones. With favorable policy support—supplementary irrigation, market security and some crop insurance policy—the migration of farmers could possibly be reversed if rice farming would be made more attractive for them. SRI provides an opportunity to address these multiple issues, as this is not only a matter of bringing yield and income benefits to farmers. In rainfed systems which cover 60% of the rice growing area and where a majority of the smallholding poor farmers reside, only a small investment is needed to make the system more productive and resilient.

This also means that the resources of irrigated areas can be freed up for other agricultural activities. There is already considerable competition for water, land and other resources. A "business as usual" scenario will make this competition even tougher and will put further pressure on environmental resources. Additionally, even with sufficient production at the macro level, food insecurity still is common in rainfed areas. Therefore, instead of trying to achieve food security through entitlement and distribution mechanisms, bringing poor smallholding farmers more directly into the process of economic growth can offer more hope at both micro and macro levels.

With respect to contribution towards food security, our results confirm that with SRI adoption, food availability, at least the staple food, can be increased substantially with increased production without an increase in the costs of cultivation, indeed with reduced cost of cultivation in most cases and with higher resource use efficiency. Greater purchasing power of poor farmers stimulated by more profitable rice production (Table 4) will improve the diversity and quality of food intake by rural households.

To cope with market uncertainty, which might threaten food stability and so security, it is important for poor farmers to save their produce for self-consumption. This is particularly relevant for Thailand's smallholder farmers. With SRI practices, they can have better yield, as evident in Table 2. A part of this increased production can be saved for self-consumption, and the rest can be sold to market.

It is to be noted that the rainfed production system is already a fragile production environment where farmers are operating with limited resources. Increasing climate change variability will further deteriorate it and will increase further pressure on them. Indeed, they need a smarter approach and low cost solution to become productive and resilient. With SRI, which is based on agro-ecological principles and practices with less input use and which encourages to feed the soil to feed the plant, production systems can become more stable and resilient.

Therefore, development efforts that raise productivity and incomes for the broad population of smallholders without further deteriorating the environment should be supported by incentives which will not only stimulate rural growth and address food security concerns, but will also influence the macroeconomic performance of the region in positive ways.

CONCLUSION

Results from over 120 sets of field-level experiments spread across 5 provinces of Cambodia and Thailand have demonstrated higher yield and resource use efficiency with SRI-T practices, as well as SRI-D management, which more fully utilized the recommended methods. The net profit achieved with SRI-T management was almost double, due to the higher yield coupled with reduced costs of inputs and also due to the higher quality of the grain produced, which commanded a higher market price. The productivity gain was even much higher with SRI-D where yield gain was 28% and 100% greater than SRI-T and CP respectively.

Surprisingly, economic gains for farmers attainable with SRI rice crop management were higher in rainfed areas compared to the irrigated areas because the cost of production was significantly lower in the rainfed areas compared to the irrigated one. Efforts to make SRI knowledge and practices available on a wider scale can raise productivity and incomes and in turn can address food insecurity of the broad population of smallholders without further deteriorating the environment, if supported by favourable and appropriate policies. Further research is warranted on the subject of higher yield and its implication on rice price either benefitting or burdening farmers and consumers. The research should take into account the current trend of increasing farmer out-migration. Finally, research is required for better location-specific adaptation of SRI practices involving smallholders.

ACKNOWLEDGEMENT

The work reported in this manuscript was conducted with funding support from the European Union for a regional project titled "Sustaining and Enhancing the Momentum for Innovation and Learning around the System of Rice Intensification (SRI) in the Lower Mekong River Basin" (Ref: EuropeAid/128500/C/ACT/Multi) which is being implemented by the Asian Centre of Innovation for Sustainable Agricultural Intensification (ACISAI) at the Asian Institute of Technology (AIT) in Klong Luang, Thailand.

REFERENCES

1. Matsuo K (2002) Development of upland cropping systems for Crop-Animal integrated farming systems in Northeast Thailand. Report of Japan International Research Centre for Agricultural Sciences (JIRCAS), Department of Agriculture (DOA) and Khon Kaen University (KKU), Thailand.
2. MAFF (2008) Report on Activities of Agriculture, Forestry and Fisheries. Workshop on National Achievement in 2007-2008 and Planning for 2008-2009. Ministry of Agriculture, Forestry and Fisheries, Phnom Penh, Cambodia, 2-3 April 2008 (in Khmer).
3. FAO (2004) FAOSTAT. Available from: http://faostat.fao.org/default.aspx
4. Kirby M, Mainuddin M (2006) Water Productivity Assessment: Mekong River Basin Approach, BFP Working Paper 4. In: *An advance edition of a working paper on the assessment of water productivity from the approach of the Mekong River Basin*. Available from: http://www.waterandfood.org/fileadmin/CPWF_Documents/

Documents/Basin_Focal_Projects/BFP_Publications/MekongBasinWaterProductivityBFPwp04Draft02.pdf

5. UN Population Division (2006) World Population Prospects: The 2006 Revision. Available from: http://www.un.org/esa/population/publications/wpp2006/wpp2006.htm
6. Nesbitt HJ (2005) Water Used for Agriculture in the Lower Mekong Basin. MRC Technical paper No 11, Mekong River Commission, Vientiane.
7. Molle F, Wester P, Hirsch P, et al. (2007) River basin development and management. In: *Water for Food Water for Life, A Comprehensive Assessment of Water Management in Agriculture*. Molden, D. (Eds.). London Earthscan. 585-625.
8. Behera D, Chaudhary AK, Vutukuru VK, et al. (2013) Enhancing agricultural livelihoods through community institutions in Bihar, India, In:*South Asia rural livelihoods*. Series 3 note no. 1. Washington D.C., The Worldbank.
9. Mishra A, Whitten M, Ketelaar JW, et al. (2006) The system of rice intensification (SRI): a challenge for science, and an opportunity for farmer empowerment towards sustainable agriculture. *Int J Agric Sustain* 4: 193-212.
10. Mishra A, Kumar P, Noble A (2013) Assessing the potential of SRI management principles and the FFS approach in Northeast Thailand for sustainable rice intensification in the context of climate change. *Int J Agric Sustain* 11: 4-22.
11. Uphoff N, Randriamiharisoa R (2002) Possibilities for reducing water use in irrigated rice production through the Madagascar System of Rice Intensification, In: *Water-Wise Rice Production*. Bouman B et al. (Eds.). Los Banos, Philippines: International Rice Research Institute. 71-88.
12. Kabir H, Uphoff N (2007) Results of disseminating the system of rice intensification with farmer field school methods in Northern Myanmar. *Exp Agric* 43: 463-476.
13. Mishra A, Salokhe VM (2008) Growing More Rice with Less Water in Asia: Identifying and Exploring Opportunities through System of Rice Intensification, In: Agricultural Systems: Economics, Technology and Diversity. Oliver, W. Castalonge (Eds.). Hauppauge, New York: Nova Science Publishers. 173-191.
14. Cohen J (1992) Statistical power analysis. *Curr Dir Psychol Sci* 1: 98-101.
15. FAO Rice Market Monitor (2015) Volume XVIII, Issue No 2, July 2015. Available from: http://www.fao.org/fileadmin/templates/est/COMM_MARKETS_MONITORING/Rice/Images/RMM/RMM_JUL15_H.pdf
16. Birdsall N, Ross D, Sabot R (1995) Inequality and growth reconsidered: lessons from East Asia. *World Bank Econ Rev* 9: 477-508.
17. Dawe D (1996) A new look at the effects of export instability on investment and growth. *World Dev* 24: 1905-1914.
18. Mishra A, Salokhe VM (2010) The effects of planting pattern and water regime on root morphology, physiology and grain yield in rice. *J Agron Crop Sci* 196: 368-378.
19. Mishra A, Salokhe VM (2011) Rice root growth and physiological responses to SRI water management and implications for crop productivity. *Paddy Water Environ* 9: 41-52.
20. Mishra A, Uphoff NT (2013) Morphological and physiological responses of rice root and shoots to varying water regimes and soil microbial densities. *Arch Agron Soil Sci* 59: 705-731.
21. Horie T, Shiraiwa T, Homma K, et al. (2005) Can yields of lowland rice resume the increases that they showed in the 1980s. *Plant Prod Sci* 8: 251-272.
22. Thakur AK, Uphoff NT, Stoop WA (2015) Scientific underpinnings of the System of Rice Intensification (SRI): What is known so far. *Adv Agron* 135: 147-179.
23. Mishra A, Kumar P (2015) Sustaining and Enhancing the Momentum for Innovation and Learning around the System of Rice Intensification (SRI) in the Lower Mekong River Basin. *Regional Review and Planning Workshop Report*. 40p. Available from: http://www.sri-lmb.ait.asia/country/doc/Regional%20Review%20and%20Planning%20Workshop-REPORT%20(02-03%20June%202015).pdf
24. Isvilanonda S, Kao-ent S (2009) Dynamics of Thailand Rice production economy and future outlook. A paper presented to Thailand Research Fund, Bangkok.

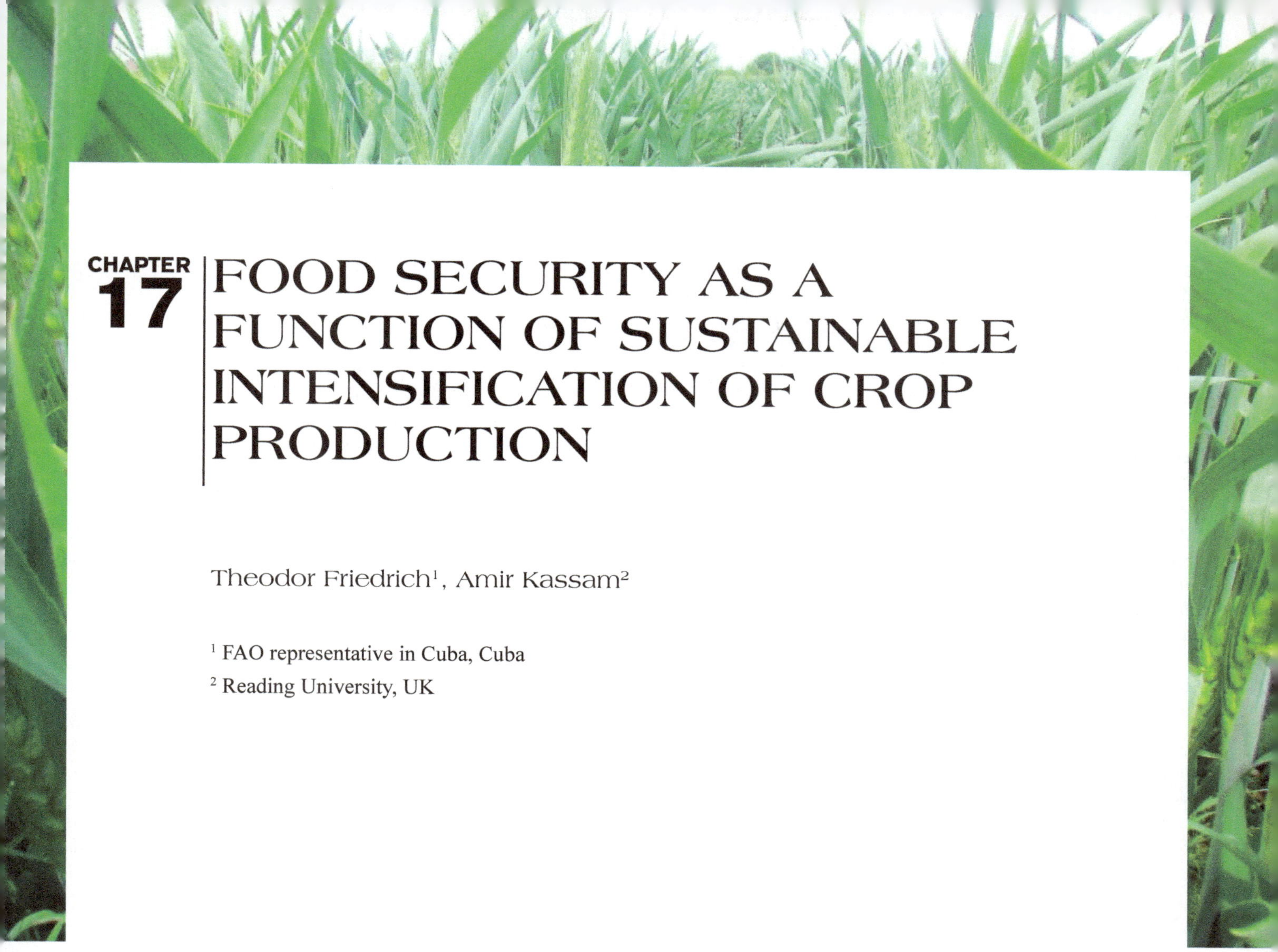

CHAPTER 17

FOOD SECURITY AS A FUNCTION OF SUSTAINABLE INTENSIFICATION OF CROP PRODUCTION

Theodor Friedrich[1], Amir Kassam[2]

[1] FAO representative in Cuba, Cuba

[2] Reading University, UK

ABSTRACT

The challenge to eradicate hunger and establish food security across all its four pillars (availability, accessibility, health and safety, and continuity) is ongoing. The actual situation in global food production leads most of the attention to improving accessibility and safety of food, particularly to vulnerable populations. However, in view of the growth in demand, which includes changes in preferences for example towards food of animal origin, availability and continuity will play larger roles in future. Food production needs to increase over the coming decades at challenging rates, while facing problems of degradation and reduced availability of natural resources for production such as soil and water, and facing increasing challenges from climate change. The actual trends in yield development suggest that a simple gradual improvement of production within the existing concepts will not provide a sustainable or feasible solution, and that more fundamental changes in the agricultural production paradigm are required to face these future challenges. The Sustainable Intensification represents such a change in paradigm in which high production levels are combined with sustainability. The concept of sustainable intensification, the rationale for it and its functional elements, represented by Conservation Agriculture, are presented in this paper.

INTRODUCTION

Food security exists when "all people have, all the time, physical, social and economic access to enough, innocuous and nutritive food for satisfying their daily energy needs and food preferences, in order to have an active and healthy life". Food security has four main pillars [1], namely:

- Physical availability of food
- Economical and physical access to food
- Appropriate and healthy use of food
- Stability of these factors over time.

The term sustainable intensification has become popular in recent years and while its definition can vary, it can be considered in both a narrow and a broader sense. The narrow definition applies to *the pursuit of the dual goals of higher yields (output) and productivity efficiency with fewer negative consequences on the environment* while building resilience, natural capital and at the same time enhancing the

Citation: Theodor Friedrich, Amir Kassam. Food security as a function of Sustainable Intensification of Crop Production. AIMS Agriculture and Food, 2016, 1(2): 227-238. doi: 10.3934/agrfood.2016.2.227.

flow of environmental services [2,3,4]. *This* combines increasing biological outputs, productivity efficiency, resilience, and ecosystem or environmental services through integrated production systems development and landscape management in rainfed and irrigated landscapes. A whole range of ecosystem services including; maintaining soil health, the quality of drinking water and air, control of erosion and other forms of land degradation, protection of water, nutrient and carbon cycling, pollination services and issues such as the protection of landscapes, habitats, biodiversity for ecosystem functioning and resilience, are considered [5,6].

In the broader context, sustainable intensification would encompass the above production and ecological dimensions, the biological products provided to and utilized by consumers and with minimum food wastage, as well as human and economic dimensions of socio-cultural aspirations, organizations and social equity, land tenure, employment and economic growth. It also implies improving the capacities of people (including farmers) and their informal and formal institutions to deliver and use inputs efficiently, manage systems, distribute and use outputs efficiently so as to avoid excessive wastage, and harness large-scale ecosystem services that benefit producers and consumers alike [2,3,7,8,9].

However sustainable intensification is defined, it is necessary to achieve increased yields and output in ways that do not harm the resource base and the environment, and even improve them. This review outlines the current situation of food security in the world and some of the challenges that can affect it. The review proposes an alternate paradigm of a "sustainable production intensification" to address the challenges. While the elements leading to this new paradigm, such as conservation agriculture, are not really new, the understanding that these concepts allow to achieve the production levels required without compromising the natural resource base, is fairly recent. This allowed the proclamation of sustainable intensification as a new paradigm, since it is achievable in practice. So far, high production levels were always combined with some sacrifices on sustainability, and so called "sustainable" approaches to agriculture sacrificed production levels. In this sense, sustainable agriculture was always a theoretic concept out of practical reach. The understanding of the impact of conservation agriculture has led to the term "sustainable intensification" which therefore can be considered a new paradigm.

THE CURRENT SITUATION OF FOOD SECURITY IN THE WORLD

Nowadays, the world produces more than enough food for the entire population, although it is unevenly distributed. Around 30% of the food produced is even wasted, and a significant amount is fed to livestock and some used for producing biofuels. Nevertheless, with population growth and its increasing food demand, which includes the changes in food preferences towards products of animal origin due to rising incomes and more people living in the urban areas, global production needs to increase by 60-70% over the next 30 years or so [7,10].

The current problem of hunger in the world is mainly due to lack of economic access to food caused by inadequate integration of developing world population into national economies which themselves are weak; however, there has been an improvement during the last three decades, as a result of the reduction in the extent of extreme poverty but also the absolute number of poor people remaining constant or decreasing slightly while the world population more than doubled.

There are also problems of undernourishment due to lack of balanced nutrients because of the inadequate use of food, but there is prevalence of obesity too in some countries at the same time. Both are signs of unhealthy and unbalanced nutrition which can be caused by problems of economic access, acquired preferences, lack of knowledge or culture.

Another challenge is the stability of food availability and accessibility. In the last decade, higher variability in availability and access has been noticed due to physical lack of food, and price volatility because of, for example, extreme climatic events [11,12].

The number of hungry people in the world has slightly decreased to 795 million in 2015. Considering the population increase, the first of the millennium goals, which was to reduce to half the percentage of hungry people between 1990 and 2015 was reached in Asia and Latin America, but missed by a little at global level [14] (Figure 1). Nevertheless, the goal of the Food World Summit, held in 1996, which suggested reducing the number of hungry people by 50% at 2015 was only reached in Latin America. It is also alarming that 98% of undernourished people live in developing countries and that hunger and poverty, particularly in rural areas, are closely linked [15], confirming the fact that the proximate cause of hunger is poverty. The goal to reduce and finally end hunger and poverty has been strengthened under the new agenda post 2015 with the Sustainable Development Goals, particularly SDG 1 and SDG 2 with the proposal to eradicate poverty and hunger until 2030[16].

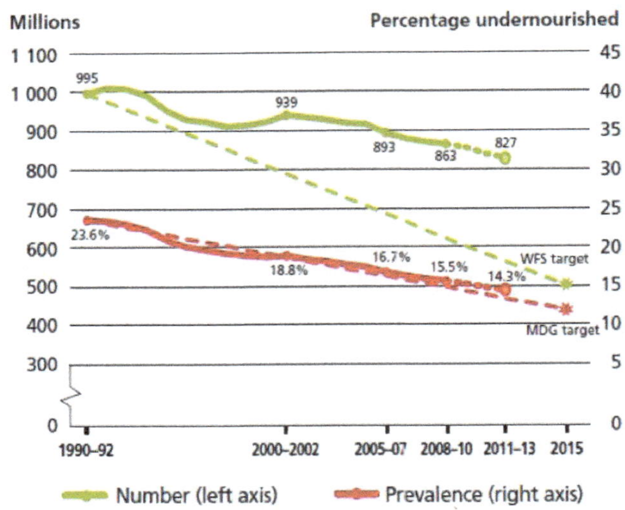

Figure 1. Trend in under nourished people in numbers and in prevalence from 1990 to 2015 (Data for 2011-13 in all graphics refer to provisional estimates; source FAO) [13]

FOOD SECURITY AND SOME OF ITS CHALLENGES

Prices

The first challenge is food prices. Since 2000 food prices generally tended to increase (Figure 2). This could be good news for farmers and food producers, but it makes it difficult for poor and vulnerable populations to access food. Besides, it causes problems to the economies of countries whose food security heavily depend on food imports. Also there has been a higher volatility of food prices in the last years. The soaring food price crisis of 2007/08 and 2011 directly affected the food security of 105 million and 44 million of people, respectively. This particular crisis has alerted many countries, particularly the net food importing countries, many of whom have since then declared the enhancement of national food production as a strategic objective for national security [17]. High price volatility seems to be a phenomenon which may not disappear soon because it results from the circumstances that characterize our world, namely [17,18]:

- Incidences of droughts, floods, plagues and diseases as a consequence of higher climatic variability due to climate change and global warming respectively.
- Strong and fast growing demands of emerging markets—many growing economies have large internal markets, and every change in them affects the world market.
- Measures of trade and agricultural policies, many times as a response to food price volatility.
- Energy price volatility, which has its direct effects on costs of food transportation and production, but also directly on food prices through the markets of biofuels.
- Speculation with food prices on the stock market.

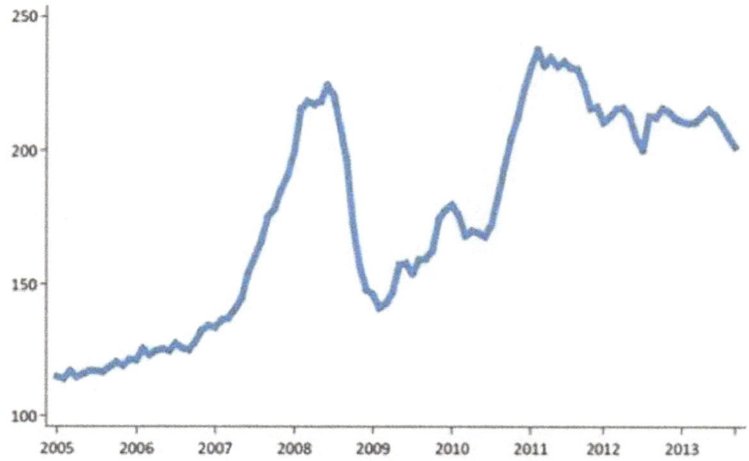

Figure 2. FAO price index for food from 2005-2013 [17]

This volatility of prices affects food security. It reduces the security of planning in the agricultural production, and this leads to a reduction of investments in the market. Agriculture can no longer be considered as a safe and calculable investment. However, there are also direct effects of price volatility on food security, like a sudden rise of food prices in the world market could endanger the food security of food importing countries with weak economies, while a drop in food prices could risk the subsistence of small producers that depend on the incomes from food sales for their survival [11,19].

Natural Resources

Food production depends on natural resources, such as water and soils, which can be limited. Nowadays, agriculture uses 70% of the fresh water of the world, and, in many regions, water is a limiting factor for production. For doubling the production, agriculture should not use more water, since the limited resources have to be shared with other sectors with increasing demands, such as domestic water for the population. An increase of agricultural production has to result from the efficiency savings of water use, since many major watersheds in the world are already overexploited and in danger of running dry [20,21,22].

Another resource in danger is land. Globally, land resources are finite but agricultural lands are constantly lost due to other uses. Besides, lands used for agricultural production suffer from soil degradation, which is a global problem [3,23,24]. Soil degradation, which is a gradual process during which the soils lose their original production potential expresses itself at later stages in erosion and finally in desertification. It is actually affecting practically all soils under agricultural production and as a result many important cropping areas are facing decreased agro-ecological yield ceilings which do not allow further production increases despite application of additional inputs.

Soil degradation can be traced back in human history [23] until the very origins of sedentary agriculture. This suggests that the causes commonly listed as reasons for soil degradation, such as compaction by heavy machinery, overuse of fertilizers and chemicals or even climate change, might not be the original root cause of the problem. Instead it suggests that mechanical soil tillage, one of the first operations applied in agriculture, is the main cause of soil degradation. Soil tillage leads to a loss of soil biodiversity and organic matter, which results in the destruction of soil structure and porosity, reduction in the water and nutrient holding capacity and retention, increase in greenhouse gas emissions, and brings about problems due to droughts and flooding, soil losses due to erosion, and, finally, desertification [6,25,26].

Climatic Change

Agricultural production is also, more recently, affected by climatic change, mainly by the increased frequency of extreme climatic events, which have their effect on food production. Once again, the fragile ecosystems, for example in the tropics, many times with vulnerable populations, are the most affected. Further, the impact of these extreme events is exacerbated by the ongoing soil degradation, which reduces the resilience of the soil and crops to respond to these extreme events. However, conventional tillage agriculture not only suffers from climate change, it also contributes to greenhouse gas emissions with the release of carbon dioxide, methane and nitrogen oxides. Carbon dioxide emissions are increased when soil organic matter is oxidized which occurs at a much faster rate when the soil is intensively disturbed by mechanical means, whether by the use of various types of ploughs, harrows, hoes or spades (Figure 3). Methane and nitrogen oxides come, for instance, from cattle production and flooded rice fields or compacted soils, i.e. conditions, were these greenhouse gases develop as a result of anaerobic fermentation processes. Agriculture also, indirectly, contributes to climatic change with the changes in the land use, as in the case of deforestation. Cutting down trees is not the only problem, but the agricultural use of lands that follows this deforestation with burning of biomass and the mechanical soil tillage, which leads to mineralization of soil organic matter, turning it into carbon dioxide emissions. Including land use change, agriculture contributes to 30% of the emissions causing the climate change [18,27,28].

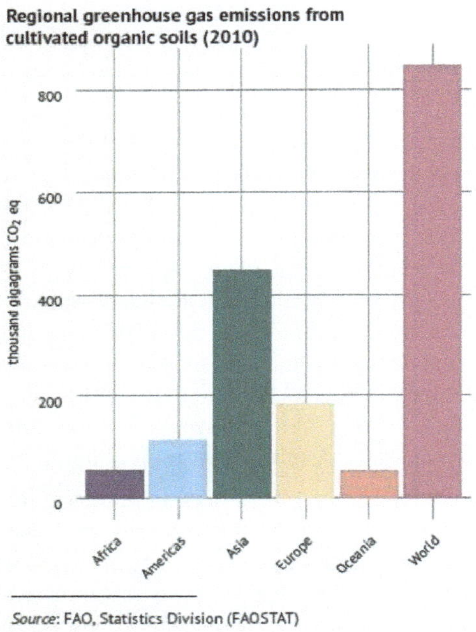

Figure 3. Carbon dioxide emissions from cultivated soils by continent [18]

Biofuel

Another factor that has affected food security during the last decade is the increasing market for biofuel. Production of biofuel crops increased up to 500% between 2001 and 2011 as a consequence of the increase in fuel prices and environmental considerations (Figure 4). In some cases, food commodities are destined directly to biofuel production. In other cases, biofuel is produced from non-food crops [29]. Nevertheless, all biofuel production competes with the food production for resources such as water, soil, workers, and technology. This way, biofuels can reduce food availability and access to food due to price increase. In both cases, food security is affected. As a response to climate change and sustainable land use biofuels also provide a questionable contribution: the carbon which is burned in biofuels for energetic purposes cannot be used for improving soil health, structure and for increasing the soil carbon pool [30,31].

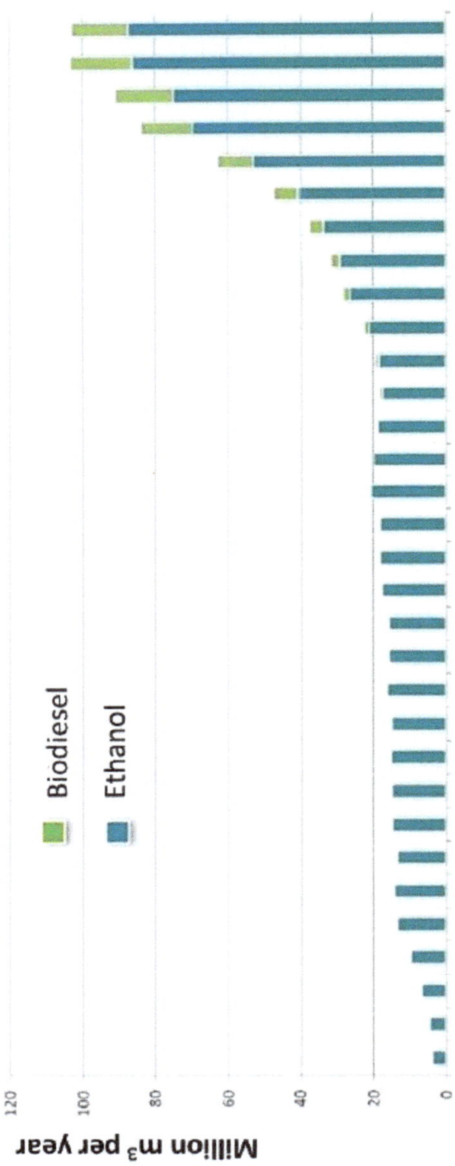

Figure 4. Biofuel production trend from 1990 to 2011 for Ethanol and Biodiesel [27]

Food Preferences

With the increase of income level, there is a change of food preferences towards products of animal origin, like meat and dairy products. This phenomenon mainly appears in growing and diversifying economies, such as China and India. With the increase of population in these countries, the increase in income, urbanization tendencies and change of food preferences, it is estimated that, until 2022, 80% of the market for meat production will be in developing countries. The biggest problem is the availability of natural resources: to assure the same nutrition with a diet of animal origin, it would take seven times the resources that would be necessary for a vegetable diet. Considering that the necessary food production increase to feed the growing population is a challenge due to the limited resources, the change of food preferences increases this challenge [12].

PROPOSITIONS FOR FACING THE CHALLENGES

As a strategy for facing the challenges of food security, countries should try to intensify production of the necessary food staples in their own territory, in order to reduce the import dependency and market volatility.

In this sense, the family farming by small- and large-scale farmers is a very useful agricultural production form, which is managed by the family unit carrying out the farm production relying mainly on family labor. This production form contrasts to investment or corporate type farm enterprises, with the main purpose of investment recovery and profit maximization. Nowadays, family farming is not only the highest contributor to food production at a global level, but also the predominant way of making a living in rural areas of many countries. Family farming, apart from assuring an economic income and improving the livelihood of the farmer's family, also has a long term interest. Therefore, it may care more about natural resources for future generations rather than only focusing on the increase of a short term economical profit, but also on the reduction of risks, for instance, with a diversified production. In times of climate change, this is a better coping strategy than specialized production and mono-cropping [32].

However, due to major weaknesses in the conventional tillage agriculture as was explained in the earlier sections, continuing to rely on 'business as usual' for further intensification of production is not an option for meeting future food needs sustainably. An alternate paradigm based on an ecosystems approach must contribute to meeting future food needs as explained in the following sections.

Sustainable Intensification Paradigm

In order to face the challenges of intensifying food production, necessary for feeding future generations, like saving and protecting limited natural resources, the United Nations' Food and Agriculture Organization (FAO) suggests a change of paradigm: from an agricultural production that follows the model of the conventional Green Revolution agriculture to a sustainable intensification model [4,33]. The Green Revolution paradigm defines a production intensity through the use of inputs and considers the environmental damage as inevitable collateral effects of this production, which can be reduced but not completely avoided. This model does not seem to respond to future challenges, as the increasing level of global agricultural soil degradation keeps demonstrating.

The sustainable intensification paradigm uses natural ecosystem processes for reaching a high and intensive production, and uses technologies and inputs, only in a way that they do not interfere permanently with nature. While equal or higher levels of production are reached, compared to the Green Revolution agriculture, the environmental impacts of the new paradigm can be kept below the threshold of the natural recovery of ecosystems. In this way, besides achieving a highly intensive production, the protection of natural resources and the improvement of environmental or ecosystem services are possible at the same time and in the same geographic area [6,34].

The main focus in this alternate system is on soil and soil functions, being the most vulnerable components of the agricultural ecosystem because of their extremely long time for recovery. Therefore, the sustainable intensification is based on an agricultural system of no tillage, known as Conservation Agriculture (CA). It is defined by three interrelated principles with synergetic effects that are permanently and simultaneously applied [15,35] using locally formulated practices:

- No or minimum disturbance of soil permanently with no tillage and direct sowing.
- Constant cover of soil surface with organic matter, for instance, plants, crop residues or stubbles.
- Crop diversity in sequence, rotation or association with annuals and perennials including legumes.

The above three principles of Conservation Agriculture can be applied in practice, along with other good agricultural practices of crop, nutrient, water, pest and energy management, to all crops in all land-based production systems, and at all operation scales, from the small farmer that uses manual implements up to the great extensions of land where big machinery is used. This type of agriculture, when adequately established, results in an increase of production, at reduced cost and input use, and in the increase of the biodiversity within the systems. It keeps soil and water in the watersheds, facilitates the competitiveness of agricultural production and adaptation to climate change [36,37,38]. Compared to tillage based systems it also facilitates the integration of other production sectors, such as livestock production and agro-forestry systems, orchards and plantations, and mixed systems. It also reduces environmental impacts and vulnerability of production systems during times of climate change, compared to specialized systems, and increases agricultural production and family income of producers [34].

Conservation Agriculture characterizes a paradigm change towards a true agro-ecology and a sustainable intensification. It is actually the core concept for this new paradigm and the reason which makes the combination of high production intensity, high delivery on ecosystem services and low ecological footprint possible [30,34,39]. CA is promoted by FAO as can be seen in the publication "Save and Grow" [11,33]. There is extensive empiric evidence for the benefits of Conservation Agriculture in the actually accelerated exponential adoption rates in many parts of the world, happening mostly under the leadership of farmers' organizations. Conservation Agriculture was practiced in 2015 on 157 Mha, corresponding to 11% of the global cropland with an annual growth rate of nearly

10 Mha. It is practiced in all land-based agro ecologies, with 50% of the CA areas distributed in developing and 50% in industrialized regions [40,41].

Changes of Paradigm in Agriculture and in the Economic Models

These changes applied in the agricultural production system do not only represent the inclusion of some good practices, but rather represent a change of paradigm, including of economic models sustaining the food and agriculture systems. An intensive production should no longer be defined by the use of inputs alone, but by its production performance and sustainability. Many economic concepts that are currently accepted, may change, such as in the case of economics of scale or the preference for a specialized production, in single crop or livestock production.

A diversified no-till CA production, and the greater national and international food and agriculture systems built upon it, could be more interesting economically in the longer term, reducing the production costs, as well as the overall production risks and negative externalities. Such a diversified production model can also improve food and nutritional security and protect natural resources with lower costs and higher profit [42]. With an integration of different production sectors, particularly the crop-livestock integration, ideally complemented by tree components, such diversified productions systems might also respond better to the consumer preferences regarding products of animal origin, avoiding the environmental challenges created by concentrated and specialized livestock production systems [43].

Thus the economics equation of the new paradigm does not only include yield that is produced with maximum efficiency and resilience, but also minimum environmental cost and with sustainability, in order to achieve the desired level of intensification. However, without the support of science, policy and institutions, this paradigm change is not possible [42].

CONCLUSIONS

The world is now facing a serious challenge of moving away from the degrading conventional tillage agriculture to a different paradigm of Conservation Agriculture to reverse the trends of land degradation and losses on productivity, re-establish sustainable production systems for intensification, and build efficient and resilient food and agriculture systems for the future to meet food security and industrial needs. Conservation Agriculture, which combines the ability to intensify sustainably with the delivery of ecosystem services is now (2013 update) spread across some 157 Mha globally, and is increasing at the annual rate of 10 Mha. Food security in all countries stands to gain from the adoption of Conservation Agriculture. In the developing regions, CA offers greater yields and profit with less resources and decreased land degradation. In industrialized regions, CA can improve profit, sustainability and efficiency at high yields with less land degradation and environmental pollution. Policy and institutional support (including science and education) is needed to advance the spread of CA worldwide and establish food security as a function of sustainable production intensification.

REFERENCES

1. FAO (1996) The Rome Declaration on World Food Security; The World Food Summit. FAO, Rome. Available from: http://www.fao.org/wfs/
2. Royal Society (2009) *Reaping the benefits: Science and the sustainable intensification of global agriculture*. RS Policy Document 11/09. London.
3. Foresight (2011) The Future of Food and Farming. The Government Office for Science, London.
4. FAO (2011) Save and Grow: a policymaker's guide to the sustainable intensification of smallholder crop production. FAO, Rome. 102.
5. MEA (2205) *Millennium Ecosystem Assessment Reports*. World Resource Institute. Washington D.C., Island Press.
6. Kassam A, Basch G, Friedrich T, et al. (2013) Sustainable soil management is more than what and how crops are grown. In: R. Lal and R. A. Stewart (Eds.). *Principles of Soil Management in Agro-ecosystems*. Advances in Soil Science. CRC Press.
7. WDR (2008) Agriculture for Development. World Development Report. Washington, DC: World Bank.
8. IAASTD (2009) *Agriculture at the crossroads*. International assessment of agricultural knowledge, science and technology for development. Island Press, Washington D.C.
9. Godfray C, Beddington JR, Crute IR, et al. (2010) Food security: The challenge of feeding 9 billion people. *Science* 327: 812-818.
10. FAO (2012) World Agriculture towards 2030/2050. The 2012 revision. N. Alexandratos and J. Bruinsma (eds). FAO, Rome.
11. FAO (2012) State of Food and Agriculture (SOFA) 2012 - Investing in agriculture for a better future. FAO, Rome.
12. FAO (2013) State of Food and Agriculture (SOFA) 2013 - Food systems for better nutrition. FAO, Rome.
13. FAO (2013) State of Food Insecurity (SOFI) 2013 – The multiple dimensions of food security, FAO, Rome.
14. FAO (2015) State of Food Insecurity (SOFI) 2015 - Meeting the 2015 international hunger targets: taking stock of uneven progress. FAO, Rome.
15. FAO (2015) State of Food and Agriculture (SOFA) 2015 - Social protection and agriculture: breaking the cycle of rural poverty. FAO, Rome.
16. UN (2015) Open Working Group proposal for Sustainable Development Goals, Full report of the Open Working Group of the General Assembly on Sustainable Development Goals, A/68/970. Available from: http://undocs.org/A/68/970
17. FAO (2014) Panorama de la Seguridad Alimentaria y Nutricional en América Latina y el Caribe 2014, FAO, Santiago de Chile
18. FAO (2014) Manual de Agricultura Climaticamente Intelligente. FAO, Rome.
19. The Global Food Security programme (2015) Extreme weather and resilience of the global food system Final Project Report from the UK-US Taskforce on Extreme Weather and Global Food System Resilience, The Global

Food Security programme, UK. 20.

20. Winterbottom R, Reij C, Garrity D, et al. (2013) "Improving Land and Water Management." Working Paper, Installment 4 of Creating a Sustainable Food Future. Washington, DC: World Resources Institute. 44.
21. Friedrich T (2007) Degradation of Natural Resources and Measures for Mitigation; in: APCAEM/MoA-PRC (Eds.): Handbook of the International Seminar on Enhancing Extension of Conservation Agriculture Techniques in Asia and the Pacific, Asian and Pacific Centre for Agricultural Engineering and Machinery (APCAEM) and Ministry of Agriculture of the Peoples Republic of China, Zhengzhou China, 24-26 Oct. 2007. 47-61.
22. SAI (2010) Water Conservation Technical Brief, TB 9 - Use of a conservation tillage system as a way to reduce water the footprint of crops; The SAI Platform, 22.
23. Montgomery D (2007) Dirt: The erosion of civilizations. University California Press, Berkeley and Los Angeles. 287.
24. Gibbs HK, Salmon JM (2015) Mapping the world's degraded lands. *Appl Geogr* 57: 12-21.
25. Friedrich T, Kassam AH, Shaxson F (2009) Conservation Agriculture. In: Agriculture for Developing Countries. Science and Technology Options Assessment (STOA) Project. European Parliament. European Technology Assessment Group, Karlsruhe, Germany.
26. Kassam A, Friedrich T, Shaxson F, et al. (2009) The spread of Conservation Agriculture: Justification, sustainability and uptake. *Int J Agric Sust* 7: 292-320.
27. HLPE (2012) Climate change and food security. A report by the High Level Panel of Experts on Food Security and Nutrition of the Committee on World Food Security. Rome.
28. IPCC (2014) Climate Change 2014: Impacts, Adaptation, and Vulnerability. Part A: Global and Sectoral Aspects. Contribution of Working Group II to the Fifth Assessment Report of the Intergovernmental Panel on Climate Change Field, C.B., V.R. Barros, D.J. Dokken, K.J. Mach, M.D. Mastrandrea, T.E. Bilir, M. Chatterjee, K.L. Ebi, Y.O. Estrada, R.C. Genova, B. Girma, E.S. Kissel, A.N. Levy, S. MacCracken, P.R. Mastrandrea, and L.L. White (eds.). Cambridge University Press, Cambridge, United Kingdom and New York, NY, USA, 1132.
29. FAO (2008) State of Food and Agriculture (SOFA) 2008 - Biofuels: prospects, risks and opportunities. FAO, Rome.
30. Haugen-Kozyra K, Goddard T (2009) Conservation agriculture protocols for green house gas offsets in a working carbon markets. Paper presented at the IV World Congress on Conservation Agriculture, 3-7 February 2009, New Delhi, India.
31. Corsi S, Friedrich T, Kassam A, et al. (2012) Soil Organic Carbon Accumulation and Greenhouse Gas Emission Reductions from Conservation Agriculture: A literature review, Integrated Crop Management Vol.16. FAO, Rome. 101.
32. FAO (2014) Agricultura Familiar en América Latina y el Caribe: Recomendaciones de Política; S. Salcedo y L. Guzmán (eds.). FAO, Chile.
33. FAO (2015) Save and Grow in practice: maize, rice and wheat. FAO, Rome.
34. Kassam A, Friedrich T (2012) An ecologically sustainable approach to agricultural production intensification: Global perspectives and developments. *Field Actions Sci Rep* Special Issue 6.
35. FAO. What is CA? Conservation Agriculture Website of FAO. 2013. Available from: www.fao.org/ag/ca/1a.html
36. Saturnino HM, Landers JN (2002) The Environment and Zero Tillage; APDC-FAO, Brasilia, Brazil UDC 504:631/635, CDD 631.521
37. Baig MN, Gamache PM (2009) The Economic, Agronomic and Environmental Impact of No-Till on the Canadian Prairies. Alberta Reduced Tillage Linkages. Canada. 134.
38. Lindwall CW, Sonntag B (Eds) (2010) Landscape Transformed: The History of Conservation Tillage and Direct Seeding. Knowledge Impact in Society. Saskatoon: University of Saskatchewan. 219.
39. Mello I, van Raij B (2006) No-till for sustainable agriculture in Brazil. *Proc World Assoc Soil Water Conserv* P1: 49-57.
40. Kassam A, Friedrich T, Derpsch R, et al. (2015) Overview of the Worldwide Spread of Conservation Agriculture. *Field Actions Sci Rep* 8.
41. FAO. AquaStat: CA Adoption Worldwide. 2016. Available from: www.fao/ag/ca/6c.html
42. Gonsalves J, Campilan D, Smith G, Bui VL, Jimenez FM (Eds.) (2015) Towards Climate Resilience in Agriculture for Southeast Asia: An overview for decision-makers. Hanoi, Vietnam: International Center for Tropical Agriculture (CIAT). CGIAR Research Program on Climate Change, Agriculture and Food Security (CCAFS). 450.
43. FAO (2010) An international consultation on integrated crop-livestock systems for development - The way forward for sustainable production intensification; Integrated Crop Management Vol.13. FAO, Rome. 63.

CHAPTER 18

DEVELOPMENT AND USE OF MATHEMATICAL MODELS AND SOFTWARE FRAMEWORKS FOR INTEGRATED ANALYSIS OF AGRICULTURAL SYSTEMS AND ASSOCIATED WATER USE IMPACTS

K.R. Fowler[1], E.W. Jenkins[2], M. Parno[3], J.C. Chrispell[4], A.I. Colón[1], R.T. Hanson[5]

[1] Dept. of Mathematics, Clarkson University, 8 Clarkson Avenue, Potsdam, NY 13699, USA

[2] Dept. of Mathematical Sciences, Clemson University, O105 Martin Hall, Clemson, SC 29634, USA

[3] Dept. of Aeronautics and Astronautics, Massachusetts Institute of Technology, 77 Massachusetts Avenue, Cambridge, MA 02139, USA

[4] Dept. of Mathematics, Indiana University of Pennsylvania, 210 South 10th St., Indiana, PA 15705 USA

[5] California Water Science Center, U.S. Geological Survey, 4165 Spruance Road, San Diego, CA 92140, USA

ABSTRACT

The development of appropriate water management strategies requires, in part, a methodology for quantifying and evaluating the impact of water policy decisions on regional stakeholders. In this work, we describe the framework we are developing to enhance the body of resources available to policy makers, farmers, and other community members in their e orts to understand, quantify, and assess the often competing objectives water consumers have with respect to usage. The foundation for the framework is the construction of a simulation-based optimization software tool using two existing software packages. In particular, we couple a robust optimization software suite (DAKOTA) with the USGS MF-OWHM water management simulation tool to provide a flexible software environment that will enable the evaluation of one or multiple (possibly competing) user-defined (or stakeholder) objectives. We introduce the individual software components and outline the communication strategy we defined for the coupled development. We present numerical results for case studies related to crop portfolio management with several defined objectives. The objectives are not optimally satisfied for any single user class, demonstrating the capability of the software tool to aid in the evaluation of a variety of competing interests.

Citation: K.R. Fowler, E.W. Jenkins, M. Parno, J.C. Chrispell, A.I. Colón, R.T. Hanson. Development and Use of Mathematical Models and Software Frameworks for Integrated Analysis of Agricultural Systems and Associated Water Use Impacts. AIMS Agriculture and Food, 2016, 1(2): 208-226. doi: 10.3934/agrfood.2016.2.208.

INTRODUCTION

Groundwater resources in the U. S. are withdrawn and used for a variety of purposes. However, based on data reported by the U. S. Geological Survey (USGS) [1], the majority of groundwater withdrawals, especially in states with a heavy agriculture sector, are used for agricultural irrigation. Recent studies of aquifer levels in these heavily agricultural states show a marked decrease primarily due to over pumping[2,3,4,5,6,7,8,9]. In several regions, the precipitous drop in levels has been exacerbated by extended periods of arguably historic droughts[10,11,12]. Farming entities in these regions have been particularly stressed as they are limited with regards to water availability options they can pursue. Many farmers must now operate under restrictions on their water use, either voluntarily or by law [13,14]. The water restrictions imposed on California farmers have made national headlines; the impact of these restrictions reverberates throughout the country as California is one of the leading producers, both in terms of quantity and selection, of fresh fruits and vegetables in the U. S. California also has some of the largest acreage dedicated to agricultural production in the U. S., and more than half of the vegetables are grown on irrigated land [15]. While recent economic data suggest California farmers have had a record year in terms of profitability [16], the penalty has been the increased use of water resources for the high-revenue, high-water consumption crops that have fed this revenue stream [17].

Our objective in this work is to aid in the decision-making process of farmers and water management agencies via the coupling of mathematical modeling and simulation software with optimization algorithms. Towards this end, we have been developing a flexible software framework to enable farmers and water management agencies to better evaluate the effectiveness of water management strategies relative to objectives connected with stakeholders in an agricultural region.

Gomaa, et.al., provide a brief survey of related work [18], where many of the referenced studies occurred outside the U. S. and covered a range of objective considerations. A more extensive review of works related to the development of models supporting crop management decisions is given in the work by Dury, et.al. [19]. This review paper includes summaries of works where the crop decisions are made based on overall acreage devoted to a single crop versus those works where the crop decisions incorporate spatial considerations, including information on properties such as soil nutrient levels.

More closely related work to this include those that consider evolutionary algorithms and multiobjective approaches to quantify relationships between often competing objectives. Evolutionary algorithms were used to understand trade-offs between sustainability and crop planning in deVoil, et.al., [20], where sustainability was used in both an economic and resource allocation context. Multiobjective optimization of farming practices, including objectives based on crop rotation and environmental farm planning, has also been the focus of several related works[21,22,23,24,25].

Our approach is the first to incorporate the USGS One-Water Hydrologic Flow Model (MF-OWHM, [26]) as a simulation tool coupled to external optimization algorithms. Our previous analysis utilized the predecessor to MF-OWHM, the MODFLOW Farm Process Model (MF-FMP2, [27]). MF-FMP2 and MF-OWHM are agriculturally-focused water management softwares, with the latter offering extended support for the analysis of a wide-range of conjunctive-use issues within a given region. Details on MF-OWHM can be found in Section 3. We chose this software as our simulation workhorse for several reasons. First, the USGS MODFLOW water simulation software is widely used and well respected. Water management was the primary concern of our farming partners in California who were responsible for the genesis of our study. Second, both MF-OWHM and MF-FMP2 have been used extensively in a variety of contexts to study water resource management in heavily farmed areas, where conjunctive use analysis is required to represent the interests of all of the stakeholders in the region [28–35]. Finally, MF-OWHM supports a range of mechanisms for predicting water usage based on climate and plant growth characteristics, all of which enables us to easily evaluate increasingly complicated objective functions with physically realistic parameter spaces.

We utilize the suite of optimization tools available within the DAKOTA optimization package [36], developed and maintained by researchers at the U. S. Department of Energy Sandia National Laboratory. DAKOTA was chosen because of its capabilities in handling simulation-based optimization problems. Users only need to supply subroutines to evaluate the objective functions and constraints without providing any gradient information. Moreover, DAKOTA has a variety of different optimization algorithms allowing for future studies to analyze algorithm performance and suitability for these types of problems.

The objective functions used in this work use output from MF-OWHM. That is, each time the objective function is evaluated for a set of possible crop distributions, new data files must automatically be generated during the course of the optimization using the points the algorithm has chosen. A major contribution of this work is a framework consisting of a set of wrappers, written in Python, that facilitate communication between the simulator and the optimizer. We demonstrate the capability of the framework to choose appropriate crop distributions for planting over a two year time horizon using only ground water as the source of irrigation. This work is a significant improvement from previous efforts in that planting decisions can be made dynamically once a crop is harvested within that two year period [37]. We explain the decision making model, which is governed by the optimization algorithm, in detail below.

We proceed by describing the modeling approach to represent the stakeholders in the agricultural setting, the software framework including the hydrological simulator and the optimization suite, and then provide numerical results. We end with conclusions and describe some future work.

VIRTUAL FARMING MODEL

We use mathematical modeling approaches to represent various stakeholders in an agricultural context [20,23,37,38,39]. In this work, our modeling focus is on an individual "virtual" farmer who needs to evaluate a potential crop portfolio based on his need to generate a profit as well as meet a demand specified from industrial partners or membership in a farming cooperative. We emphasize, however, this is a particular choice made for this presentation and the framework can easily be manipulated to consider decisions from other perspectives (e.g., industrial partners, residential community members, water management agencies, or policy makers). The only requirement is the ability to quantify and appropriately model the primary goal of the represented stakeholder. The modeling effort requires identification of appropriate decision variables, objective functions, and constraints. One must also determine the number of crops for which decisions should be made along with the times at which the crops are planted and harvested. Note planting decisions are seasonal, depending on the calendar month and the growing cycle, making it necessary to find the minimum number of decision variables required to simulate a multi-year planting horizon. From a farming perspective, the purpose of the optimization problem may be to select crops based on trade-offs, e.g., between revenue generation and demand or between revenue generation and water usage. In this case, the optimization tool must be able to make decisions in the same time frame as the farmer.

The objective functions aim to provide a farmer with a suite of feasible farm states that allow for crop selection based on individual farming goals. To this end, let N_{crop} be the number of different crops under consideration. We consider changing the crop distribution after a harvest, which drives the number of decision variables in the model in that each time a crop is harvested and land thus becomes available for planting, a decision must be made as to what to do with that land.

The simulation tool used in this work evolves the farming environment over time using *stress periods*. The key idea is that over a given stress period, model parameters are kept constant, inflows and outflows are specified, and assignments for land use, such as cropping, are also held constant. The actual use and movement of water are changing over smaller time periods. We define the variable $x_i^t \in [0,1]$ to be the fraction of the farm area devoted to crop ii during stress period t. This variable is used in our framework to create input files needed by MF-OWHM (specifically the CID files).

We formulate constraints based on the amount of each crop planted at the beginning of each stress period. Let $p_i^t \in [0,1]$ be the fraction of each farm's area *planted* with crop i at the start of stress period t. Let L_i be the time between planting and harvesting for each crop. Thus, if crop i is planted at the beginning of stress period t, it will be harvested at the beginning of stress period $t+L_i$. We assume that the crops can be planted and harvested instantaneously, so it is possible to both harvest and plant a crop at the beginning of a stress period.

The planting variables pti are related to the fraction of each crop in the ground x_i^t by

$$x_i^t = x_i^{t-1} + p_i^t - p_i^{t-L_i}. \tag{1}$$

Assume that the simulation starts with t=1 and that x_i^t =0 for t<1 and p_i^t =for t<1. Thus, given p_i^t for each stress period, x_i^t can be calculated for each stress period. The advantage of framing the problem in terms of p_i^t instead of x_i^t lies in the constraint definitions.

Note that each crop has a specified month it can be planted. Let $P_i \subseteq \{1,2,3,\ldots,12\}P_i \subseteq \{1,2,3,\ldots,12\}$ be the months that crop ii can be planted. For example, if crop ii can only be planted in May or June, $P_i=\{5,6\}$Pi$=\{5,6\}$. These sets define constraints on the planting variables p_i^t. In particular, we have

$$p_i^t = 0 \quad \text{for} \quad t \notin P_i. \tag{2}$$

The constraints in both (1) and (2) can be generated automatically for all ii and t if Pi and Li are known for each of the N_c crops. Thus, in the framework presented here, these variables are included in the crop class.

Our objectives are defined in terms of farm revenue and an industrial demand for specific crops. We seek to maximize the total revenue over the entire time horizon as we simultaneously minimize deviation from industry demand. The demand objective can be in competition with the revenue objective in scenarios where high-demand crops are water-intensive. The simultaneous optimization means each objective function is considered separately, instead of combining them in a weighted, single-objective approach. As the objectives are competing, there is no single solution that simultaneously optimizes each objective. Our multi-objective approach instead provides a set of points, each optimal in a given context, giving stakeholders the ability to analyze trade-offs between possible solutions.

We calculate the revenue generated from a crop portfolio as the sales price times the yield of each crop minus the cost of water required for irrigation. For simplicity we drop the super scripts and use xi to denote the fraction of crop ii planted at a given time period. The revenue generated from that planting can be calculated using

$$\text{Maximize } R_i = [x_i \times Y_i \times C_i \times A] - C_w \times W_{gw}, \tag{3}$$

where Yi is the total yield for crop ii (Weight/L^2), C_i is the sales price of crop ii (\$/Weight), AA is the acreage of the farm (L^2), C_w is the cost of groundwater pumping (\$/L^3) based on volume only, and Wgw is the volume of water extracted from the aquifer (L^3). Alternative revenue models could incorporate seed and labor costs for the distribution of different crops, varying water prices [41], the lift cost associated with groundwater pumping from shallow wells, or linear components that represent penalties or incentives that reflect policy or governance.

We model the demand objective using the l212 norm of the deviation from a specified demand

$$\text{Minimize } D = \|Y_a - Y_d\|_2 = \sqrt{\sum_{i=1}^{N_c} ((Y_a)_i - (Y_d)_i)^2}, \tag{4}$$

where $(Y_a)_i$ is the actual yield (weight) and $(Y_d)_i$ is the demand yield (weight) for crop ii dictated, for example, by an industrial partner or market trend. For any crop, the yield, Y_a, is not calculated explicitly by the simulation tool for this work but can be approximated based on the actual evapotranspiration data provided as output by from the MF-OWHM hydrologic model. To derive Y_a, we use a model given by the Food and Agriculture Organization of the United Nations (FAO) [41]

$$\left(1 - \frac{Y_a}{Y_m}\right) = K_y\left(1 - \frac{ET_a}{ET_m}\right), \quad (5)$$

where Y_m is the maximum yield in unstressed conditions (weight), ET_a is the actual crop evapotranspiration (L/T), K_y is the crop water production response coefficient, and ET_m is the maximum crop evapotranspiration in unstressed conditions (L/T). The values of Y_m for most crops are available in the literature.

We can find the maximum crop evapotranspiration ET_m given a known reference evapotranspiration (ET_0) and crop coefficient (K_c) using

$$ET_m = ET_0 \times K_c. \quad (6)$$

In practice, given a reasonable estimate on the unstressed yield for a given crop and the actual crop evapotranspiration, we can estimate the actual yield for a given crop in both stressed and unstressed conditions, extending the robustness of our results to account for drought scenarios.

Since our decision variables are the fraction of each type of crop, we require

$$\sum_{i=1}^{N_c} x_i \leq 1 \quad (7)$$

at each planting decision. Note this also allows land to go fallow.

COMPUTATIONAL FRAMEWORK

The main contribution of this study is the implementation of optimization algorithms in conjunction with hydrological models and mathematical representations of stakeholders to guide decisions in an agricultural setting. We proceed by describing the USGS MF-OWHM simulation code used to represent the agricultural setting. We then describe the optimization software suite and the framework developed to create the virtual farming model, which can account for dynamical planting decisions. This is achieved by creating a suite of subroutines that assigns the appropriate input (decision variables) and directs the output from the simulation tool to the objective functions during the course of the optimization.

Hydrological and Farm Processing Simulation

The One-Water Hydrologic Flow Model (MF-OWHM, [26]) is a MODFLOW-based (MF-05, [42]) integrated hydrologic flow model that is the most complete version, to date, of the MODFLOW family of hydrologic simulators needed for the analysis of a broad range of conjunctive-use issues. MFOWHM fully links the movement and use of groundwater, surface water, and imported water for consumption by agriculture and natural vegetation on the landscape, and for potable and other uses within a supply-and-demand framework. MF-OWHM is based on the Farm Process for MODFLOW (MF-FMP) [27,43] combined with local grid refinement, streamflow routing, surface-water routing process, seawater intrusion, and riparian evapotranspiration.

MF-OWHM allows not only for head-dependent flows of a traditional groundwater model but also flow-dependent and deformation-flows for a more complete coupling within the hydrosphere. By retaining and tracking the water within the hydrosphere, MF-OWHM accounts for "all of the water everywhere and all of the time." This approach provides more complete water accounting and provides a platform needed to address wider classes of problems such as evaluation of conjunctive-use alternatives, including sustainability analysis, potential adaptation and mitigation strategies, and development of best management practices [44]. MF-OWHM's broader ability to simulate more of the hydrosphere has served as a valuable tool for multiple research and applied modeling projects.

As research tools, both MF-FMP and MF-OWHM have been modified to investigate mathematical techniques, including subsidence feedback on conjunctive use [45], effects of climate change [35,46], crop optimization [37,43,47], water-rights driven surface water allocations [48], and proper orthogonal decomposition model reduction [49,50,51]. MF-FMP and MF-OWHM have also been used to evaluate many applied projects within the U. S. Geological Survey and the private sector [28,29,31,34,49,52,53,54,55,56].

The allocation of crops is input for an MF-OWHM simulation and evapotranspiration and water usage values are extracted at the end of a simulation. These values are then used to calculate revenue and yield using Equations (3) and (4) above. The details of the crops and the physical description of the farm for this work are described later.

Optimization

DAKOTA provides a suite of optimization strategies for a range of simulation-based scenarios and is an ideal framework for this study since it is open-source and flexible [36]. Understanding the applicability of optimization algorithms, in particular evolutionary algorithms, to real-world problems is an active area of research [57]. Since the objectives are competing, there is no single solution that simultaneously optimizes each objective. A multi-objective approach instead provides a set of points, giving stakeholders the ability to analyze trade-offs between points. In general, genetic algorithms move through "generations" of evaluation points by assessing the fitness of members of the generation and selecting members to continue to the next generation (through mutation or cloning), parent offspring for the next generation, or die (i.e. removing those points from the population). In a multiobjective setting the population evolves towards a set in which the points are non-dominated, known as a Pareto set. A non-dominated point has the property that its fitness cannot improve with respect to one objective without degrading the value of another. The basic steps of the algorithm are to initialize a population, evaluate the objective function and constraints, perform crossover and mutation, evaluate the new population members and assess the fitness of each population. Population members are then replaced to continue to the next generation. Termination of the optimization can be based on a maximum number of function evaluations (or iterations) or on performance metrics. The performance metrics track changes in the population from generation to generation. Finally, a post processing step reduces the final solution set so that a minimum distance exists between any two design points.

We use the multi-objective genetic algorithm (MOGA) developed by Eddy [58] as implemented in DAKOTA. This method was shown to perform well on a simplified farming model [37].

Implementation

For this work, DAKOTA and MF-OWHM are linked via objective function evaluations facilitated by a framework of Python wrappers. Figure 1illustrates the configuration of the framework. Initial set up of the physical setting, generation of the data files for the simulator, and assignment of objective function parameters are done with an IPython notebook. We list the specific model parameters in Figure 1 in the upper left box. The objective functions and constraints are defined via a setup script and DAKOTA is used to solve the optimization problem with repeated calls to MF-OWHM for each objective function evaluation. Each simulation requires a new crop distribution, thereby requiring new data files which are created automatically by additional Python scripts. This is demonstrated in the upper right box in Figure 1.

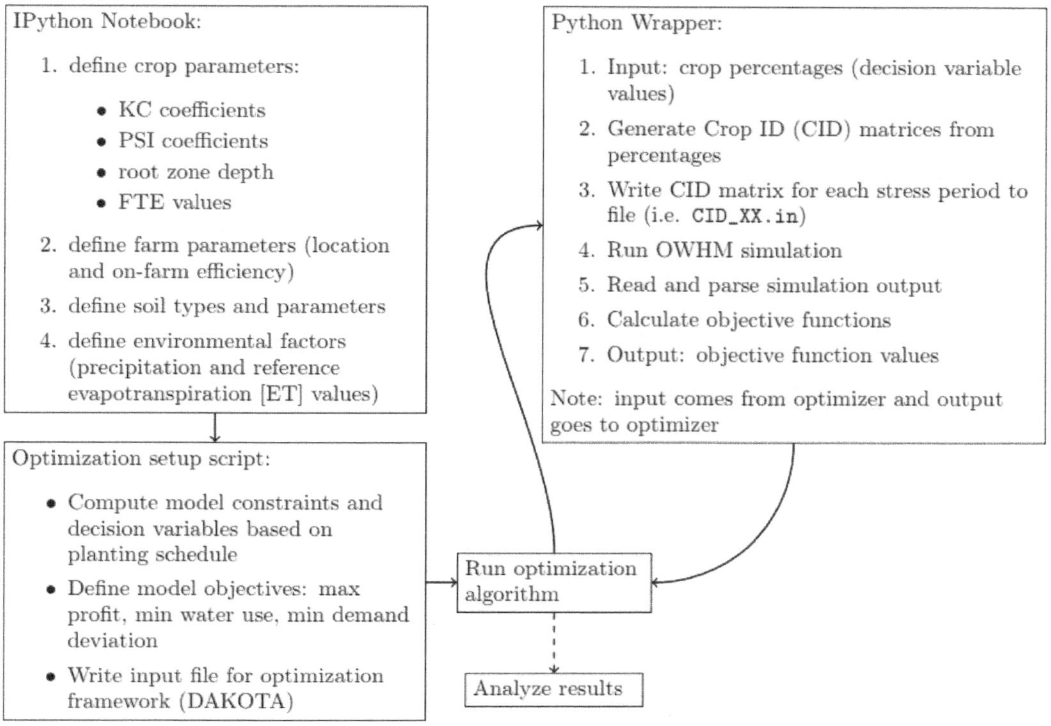

Figure 1. Graphical depiction of the workflow for the software coupling. Preprocessing and communication are handled using Python based tools.

DESCRIPTION OF EXAMPLE PROBLEM

We demonstrate our modeling approach by considering a farm originally implemented in MF-FMP2 and modified for use in MF-OWHM and the virtual farmer as modeled by Equations (3) and (4) above. We consider three crops with competing properties and varying planting and growing schedules to demonstrate how the optimization algorithm can guide the planting decisions.

Agricultural Setting

The problem formulation uses a dataset to represent a farm originally included as a test problem with the suite of MF-FMP2 test problems and later modified for use with MF-OWHM. This farm model is a hypothetical realization of a region with multi-layered hydrology and distinct regions with different water usage requirements [27]. The model includes a 10 km by 11.5 km region with multiple farms, urban zones, riparian zones (interfaces between land and streams), and areas of natural vegetation. The water management system includes multiple wells, stream inflows and outflows, and natural precipitation. The topography slopes downward from west to east and converges from the north and south toward a riparian region along the eastern edge. The underlying geology contains four aquifer layers separated by three layers of confining material [27]. The aquifer nearest the surface is unconfined with varying depth. The remaining (confined) layers are uniformly 60 m thick, with each layer of confining material between 5 m and 15 m thick. The saturated hydraulic conductivity (a measure of the ability of the aquifer to transmit fluid) varies from 10 m/day in the aquifer nearest the surface to 0.15 m/day in the aquifer furthest from the surface.

The example region is divided by a stream flowing west to east. The stream flow into the domain is prescribed at 50,000-100,000 m^3/day. No fluid flow is allowed into the region through the northern and southern boundaries. The eastern and western boundaries are general head boundaries. These specify a head-dependent flux designed to mimic known groundwater head at a specified distance. The topography and boundary conditions dictate a west-to-east directional groundwater flow.

The model domain consists of a 20×23 grid, where each cell within the grid is associated with a specific farm type. The production farm is modeled using a 10x10 block of cells in the upper left corner of the domain. The riparian vegetation zone is comprised of a block of 25 cells lying along the east boundary, and the remaining cells are associated with native vegetation landscape. A schematic of the model domain is given in Figure 2, with the different regions represented by distinct colors. The orange cells denote the production farm, the green cells denote the riparian region, and the dark gray cells denote the native vegetation. The blue circles show the locations of the wells, while the blue line represents a stream. The production farm is the focus of the simulation, but the properties associated with the surrounding landscape also affect the dynamics of the simulation [40].

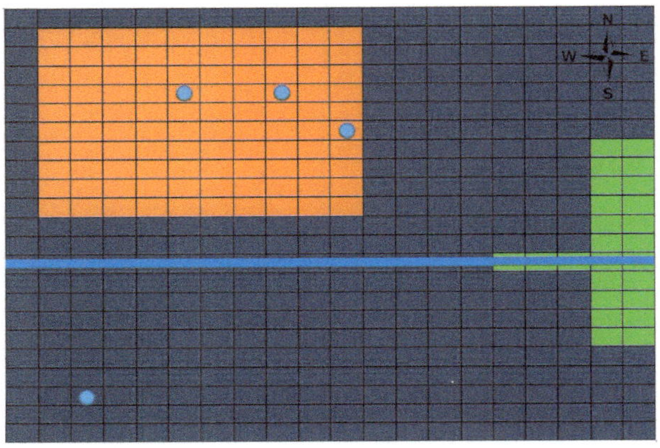

Figure 2. Diagram of test problem farm configuration. Orange cells are farm regions, dark gray cells are native vegetation, and green cells are riparian vegetation. The blue line represents a river, and blue circles denote wells.

In practice, the physical composition of the subsurface impacts the optimization as soil properties govern runoff and inefficiencies in the irrigation system. Our test problem consists of three different soil types which are predefined in MF-FMP and MF-OWHM. The soil types are silt, sandy loam, and silty clay. Their distribution is shown in Figure 3.

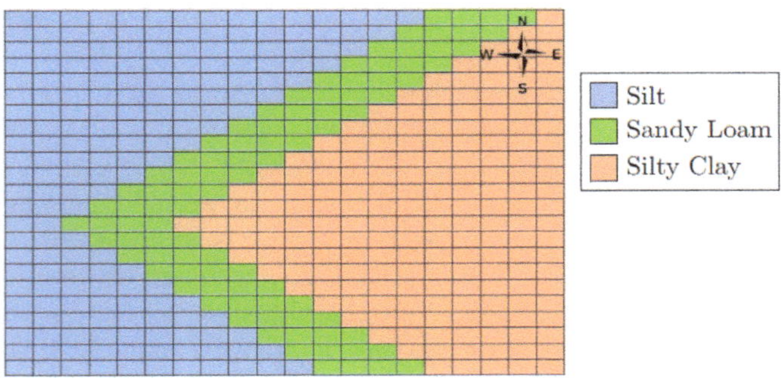

Figure 3. Distribution of soils throughout the model domain.

Several adjustments were made to allow for a more definitive assessment of the performance of the optimization algorithm and the coupling strategy. In particular, we reduced the number of farms to three; an agricultural farm, a riparian zone, and an area of native vegetation. We consider five crops: three are agricultural products of the farm (our decision variables), one is the vegetation in the riparian zone, and the final is the native vegetation. In this model, crops receive water through groundwater irrigation or precipitation only, with a uniform distribution across all crops. Four extraction wells are included in the model and all other water delivery options were not used. We incorporate precipitation data taken from a weather station in Watsonville, CA from January 2013 – December 2014. The volumetric flow rates for the precipitation are shown in Figure 6 (the dashed line) which varies significantly over the two years and includes two periods of little to no precipitation.

Crops

For this demonstration, we consider three crops over a two year time frame. We allow land to go fallow as an implicit fourth crop. That is, the production crops are not forced to occupy the total farm area, leaving any unoccupied acreage fallow. We assume that the water and sales prices do not vary significantly, but varying model parameters are a focus of future work. Further, we chose the three crops with differing properties to demonstrate the trade-off analysis for decision making. Crop 1 is a water intensive but high profit crop with low demand. Crop 1 can be planted in April or May and is in the ground for 4 months. Crop 2 uses the least amount of water, has medium demand, and is in the middle with regards to profitability. Crop 2 can be planted in January or June stays in the ground for 2 months. Finally, Crop 3 is the highest demand crop, has medium water usage, but is low profit in comparison with the other crops. Crop 3 can be planted only in June and is in the ground for 5 months. We show the planting possibilities over the two year horizon in Figure 4. Here pti refers to the fraction of farmland planted with crop ii at stress period t.

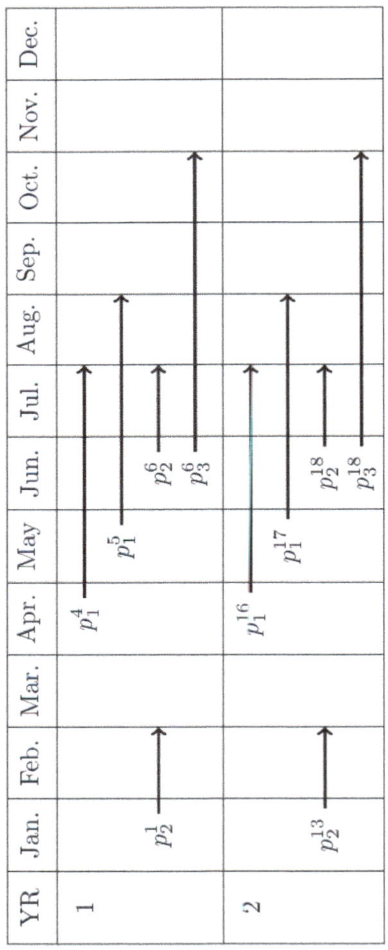

Figure 4. Planting schedule for crops 1, 2, 3 over the 2 year period.

Specific model parameters for the crops used in our simulation are provided in a database downloaded with MF-OWHM and [59,60,61] (We loosely base them on vegetable row crops, strawberries, and artichokes). The model parameters describing the attributes are shown in Table 1.

Table 1. Model parameters for each crop. The relative values between crops for each parameter give each an advantage with respect to a distinct objective

Crop	1	2	3
Max Yield (Y: Metric Ton/acre)	8.0	24.5	8.8
Sales Price (C_i: $/Metric Ton)	150.47	10.16	5.58
Demand (Y_d: Metric Tons)	3,300	19,700	53,750

NUMERICAL RESULTS

MOGA is a heuristic algorithm with randomness in its search phase; therefore, it is ideal to run the optimization multiple times. We performed 13 optimization trials using the default algorithm parameters including a mutation rate of 0.1 and a crossover rate of 0.8. The numerical experiments were done on a machine running Ubuntu 14.04.3 LTS with 16x AMD Opteron (TM) Processor 6212 and 32G of Memory. We use DAKOTA version 6.3 released November 15,2015, and OWHM 1.0 Version 1.0.11 released May 5, 2015 for all simulations. A single optimization experiment took roughly two hours.

Figure 5 shows a typical trade-off curve generated by the optimizer. Each point on the curve corresponds to a specified planting portfolio so that a farmer could select a distribution of crops based on a personal preference in balancing the demand and profit objectives. In particular, notice that the objectives are indeed competing in that deviating from the demand leads to an increase in profit. We note the demand deviation measures the difference between what the farm yields and what we have specified as a demand for the corresponding crop (i.e., Y_a -Y_d in Equation (4)). Recall with competing objectives there is no single solution that will optimize all the objective functions. DAKOTA does identify a "best" point in the Pareto set defined in terms of distance from the so-called utopia point (labelled above with a diamond and a star). The utopia point is defined as the point of extreme best values for each objective. The best point is then the point on the Pareto front closest to the utopia point. An example from the DAKOTA manual [36] highlights this concept. The best point as well as the one that gave a higher profit for one of the optimization runs are given in Table 2. The second point, (indicated with a triangle) was chosen by scaling the objective functions to both be between zero and one and then selecting the point closest to the Utopia point. A benefit of using multi-objective optimization is that a user can consider the trade-offs between the solutions on the Pareto front.

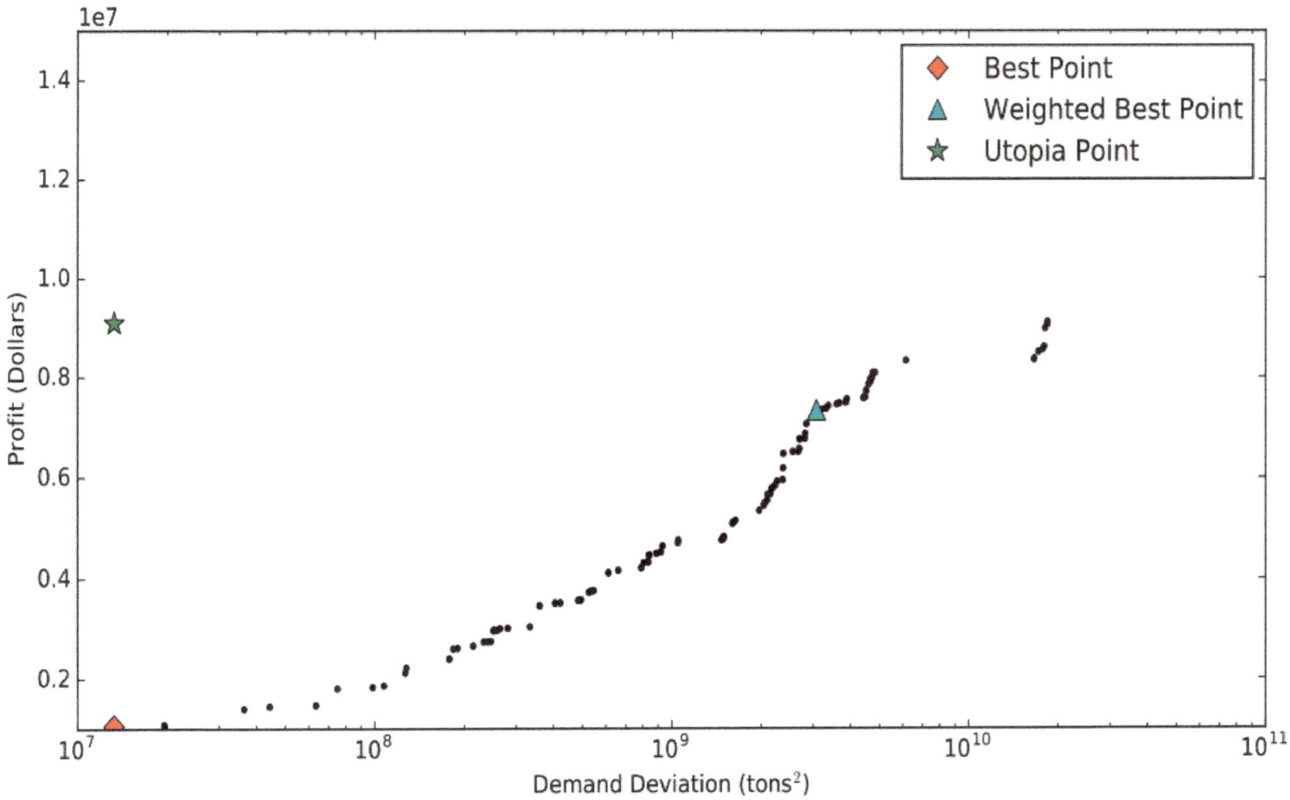

Figure 5. Representative trade-off curve from data in Table 2.

Table 2. Fractions of Crops 1, 2, 3 and the resulting profit, deviation from demand, and water usage. The superscript on p_i denotes the stress period

	Example Best Point	Weighted Best Point	Mean (Best Point)	Std. Dev. (Best Point)
p_1^4 (April Yr 1)	0.06	0.41	0.08	0.0617
p_1^5 (May Yr 1)	0.46	0.43	0.11	0.1284
p_1^{16} (April Yr 2)	0.00	0.00	0.09	0.0929
p_1^{17} (May Yr 2)	0.14	0.14	0.09	0.0548
p_2^1 (Jan Yr 1)	0.37	0.37	0.33	0.0900
p_2^6 (June Yr 1)	0.08	0.06	0.10	0.1066
p_2^{13} (Jan Yr 2)	0.02	0.74	0.56	0.3360
p_2^{18} (June Yr 2)	0.21	0.14	0.17	0.1129
p_3^6 (June Yr 1)	0.37	0.09	0.36	0.1478
p_3^{18} (June Yr 2)	0.60	0.69	0.34	0.2016
Profit ($)	1,053,365.84	7,337,207.88	786,983.68	184,460.12
Demand Deviation (Tons)	3,640	55,439	2654	1922
Water Use (m^3)	6,376,770.85	6,723,169.86	6,452,116.95	543,924.99

The crop portfolios are the same in terms of the fraction of Crop 1 for the third and fourth planting as well as the first planting of Crop 2. Note Crop 1 was a low demand, high water use crop. The precipitation during the time for the April and May plantings of Crop 1 in the second year was low (see Figure 6), requiring increased groundwater pumping and hence increased cost. Thus, the optimizer selected low percentages of this crop to appropriately balance the competing objectives.

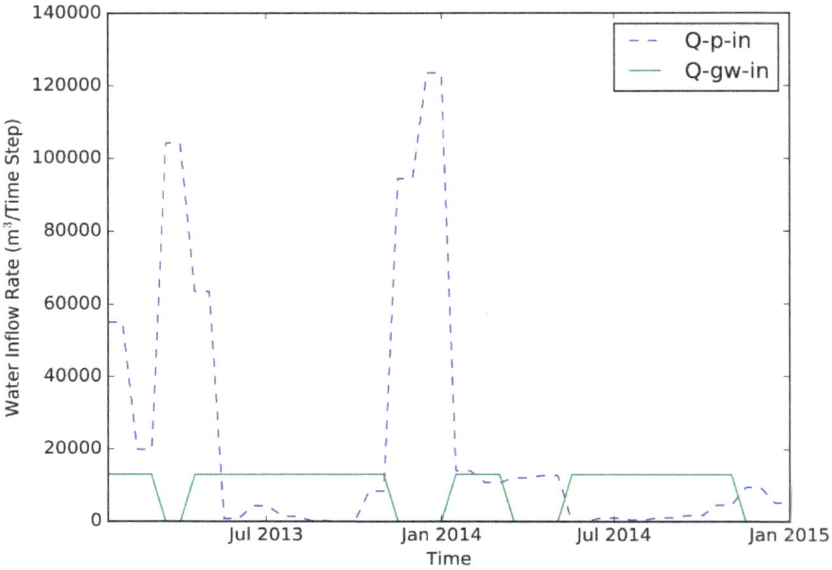

Figure 6. Volumetric water usage from wells and rainfall events over the two year horizon. Q_p denotes precipitation and Q_{gw} denotes groundwater usage.

The portfolios differ most in the amount of Crop 2 for its third planting. Crop 2 was a low water, medium demand, and high profit crop. Crop 2 is planted in January and in year 2 at that time, there is a significant precipitation event (see Figure 6). They also differ significantly in Crop 3 (the high demand crop) during its first planting. Since the best point planting portfolio is also trying to satisfy the demand objective and since the only possible planting for January is Crop 2 (leading to a profit), these differences make sense.

For any given crop portfolio, a post-processing simulation can be run to analyze water usage. Figure 6 shows the volumetric flow rate for the precipitation events over the two year period and the extraction wells for the farm. The figure highlights that during rainfall events, the wells shut down; otherwise, they are operating at their maximum capacity. The crop fractions used to generate this plot were from Column 2 in Table 2 (the high profit point).

CONCLUSIONS

This work is motivated by a need to understand, in a quantifiable way, how decisions made by stakeholders in an agricultural community impact water usage and revenue considerations for both individuals farmers and the entities purchasing their harvests. We have proposed a computational framework which uses dynamic planting rules and a simulation-based optimization approach to create a virtual farmer. The framework uses Python wrappers to manage the communication between the conjunctive-use hydrological simulator MF-OWHM and the DAKOTA optimization suite. We considered two competing objectives, minimizing deviation from industry demand and maximizing profit, and applied the MOGA to search the design space and present feasible solutions. We demonstrated the capabilities of this approach on a test problem with a realistic farming scenario and three crops with competing properties. Sets of solutions in the form of trade-off curves provide stakeholders with options when selecting a crop planning portfolio to balance their personal preferences. In particular, this work is an improvement over the initial use of DAKOTA coupled with MF-FMP2 proposed in Fowler, et.al., [37] in that a crop distribution could only be determined once at the beginning of a simulation. In that work, for example, the distribution of crops was defined at the beginning of the two-year horizon and the same crops were always planted when land became available. The approach described here is more realistic and allows the optimization algorithm to select new crops to be planted once land opens up from a harvest.

More important, however, is the ability of the framework to facilitate a wide range of future studies. More realistic objective functions that represent additional stakeholders, such as citizens or policy makers can easily be implemented. For example, the crop that is in highest demand could be weighted more heavily in that objective. Varying water prices and sales prices would also impact the solutions. In addition, DAKOTA has a suite of optimization algorithms and sensitivity analysis tools that can be applied to better understand the nature of the problems.

Finally, each component is open source, and the communication mechanism is defined in an object oriented context. These communication tools have allowed us to take advantage of the significant research contributions from the USGS and Sandia National Laboratories. The object-oriented nature of the tools allows us to easily add more components to a crop class and incorporate new parameters into a given study. Future work includes development of graphical user-interfaces enabling users to easily define distinct objectives or new crop information, and studies on larger regions with several farmers and urban entities accessing the same water sources for their needs. We also note we only use groundwater sources for irrigation in this work; consideration of all possible water delivery systems is a direction of future research.

Table 3. Table of notation. Units used in this work are in parentheses

Notation	Description	Units
x_i^t	fraction of farm area devoted to crop i during stress period t	no units
p_i^t	fraction of farm area planted with crop i at stress period t	no units
R_i	revenue from crop i	\$
Y_i	total yield for crop i	weight/L^2 (tons/acre)
C_i	sales price of crop i	\$/weight (\$/tons)
A	acreage of the farm	L^2 (m^2)
C_W	cost of groundwater pumping	\$/L^3 (\$/m^3)
W_{gw}	volume of extracted water	L^3 (m^3)
Y_a	actual yield	weight (tons)
Y_d	demand yield	weight (tons)
K_y	crop water production response coefficient	no units
ET_a	actual crop evapotranspiration	L/T (m/day)
ET_m	maximum crop evapotranspiration	L/T (m/day)
ET_0	reference evapotranspiration	L/T (m/day)
K_C	crop coefficient	no units
N_C	number of crops	no units

ACKNOWLEDGMENTS

We would like to thank the American Institute of Mathematics for facilitating and supporting this collaboration. We would also like to thank the McNair Scholars program for their support of this work. We thank Corey Ostrove of the Applied Research Laboratories, The University of Texas at Austin, and Matthew Farthing of the Engineer Research and Development Center, U. S. Army Corps of Engineers, for their initial efforts on this framework. We would also like to thank USGS internal reviewers and the anonymous referee for the journal for suggestions that improved this manuscript, and we thank Dr. Alexander Schreiber for his assistance in generating figures.

REFERENCES

1. M. Maupin, J. Kenny, S. Hutson, J. Lovelace, N. Barber and K. Linsey, (2014), Estimated use of water in the United States in 2010, Circular 1405, U.S. Geological Survey, Available from: http://dx.doi.org/10.3133/cir1405.
2. 2.I. James and S. Reilly, (2015), Pumped beyond limits, many U.S. aquifers in decline, The Desert Sun.
3. J. Thomas, G. Stanton, J. Bumgarner, D. Pearson, A. Teeple, N. Houston, J. Payne and M. Musgrove, (2013), A conceptual hydrogeologic model for the hydrogeologic framework, geochemistry, and groundwater-flow system of the Edwards-Trinity and related aquifers in the Pecos County region, Texas, Fact Sheet 2013-3024, U.S. Geological Survey.

4. F. Morris, (2013), Western Kansas farmers face dwindling water supply, Available from: http: //hereandnow.wbur.org/2013/08/19/kansas-farmers-water.

5. D. Steward, P. Bruss, X. Yang, S. Staggenborg, S. Welch and M. Apley, (2013), Tapping unsustainable groundwater stores for agricultural production in the High Plains aquifer of Kansas, projections to 2110, Proceedings of the National Academy of Sciences, 110: E3477–E3486.

6. J. Peterson and Y. Ding, (2004), Economic adjustments to groundwater depletion in the High Plains: Do water-saving irrigation systems save water?, Am. J. Agr. Econ., 87: 147–159.

7. V. McGuire, M. Johnson, R. Schie er, J. Stanton, S. Sebree and I. Verstraeten, (2002), Water in storage and groundwater management approaches, High Plains Aquifer, 2000, Circular, U.S. Geological Survey.

8. J. Musick, F. Pringle,W. Harman and B. Stewart, (1990), Long-term irrigation trends – Texas High Plains, Appl. Eng. Agric., 6: 717–724.

9. M. Sophocleous, (2010), Review: groundwater management practices, challenges, and innovations in the High Plains Aquifer, USA – lessons and recommended actions, Hydrogeol. J., 18: 559–575.

10. California Water Science Center, (2016), The California drought, Available from: http://ca. water.usgs.gov/data/drought/drought-impact.html.

11. V. McGuire, (2014), Water-level changes and change in water in storage in the High Plains Aquifer, Predevelopment to 2013 and 2011–2013, Scientific Investigations Report 2014–5218, U.S. Geological Survey.

12. 12.B. Scanlon, C. Faunt, L. Longuevergne, R. Reedy, W. Alley, V. McGuire and P. McMahon, (2012), Groundwater depletion and sustainability of irrigation in the US High Plains and Central Valley, PNAS, 109: 9320–9325.

13. J. Medina, (2015), California cuts farmers' share of scant water, The New York Times, Available from: http://www.nytimes.com/2015/06/13/us/ california-announces-restrictions-on-water-use-by-farmers.html?_r=0.

14. D. Charles, (2013), Kansas farmers commit to taking less water from the ground, All Things Considered, Available from: http://www.npr.org/blogs/thesalt/2013/10/22/230702453/ in-kansas.

15. H. Wells, (2015), Vegetables and pulses, Available from: http:www.ers.usda.gov/topics/ crops/vegetables-pulses.aspx.

16. D. Goolsby, (2015), Coachella Valley agriculture industry continues growing, The Desert Sun.

17. H. Cooley, K. Connelly, R. Phurisamban and M. Subramanian, (2015), Impacts of California's Ongoing Drought: Agriculture, Technical report, Pacific Institute, Oakland, CA, Available from: http://pacinst.org/publication/ impacts-of-californias-ongoing-drought-agriculture/.

18. W. Gomaa, N. Harraz and A. el Tawil, (2011), Crop planning and water management: A survey, in Proceedings of the 41st International Conference on Computers & Industrial Engineering, Los Angeles, CA, 319–324.

19. J. Dury, N. Schaller, F. Garcia, A. Reynaud and J. Bergez, (2012), Models to support cropping plan and crop rotation decisions: a review, Agronomy Sust. Developm., 32: 567–580.

20. P. deVoil, W. Rossing and G. Hammer, (2006), Exploring profit – sustainability trade-o s in cropping systems using evolutionary algorithms, Environ. Modell. Softw., 21: 1368–1374.

21. J. E. Annetts and E. Audsley, (2002), Multiple objective linear programming for environmental farm planning, J Oper. Res. Soc., 53: 933–943.

22. 22.R. Beneke and R. Winterboer, (1984), Linear Programming. Applications to Agriculture, Aedos.

23. J. Groot, G. Oomen and W. Rossing, (2012), Multi-objective optimization and design of farming systems, Agr. Syst., 110: 63–77.

24. B. Sahoo, A. Lohani and R. Sahu, (2006), Fuzzy multiobjective and linear programming based management models for optimal land-water-crop system planning, Water Resour. Manag., 20: 931–948.

25. R. Sarker and T. Ray, (2009), An improved evolutionary algorithm for solving multi-objective crop planning models, Comput. Electron. Agric., 68: 191–199.

26. R. Hanson, S. Boyce,W. Schmid, J. Hughes, S. Mehl, S. Leake, T. Maddock III and R. Niswonger, (2014), One-Water Hydrologic Flow Model (MODFLOW–OWHM), Techniques and Methods 6-A51, U.S. Geological Survey, Available from: http://dx.doi.org/10.3133/tm6A51.

27. W. Schmid and R. Hanson, (2009), The Farm Process Version 2 (FMP2) for MODFLOW-2005 — Modifications and Upgrades to FMP1, Techniques in Water Resources Investigations 6-A32, U.S. Geological Survey.

28. C. Faunt, R. Hanson, K. Belitz and L. Rogers, (2009), California's Central Valley Groundwater Study: A powerful new tool to assess water resources in California's Central Valley, Fact Sheet 2009-3057, U.S. Geological Survey.

29. C. Faunt, C. Stamos, L. Flint, M. Wright, M. Burgess, M. Sneed, J. Brandt, A. Coes and P. Martin, (2015), Hydrogeology, hydrologic e ect of development, and simulation of groundwater flow in the Borrego Valley, San Diego County, California, Scientific Investigations Report 2015–5150, U.S. Geological Survey.

30. R. Hanson,W. Schmid, J. Lear and C. Faunt, (2008), Simulation of an aquifer-storage-and-recovery (ASR) system for agricultural water supply using the farm process in MODFLOW for the Pajaro Valley, Monterey Bay, California, in MODFLOW and More 2008: Groundwater and Public Policy, International Ground Water Modeling Center, Colorado School of Mines, 501–505.

31. R. Hanson, W. Schmid, J. Knight and T. Maddock III, (2013), Integrated hydrologic modeling of a transboundary aquifer system - Lower Rio Grande, in

MODFLOW and More 2013: Translating Science into Practice, International Ground Water Modeling Center, Colorado School of Mines, Golden, CO, 5 p.

32. 32.R. Hanson,W. Schmid, C. Faunt and B. Lockwood, (2010), Simulation and analysis of conjunctive use with MODFLOW's farm process, Ground Water, 48: 674–689.

33. W. Schmid, J. P. King and T. Maddock III, (2009), Conjunctive Surface-water/groundwater Model in the Southern Rincon Valley Using MODFLOW-2005 with the Farm Process, Technical report, New Mexico Water Resources Research Institute.

34. R. Hanson, W. Schmid, C. Faunt, J. Lear and B. Lockwood, (2014), Integrated hydrologic model of Pajaro Valley, Santa Cruz and Monterery Counties, California, Scientific Investigations Report 2014-5111, U.S. Geological Survey.

35. R. Hanson, L. Flint, A. Flint, M. Dettinger, C. Faunt, D. Cayan and W. Schmid, (2012), A method for physically based model analysis of conjunctive use in response to potential climate changes, Water Resour. Res., 48: 23.

36. B. Adams, L. Bauman, W. Bohnho , K. Dalbey, M. Ebeida, J. Eddy, M. Eldred, P. Hough, K. Hu, J. Jakeman, L. Swiler and D. Vigil, (2009), DAKOTA: A Multilevel Parallel Object-Oriented Framework for Design Optimization, Parameter Estimation, Uncertainty Quantification, and Sensitivity Analysis: Version 5.4 User's Manual, Technical report SAND2010-2183, Sandia National Laboratories (updated April 2013).

37. K. Fowler, E. Jenkins, C. Ostrove, J. Chrispell, M. Farthing and M. Parno, (2014), A decision making framework with MODFLOW-FMP2 via optimization: Determining trade-o s in crop selection, Environ. Modell. Softw., 69: 280–291.

38. N. Lehmann, R. Finger, T. Klein, P. Calanca and A. Walter, (2013), Adapting crop management practices to climate change: Modeling optimal solutions at the field scale, Agr. Syst., 117: 55–65.

39. S. Lautenbach, M. Volk, M. Stauch and G. Whittaker, (2013), Optimization-based trade-o analysis of biodiesel crop production for managing agricultural catchment, Environ. Modell. Softw., 48: 98– 112.

40. J. Bokhiria, K. Fowler and E. Jenkins, (2014), Modelling and optimization for crop portfolio management under limited irrigation strategies, Journal of Agriculture and Environmental Sciences, 2: 1–13.

41. P. Steduto, T. Hsiao, E. Fereres and D. Raes, (2012), Crop Yield Response to Water, FAO irrigation and drainage, Paper 66, Food and Agriculture Organization of the United Nations.

42. 42.A. Harbaugh, (2005), MODFLOW-2005: The U.S. Geological Survey modular ground-water model: The groundwater flow process, Techniques and Methods 6-A16, U.S. Geological Survey.

43. W. Schmid, R. Hanson, T. M. III and S. Leake, (2006), User guide for the farm process (FMP1) for the U.S. Geological Survey's modular three-dimensional finite-di erence ground-water flow model, MODFLOW-2000, Techniques and Methods 6-A17, U.S. Geological Survey.

44. R. Hanson and W. Schmid, (2013), Economic Resilience through "One-Water" Management, Open File Report 2013-1175, U. S. Geological Survey.

45. W. Schmid, R. Hanson, S. Leake, J. Hughes and R. Niswonger, (2014), Feedback of land subsidence on the movement and conjunctive use of water resources, Environ. Modell. Softw., 62: 253–270.

46. I. Ferguson and D. Llewellyn, (2015), Simulation of Rio Grande project operations in the Rincon and Mesilla Basins: Summary of model configuration and results, Technical Memorandum 86– 68210–2015–05, U.S. Bureau of Reclamation.

47. G. Schoups, C. Addams, J. Minjares and S. Gorelick, (2006), Sustainable conjunctive water management in irrigated agriculture: Model formulation and application to the Yaqui Valley, Mexico, Water Resour. Res., 42: 19.

48. W. Schmid and R. Hanson, (2007), Simulation of intra- or trans-boundary water-rights hierarchies using the farm process for MODFLOW-2000, J. Water Res. Pl. – ASCE, 133: 166–178.

49. S. Boyce and R. Hanson, (2015), An integrated approach to conjunctive–use analysis with the one–water hydrologic flow model, MODFLOW-OWHM, in MODFLOW and More 2015: Modeling a Complex World – Integrated Modeling to Understand and Manage Water Supply, Water Quality, and Ecology, Colorado School of Mines, Golden, CO, 6–10.

50. S. Boyce, (2015), Model Reduction via Proper Orthogonal Decomposition of Transient Confined and Unconfined Groundwater Flow, PhD thesis, University of California Los Angeles.

51. S. Boyce, T. Nishikawa and W. Yeh, (2015), Reduced order modeling of the Newton formulation of MODFLOW to solve unconfined groundwater flow, Adv. Water Res., 83: 250–262.

52. 52.C. Faunt, (2009), Groundwater availability of the Central Valley aquifer, California, Professional Paper 1766, U. S. Geological Survey.

53. R. Hanson, B. Lockwood and W. Schmid, (2014), Analysis of projected water availability with current basin management plan, Pajaro Valley, California, J. Hydrol., 519: 131–147.

54. R. Hanson, L. Flint, C. Faunt, D. Gibbs and W. Schmid, (2014), Hydrologic models and analysis of water availability in Cuyama Valley, California, Science Investigations Report 2014-5150, U.S. Geological Survey, Available from: http://dx.doi.org/10.3133/sir20145150.

55. R. Hanson and D. Sweetkind, (2014), Water availability in Cuyama Valley, California, Fact Sheet FS2014-3075, U.S. Geological Survey.

56. T. Russo, A. Fisher and B. Lockwood, (2014), Assessment of managed aquifer recharge site suitability using a GIS and modeling, Ground Water, 53: 1–12.

57. H. Maier, Z. Kapelan, J. Kasprzyk, J. Kollat, L. Matott, M. Cunha, G. Dandy, M. Gibbs, E. Keedwell, A. Marchi, A. Ostfeld, D. Savic, D. Solomatine, J. Vrugt, A. Zecchin, B. Minsker, E. Barbour, G. Kuczera and F. Pasha, (2014),

Evolutionary algorithms and other metaheuristics in water resources: Current status, research challenges and future directions, Environ. Modell. Softw., 62: 271–299.

58. J. Eddy and K. Lewis, (2001), E ective generation of Pareto sets using genetic programming, in Proceedings DETC '01: ASME 2001 Design Engineering Technical Conferences, Pittsburgh, PA, 1–9.

59. K. Demchak, J. Harper and L. Klime, Strawberry production, Available from: http://extension.psu.edu/business/ag-alternatives/horticulture/fruits/ strawberry-production#section-4.

60. R. Smith, A. Baameur, M. Bari, M. Cahn, D. Giraud, E. Natwick and E. Takele, Artichoke production in California, Available from: http://anrcatalog.ucanr.edu/pdf/7221.pdf.

61. 61 M. LeStrange, M. Cahn, S. Koike, R. Smith, O. Daugovish, S. Fennimore, E. Natwick, S. Dara, E. Takele and M. Cantwell, Artichoke production in California, Available from: http://anrcatalog.ucanr.edu/pdf/7211.pdf.

INDEX

O

P

R

S

T

U

V

W

Z